Ethnographic research

ASSOCIATION OF SOCIAL ANTHROPOLOGISTS

ASA Research Methods in Social Anthropology

Panel of Honorary Editors

A. L. Epstein   James F. Fox   Clifford Geertz
Adam Kuper   Marilyn Strathern

Series Editor

R. F. Ellen, University of Kent at Canterbury

Ethnographic research

A guide to general conduct

edited by

R. F. Ellen

Faculty of Social Sciences
University of Kent at Canterbury
Canterbury, England

With a Foreword by
Professor Sir Raymond Firth

Academic Press
Harcourt Brace & Company, Publishers

London · San Diego · New York
Boston · Sydney · Tokyo · Toronto

ACADEMIC PRESS LIMITED
24–28 Oval Road, London NW1 7DX

United States edition published by
ACADEMIC PRESS INC.
San Diego, CA 92101

British Library Cataloguing in Publication Data
A Guide to the general conduct of ethnographic
research. – (ASA research methods in social
anthropology; 1)
1. Ethnology – Field work
I. Ellen, R. F.   II. Series
306′.028   GN346
ISBN 0-12-237181-X

Typeset by Oxford Verbatim Limited
and printed in Great Britain by
St Edmundsbury Press Limited, Bury St Edmunds, Suffolk

Contributors

A. V. Akeroyd
Department of Sociology
University of York
York YO1 5DD, UK

S. G. Ardener
Queen Elizabeth House
21 St Giles
Oxford OX1 3LA, UK

M. J. Auckland
British Library of Political and
    Economic Science
10 Portugal Street
London WC2A 2HD, UK

J. A. R. Blacking
Department of Social Anthropology
The Queen's University of Belfast
Belfast BT7 1NN
Northern Ireland

J. M. H. du Boulay
Department of Sociology
University of Aberdeen
Edward Wright Building
Dunbar Street
Old Aberdeen AB9 2TY, UK

P. Brown Glick
Department of Anthropology
State University of New York
Stony Brook, New York
New York 1174, USA

J. R. Clammer
Department of Sociology
National University of Singapore
Kent Ridge, Singapore 0511
Singapore

A. P. Cohen
Department of Social Anthropology
University of Manchester
Roscoe Building
Brunswick Street
Manchester M13 9PL, UK

J. Davis
Faculty of Social Sciences
Eliot College
The University of Kent at Canterbury
Canterbury
Kent CT2 7NS, UK

Y. Dhooge
Resource Options Unit
London School of Economics and Political
    Science
Houghton Street
London WC2A 2AE, UK

R. F. Ellen
Faculty of Social Sciences
Eliot College
The University of Kent at Canterbury
Canterbury
Kent CT2 7NS, UK

N. J. Goward
Faculty of Social Sciences
Eliot College
The University of Kent at Canterbury
Canterbury
Kent CT2 7NS, UK

D. Hicks
Department of Anthropology
State University of New York
Stony Brook, New York
New York 1174, USA

L. Holy
Department of Geography
The University
St Andrews
Fife KY16 9AL, UK

J. H. Kemp
Centre for Southeast Asian Studies
Eliot College
The University of Kent at Canterbury
Canterbury
Kent CT2 7NS, UK

J. C. Mitchell
Fellow and Senior Tutor
Nuffield College
Oxford OX1 1NF, UK

A. Raspin
British Library of Political and Economic
  Science
10 Portugal Street
London WC2A 2HD, UK

J. G. Sarsby
School of Continuing Education
Rutherford College
The University of Kent at Canterbury
Canterbury
Kent CT2 7NT, UK

J. E. A. Tonkin
Centre of West African Studies
University of Birmingham
PO Box 363
Birmingham B15 2TT, UK

J. Urry
Department of Anthropology
Victoria University of Wellington
Wellington
New Zealand

S. S. Wallman
Resource Options Unit
London School of Economics and Political
  Science
Houghton Street
London WC2A 2AE, UK

R. Williams
MRC Medical Sociology Unit
University of Aberdeen
Westburn Road
Aberdeen, UK

Foreword to the series

Writing about research methods in social anthropology may have several functions. An anthropologist presents a mass of information about some issue, with reflections upon its significance; it may be of prime importance for his colleagues to know how the information was collected, its representativeness, the manner of communication between all the parties, the type of systematic record, in order that the author's interpretation may be fairly judged. An account of research methods can also stimulate comparative interest, and aid in the evaluation of similar or contrasting views on a problem. It can perhaps serve a most useful function in suggesting to a prospective or actual fieldworker more fruitful ways of tackling an enquiry. This *Research Methods* series is designed to cover a wide range of interest, including that of colleagues outside the discipline of anthropology who legitimately want to know what is the justification for what we so broadly and confidently say about the human condition. But understandably, it is directed strongly towards fellow anthropologists who are grappling with problems of this fascinating, exasperating, rewarding, always unique experience known vaguely as "the field".

Writing about research methods in anthropology is a bold undertaking. While it may be gratifying to the ego either to explain one's own procedures or to tell other people how to improve theirs, an account of method does mean often an exposure of the self from which many anthropologists have shrunk. Social anthropology in its fieldwork phase demands a commitment of the self to personal relationships of a wide variety, with qualities hard to define, often ranging from casual encounter to prolonged and serious communication, perhaps of an emotional character which can be characterized properly as friendship. Data are collected by observation and discussion, not in neutral situations of mechanical action, but in situations of vivid human interaction in which the anthropologist is often directly involved in a complex interplay of judgement, sympathy and personal statement. To generalize from this in terms of more abstract procedures or even practical rules for conduct is not easy. Nevertheless the attempt at generalization has to be made if anthropologists are to retain credibility in the face of charges that their work is primarily a series of aesthetic constructs, a set of ego-trips into the exotic.

In social anthropology we are faced by a fundamental dilemma: we work at three levels or in three dimensions. The declared object of our study is some aspect of the ideas and institutions of society – often exotic but not necessarily so – and there is a professional obligation upon us to describe, represent or interpret such ideas and institutions as accurately as possible. Whatever be the talk about models or the social conditioning of perception, whatever the obscurity in relation between portrayal and portrayed, we are trying to perceive and make statements about the order of an actual, not an invented society. But our perceptions and statements must inevitably pass through a personal lens, with all the individual assumptions and mode of thought which our specific upbringing and our general literate training have helped to establish. The individual component in the study of the actual society is involved in an historical time-flow, a series of particular incidents which in themselves may be trivial and unilluminating. Yet anthropologists deem themselves to be concerned with more than a record of past events; their work should relate to more general human issues. So while a contribution to social anthropology can be rendering of personal experience in an identifiable social milieu, it should not be purely idiosyncratic; it must carry an implication of generality. As I see it, this dilemma is ultimately not resolvable. But it is not unique – there is something akin to it for example in the efforts of Asian or African novelists to write about their own natal societies. It is inherent indeed to some degree in any relationship of the Self to the Other. In all human relationships we operate with sets of assumptions if not about the reality at least about the viability of our observations and inferences, and with convictions that however approximate our models of other people's character and behaviour may be, at least some fair degree of consistency in relations with the Other is predictable.

This *ASA Research Methods* series has been and is being composed in clear realization of the difficulties faced by anthropologists in these and allied sensitive areas. With sophistication and without dogmatism the volumes aim to face the problems and indicate where some solutions may be found. In this way they attempt to reduce the burden of uncertainty that often lies on anthropologists facing the fieldwork situation. Each book sets out in varying detail the framework upon which our work has been built so far, and indicates the kinds of lines on which improvement is needed if our conclusions are to hold their validity. One can always argue about the value of such guides, which have a long history in our discipline. Some anthropologists of great experience have argued that it is useless to try and teach the craft, that each neophyte must find his or her own solution. I myself have never held this opinion. Not only have I tried to publish at least some brief account of my field condition and methods in Tikopia and again in Kelantan (and Rosemary Firth has given more extended and more penetrating examination of some aspects of our joint work in the latter field), I have also collaborated with Jane Hubert and Anthony Forge in a systematic exposé of the methods adopted in a team research project in North London. Again, in cooperation with colleagues at the London School of Economics and University College, London, I was involved for upwards of 20 years in seminars on fieldwork methods. Anthropologists in other institutions have done likewise. But I think we have all shared the view of the uniqueness and intensely personal character of a field experience, and that any training for the field is therefore not injunction but example, not prescription but suggestion. This seems also to be the spirit of these volumes.

What I can say here is that looking back over my own experience, I am very aware of how much my own enquiries would have benefited from use of monographs such have been prepared or are planned in this series. They combine theoretical stimulus and practical suggestion in fertile ways, and though I may not agree with all their formulations I find them definitely thought-provoking. Of some vocal economic historians R. H. Tawney once said "I wish these fellows who talk so much about methods would go and *do* some of it". The contributors to volumes in this series have already completed a great deal of field research, and the methods they advocate are a distillation from a very wide experience.

I look at these research method volumes much as a cook looks at a recipe book. They are no substitute for individual ingenuity and skill, but they can offer a range of alternative suggestions about procedure, and warnings about pitfalls, that are most unlikely to have occurred in such fullness to the practitioners. To change the analogy, if I can conclude on a light-hearted note, their role is parallel to that ascribed by a Kelantan religious man to his ritual and formula for securing fish. In terms almost reminiscent of Malinowski he said he could not turn an unskilled fisherman into a skilled one, but through the bounty of Allah he could "help" a competent fisherman to get a better catch. Fisherman or cook – and figuratively an anthropologist is sometimes both – any one in our profession should keep these volumes to hand.

Raymond Firth

Preface

This series, of which the present volume is the first, was first proposed in 1981 at the Edinburgh meeting of the Association of Social Anthropologists. Its appearance is a response to the absence of suitable, systematic and thorough reference guides to research methods employed in the discipline of social anthropology. It is not the outcome of a perceived lack of general interest in most questions pertaining to methods, since this is obviously an area of current concern. A detailed justification for the series as a whole, and for this volume in particular, is presented in Chapter 1, where the reader can also find a statement of our aims and scope. Although some acknowledgements are set out in the conventional manner following this preface, it will be clear to many that the full indebtedness of the various contributors extends beyond this to several generations of scholars and fieldworkers upon whose work our contemporary research practices rest.

Canterbury
November, 1983   R. F. Ellen

The trunks of notes up in his attic, which he had never even sorted out, were a constant reproach to him.

p.58, Barbara Pym, *Less than Angels*

Our ethnographic sources provide our indispensible claims to legitimacy. . . . This ultimate reference of all our ideas to the ethnographic 'facts' elucidated by painstaking field research is now so deeply engrained in our subject and so taken for granted that we tend to forget how dependent we are on our sources.

p. 11, Ioan Lewis, *The Anthropologist's Muse*

We also planned names . . . and had, a procedure built into the plans of every true planner, to prune, to make new plans, distil and clarify old ones.

p. 32, Günther Grass, *From the Diary of a Snail*

Acknowledgements

In preparing this volume for publication an unusually large number of individuals have been consulted and whose help should be formally put on record. Extensive consultation is perhaps understandable in a book of this kind, which seeks to be a research tool of the widest possible application, cutting across personal idiosyncrasies and differences of approach. But although the book has undoubtedly benefited from the good advice, suggestions, comments and kind assistance of many, it is especially necessary to issue the general disclaimer that the final results, all opinions expressed and all deficiencies, are the responsibility of the various contributors and the editor alone. Also, while the volume may have the imprimatur of the Association of Social Anthropologists, this can in no way be said to represent official endorsement by the association or reflect the views of its membership as a whole.

We would like to thank the following for formal permission to reproduce previously published or circulated material: The Social Science Research Council (SSRC) for permission to reproduce a version of a short guide (published August, 1982) which appears here in a modified form as §7.8; Professor F. Bailey of the University of California at Santa Barbara who generously agreed to the use of material from an unpublished handbook on anthropological fieldwork produced in 1968 under his supervision at the University of Sussex.

Numerous individuals have been helpful in the compilation of particular chapters and sections. In this connection we would like to thank the following: Chapter 3 – Professor Sir Raymond Firth, Professor G. W. Stocking of the Department of Anthropology, University of Chicago, Dr Michael Young of the Research School of Pacific Studies at the Australian National University in Canberra. Chapter 5 – Dr C. W. Watson of the Centre for Southeast Asian Studies, University of Kent at Canterbury (§5.3); Nikki Goward, Rosemary Ridd, Deborah Kirkwood, Mona MacMillan, Pat Holden, Jane Khatib-Chahindi, Soraya Tremayne, Ann-Marie O'Neil-Baker, Renee Hirschon, June Starr, Renata Barber and Fiona Bowie (§5.5). Chapter 6 – Professor John Barnes, Department of Sociology, University of Cambridge. Chapter 7 – Centre for Information on Language Teaching and Research (CILT) (§7.5); Dr John Malin of the Social Affairs Committee and Dr Dugget of the Information Division of the Social Science Research Council (SSRC), Dr Anne Akeroyd of the Department of Sociology at the University of York, Mr Jonathan Benthall, Director of the Royal Anthropological

Institute (RAI), Dr Simon Charsley of the Department of Sociology at the University of Glasgow, Dr Ralph Grillo of the School of African and Asian Studies at the University of Sussex, Professor David Hicks of the Department of Anthropology, State University of New York at Stonybrook, Dr Peter Riviére of the Institute of Social Anthropology in Oxford, and Dr J. L. Watson of the Department of Anthropology and Sociology of the School of Oriental and African Studies (SOAS) in the University of London (§7.5); Dr Martin Southwold of the Department of Social Anthropology in the University of Manchester, Professor Paul Ekman of the Human Interaction Laboratory at the University of California in San Francisco, Professor Jack Goody, Dr Susan Benson and Janey Hall of the Department of Social Anthropology at the University of Cambridge, Dr Farnham Rehfisch of the Department of Sociology and Anthropology at the University of Hull, Dr J. A. Soulsby of the University of St. Andrews, and Neil Price of Bristol University (§7.8); Dr C. MacCormack and Dr J. P. Vaughan of the Ross Institute in the London School of Hygiene and Tropical Medicine, Colonel A. G. Harwood of the Army Personnel Research Establishment at Farnborough (§7.10).

A large number of people have served in a variety of additional capacities, especially as anonymous readers and sources of good sense and advice. In this respect we would like to thank: Dr Anne Akeroyd, Dr Juliet du Boulay of the Department of Sociology at the University of Aberdeen, Professor Paula Brown-Glick at the Department of Anthropology in the State University of New York at Stonybrook, Professor Abner Cohen of the Department of Anthropology and Sociology at the School of Oriental and African Studies in the University of London, Professor Scarlett Epstein and Professor A. P. Epstein of the School of African and Asian Studies at the University of Sussex, Dr G. K. Garbett of the Department of Anthropology at the University of Adelaide, Olivia Harris of the Department of Social Anthropology at Goldsmith's College in the University of London, Miss B. J. Kirkpatrick, formerly Librarian of the Royal Anthropological Institute, Professor Ioan Lewis of the Deparment of Anthropology at the London School of Economics, Dr Joe Loudon, and Dr J. Urry of the Department of Anthropology at the Victoria University of Wellington, New Zealand. At the University of Kent at Canterbury, special thanks are due to Professor J. Davis, Dr J. S. Eades, Dr N. Goward, Professor R. E. Pahl, Professor Paul Stirling and Ms Claire Wallace of the Faculty of Social Sciences, and to Dr R. Veltman of the Institute of Language and Linguistics.

We should like to thank also the following for permission to reproduce quoted material: Addison Wesley Publishing Company, Cambridge University Press, the Department of Anthropology at the London School of Economics and Political Science, Elsevier-Dutton Publishing Co. Inc., Harper and Row Publishers Inc., Jossey-Bass, MacMillan Publishing Co. Inc., Routledge and Kegan Paul Ltd., Tavistock Publications and Witwatersrand University Press. Full acknowledgement is given in the Bibliography at the end of the book.

Finally, the editor would like to thank all those who have provided information on the teaching of research methods in various institutions, and all those involved in the production of this volume at the technical level, both in the secretarial office of Eliot College in Canterbury and at Academic Press.

Contents

1 Introduction

1.1 The series

Social anthropologists have sometimes had an image of their profession
which presents them as highly atomized practitioners working with a shared
(although not always specified) set of assumptions and procedures. To the
extent that this picture is, in any sense, accurate, it has not always served the
discipline well. While it has generated lively debate and stimulated innovation, it has often mystified rather than clarified, and has encouraged a
secretiveness which is ethically suspicious and ill-suited to both its humanistic
and scientific pretensions. Research proposals which indicate that an appli-
cant intends to employ "standard anthropological methods" have proved
routinely acceptable in the past, even though techniques are not specified.
Increasingly, techniques are by no means standard, and it is rather doubtful
whether they ever have been.

It is now widely acknowledged that for a long time there was a virtual
"conspiracy of silence" (Berreman 1962) on how ethnography is produced.
Indeed, some would consider that little of any profundity has been said about
our own professional practices and methods of enquiry until relatively
recently (Diamond 1980; Scholte 1980; Crick 1982). In a modest way, it is
the intention of this series to make anthropological research practices much
more of an open book and object of critical scrutiny than they have perhaps
been in the past.

The discipline of social anthropology has sometimes been compared to a
craft (e.g. Epstein 1967) and its training to an apprenticeship. That training in
ethnographic methods approaches apprenticeship is, except in a few cases, a
quite unfounded assertion (SSRC 1968: 100). The applicability of the term
"craft" is more complicated. The image generated by this term is one of

ETHNOGRAPHIC RESEARCH
ISBN 0 12 237180 1

esoteric knowledge being absorbed rather than learned, of an oral tradition passed on from hand to mouth within a closed circle of scholars. That the view anthropologists have of themselves should be the same as that which most of them project of the societies which they study is, perhaps, hardly surprising. But while craft it may be in a specific technical sense, there is nothing mystical about how research data are produced. Results do not arise from thin air, nor from some pseudo-spiritual empathy between analyst and subject. Ethnographic research has to be worked at, and results are achieved through the painstaking application of a body of techniques, which may be either implicit or explicit, and based on certain kinds of assumptions. To say that anthropological research methods cannot, and should not, be taught but can only be acquired through absorption, intense moral sensitivity and total immersion (aided only by a notebook and pencil) is no longer tenable, along with Evans-Pritchard's (1973: 1) appeal to gentlemanly values and common sense (c.f. also Beattie 1965: 2; Beals 1970: 38).[1] Now, this is not to take a tough "positivist" stance, in contrast to a "wet" humanism which abhors the codification of rules. Neither is it to deny the very considerable value to be placed on prolonged experience, on specific contexts, on appreciating the demands of maintaining adaptive flexibility, and on using non-directive methods (Henry 1969: 46; Cohen and Naroll 1970: 7). Methods have always been taught, or at least talked about, and people do learn how to DO anthropology. What is perhaps more important is that they can learn to do it well, badly or indifferently. This series starts from the assumption that, since people will anyway be taught or learn for themselves, we should discuss and scrutinize more candidly the methods we use and learn from the detailed experience of others.

A report written by a committee of the Social Science Research Council (1968: 83) states that:

> Anthropologists have written relatively little about their field methods by comparison, for example, with sociologists and psychologists . . . partly because of the intensely personal character of much anthropological field research which makes it difficult to characterise method effectively except by extensive documentation.

Whether or not this has ever really been the case, there is in fact a long history of written accounts of research practices (Chapter 3) while the literature on methods is now considerable (§1.5, and cumulative final bibliography). Most of this is, nevertheless, relatively recent. The growth stems mainly from the

[1] Elsewhere, however, Evans-Pritchard (1954: 82) is less flippant: "he [*the anthropologist*] must also have intuitive powers . . . ability, special training, and love of careful scholarship . . . the imaginative insight of the artist . . . and the literary skill necessary to translate a foreign culture".

1960s; a period when social anthropology was on the defensive in the face of criticisms of its historical and ideological role as the hand-maiden of colonialism and neo-colonialism, and when anthropologists themselves were becoming increasingly aware of the shortcomings of their discipline and its methods (Salamone 1979: 47). This has been transformed more recently into a sustained interest in the intellectual and sociological analysis of the subject, its procedures and practitioners (§1.4).

The literature is still patchy. Some subjects are well-covered: quantitative methods and the subjective experience of fieldwork (at least in some of its aspects), for example. Other areas, such as the production of fieldnotes, the processes by which these are transformed into "analysis", and the use of language, are still poorly covered. Moreover, anthropologists trained in the British tradition have been more reticent about discussing their methods than their American colleagues. Apart from the ubiquitous *Notes and Queries*, the only book length treatment of the subject has been the Epstein edition (1967). It is some measure of current consciousness of the importance of discussing methods that, while the volume mentioned amounted to a slender 250 pages (with many aspects of enquiry excluded), in 1982 it was thought fit for a professional association of social anthropologists to embark upon an entire series.

This handbook is planned to be the first of a series on methods in social anthropology which we hope will become standard reference works. Each one will attempt to be both practical and authoritative; to provide guidance to anyone likely to be involved in the conduct of ethnographic research, to help readers interpret in a critical fashion the assumptions upon which research is based, and to evaluate research procedures resulting in published work.

By *practical* we mean that each handbook emphasizes fieldwork and research practices and provides clear and concise guidance and information on methods in a convenient form, without embedding the discussion in lengthy theoretical and methodological discourse. This does not mean that we take a simple-minded view of the relationship between research procedures, theory and methodology (§1.3, Chapter 2), or believe that its discussion has no place in research training. This, of course, must be so. It is in this sense that we hope that the notion of a "handbook" (in the best sense of this term) is an appropriate one. It stresses the practical compendious character of the work without, we hope, suggesting the extreme formalism and standardized routines of a manual, or the naive guidance for amateurs enshrined in *Notes and Queries*. There is always a danger that some will think that ethnography is simply a matter of the mechanical application of self-contained and well-worn techniques. If anything, we anticipate that a series of this kind will dispel such a false impression. We provide few firm rules; neither are we writing prescribed texts for undergraduate courses. Much of what is said will only make sense when coupled with proper attention to theoretical questions and underlying principles, and in relation to a consideration of substantive cases by those who have acquired a training or who are in the process of doing so.

Each volume, as well as the series as a whole, does not stand by itself: you cannot learn to do anthropology from a manual. What we do hope to provide are some directions that will enable workers to develop the most appropriate approach to a particular ethnographic project. For the most part, this takes the form of an overview of the major problems, some practical advice, and (most importantly) a guide to the literature.[2] The handbook has been designed for use by intelligent undergraduates, research students in training, field-workers, or any other member of the profession.

By *authoritative* we mean that each handbook will aim to set out realistic standards for the conduct of research in the light of work already undertaken, and to standardize procedures, and in some cases terminological usages, which will make comparison and evaluation easier and more productive (§1.3). This means that the methods recommended are only those which have demonstrated their effectiveness, and not those which require impossibly ideal conditions and a great deal of luck. Throughout, we work from the assumption that certain methods of enquiry are better than others for achieving particular ends, and that particular standards are desirable. This is not to adopt a high-minded tone or to lay down immutable conditions for professionally acceptable research. It will, indeed, be rare for any of those contributing to this series to have maintained consistently in their own work the high standards which they recommend and aspire to. Often they will have perpetrated the horrifying errors and misjudgements that they warn against. Moreover, we fully recognize that not all procedures mentioned or recommended are necessarily those most effective all of the time, and that their very effectiveness may (as a result of new research and thinking) become highly questionable. In a thriving discipline books of this kind may swiftly be superseded and must always be treated as provisional documents. It is not our wish to produce textbooks which imply a changeless body of techniques enshrined by past usage, although it is perfectly reasonable that we should aim to produce something which may have something in it of lasting value.

[2] In this latter respect we have tried to be thorough, although given the total number of potential sources, it would have been impossible to be scrupulously comprehensive, even if this were desirable. Where we have had to be selective, texts have been chosen which provide the most exemplary or characteristic illustrations of the matter in hand.

While we have tried to be authoritative and (we might add) comprehensive, we have eschewed the idea of handbooks written by anonymous committees. These are seldom a recommendation for clarity and excellence. Each volume is edited or authored by selected individuals who are directly responsible for the content to the series editor and a panel of honorary editors. At the same time, some of the handbooks, such as Volume 1, are the product of a large number of consultations, a great deal of correspondence and many meetings. Many of the ideas and individual sections have been the subject of discussion at conferences, seminars and workshops. In some cases, as in Volume 1, the compilation and writing of particular chapters and sections has been delegated. Each volume is, therefore, very definitely, a collaborative exercise. For this reason, it can legitimately be claimed that a large measure of agreement has been reached in formulating those recommendations which have been made. Nevertheless, it is unlikely that we will be able to please everybody.

1.2 The present volume

The present volume, the first in the series, is intended to provide basic outlines on general practical matters of concern to all those engaged in ethnographic research, to introduce the series as a whole, and to serve as a guide to existing literature on issues not specifically covered by the more specialized volumes which are to follow. The book is about ethnographic research, about the production of data. It is not about other non-fieldwork methods of anthropological research, such as cross-cultural survey, comparison and generalization. These entail the manipulation of data at a higher level of abstraction. The various levels of anthropological research are, admittedly, not easily separated, particularly when we begin to analyse the research process itself (§1.4). However, such distinctions must be made for pragmatic purposes. It is hoped that such methods of secondary analysis will be tackled in a later volume in the series.

Certain sections of the present volume provide only a brief summary coverage of their subject-matter, usually in the form of a bibliographical guide. There are several reasons for this. First, the volume is intended to be general and introductory. Secondly, we anticipate some overlap with the content of the more specific handbooks to follow. For example, the use of archival and other written materials, and the subjects of quantitative and comparative methods deserve quite separate and detailed treatment. A volume on quantitative methods is currently in preparation, while plans are in hand for volumes on the other two subjects. Handbooks will also follow dealing with research methods as they apply to particular substantive fields, such as subsistence, oral literature and material culture. Among those subjects to be dealt with in forthcoming volumes are kinship, ethnomusicology and applied anthropology. Thirdly, the existing literature on many aspects of research methods is now extensive and it would be impossible to put into one book all that might be said (or all one should know). Rather than reproducing or summarizing what has already been written by others, we have preferred to provide a general outline, a few signposts, and refer the reader to relevant books and articles. Of course, there are some areas, such as the compilation of field notes, which have been discussed little in the literature, or which it is felt have been dealt with inadequately. In such cases the subject has been tackled at greater length.

This volume has been designed as a handbook. By this is meant that we hope it will be suitable for rapid occasional reference. Chapters 2, 3, 4 and 6 are in essay format, written by single authors, and provide an extended backdrop and context for all the other sections. The remaining chapters deal with specific, and for the most part, practical aspects of research practice. Each chapter, section, and sometimes sub-section, has been numbered and coded to facilitate easy reference (e.g. "§7.1.2"). Footnotes have been avoided wherever possible, although long lists of references have sometimes been placed at the bottom of the page to avoid breaking-up the text. There is a full cumulative bibliography at the end of the book.

The chapters, and often sections within them, are written by different individuals. We have tried to divide the labour in a way which draws upon the particular interests, experiences and specialisms of the contributors. Because of the holistic character of much ethnographic research, there is inevitably some overlap and repetition. Issues discussed in more than one place are cross-referenced. Cross-references are to section and sub-section codes and appear in parentheses with the section symbol (§). Qualifiers such as "see above" or "following" have been eliminated to save space since cross-references are so frequent. A cross-reference, therefore, is generally indicated thus: (§6.5.2). Apart from a certain amount of judicious editorial pruning, overlap has not been discouraged. It would, anyway, have been difficult to avoid in practice, while rigorous attempts to eliminate it would have resulted in a grossly misleading impression of the relationship between different aspects of the ethnographic enterprise. Much the same might be said of certain disagreements and contradictions in the methods outlined and advice given. This is inevitable, and arises directly from the complexity of the objects of our analysis. It should not be resolved simply in the interests of editorial harmony.

Many inaccuracies will have crept into the text, or have been overlooked, during the editing and redrafting of material for inclusion. For this we would like to apologize in advance. In its present form, the volume should be regarded as highly provisional. We hope that the very existence of the series, and in particular this volume, will generate a need for rapid revision, as people respond to its inadequacies as they become apparent through their own experience. If this is so, then we will have succeeded in one of our major aims: that of encouraging a new degree of consciousness about the research practices which social anthropologists engage in. All suggestions and criticisms are welcome in the hope that they will result in an improved revised edition at some stage in the future.

1.3 Precision in the use of words

It is generally agreed that jargon should be avoided wherever possible in ethnographic description and analysis. Not only can it obscure, but it may also stultify creativeness through the imposition of inflexible terminologies and meanings. On the other hand, much confusion arises from the lack of agreement on the use of certain terms and their definitions, and from the blatant misuse and misunderstanding of those terms and concepts which are generally assumed to be widely shared (Thouless 1939). Clearly there is a problem; for while it may be true that on occasions terminological squabbles have generated more heat than light, it is equally the case that similar disputes serve as icons of competing methodological and theoretical systems, the differences between which should not be concealed. Consider, for example, the word "ethnography". This word is used regularly to refer to empirical accounts of the culture and social organization of particular human populations (as in "an ethnographic monograph", "an ethnography"). The implication is that of a completed record, a product. But then the sense alters somewhat if we speak of "ethnography" as opposed to "theory", or of "an ethnographic account" (meaning living people) as opposed to an historical or archaeological account. Different again from all of these is the use of the term to indicate a set of research procedures, usually indicating the intensive qualitative study of small groups through "participant-observation" (§8.2). Finally, "ethnography" may refer to an academic subject, a discipline in the wider sense involving the comparative study of ethnic groups, or as the Concise Oxford Dictionary rather unhelpfully puts it, the "scientific description of the races of men" (1964: 415). At this point, "ethnography" appears to be equivalent to at least one sense of the word "ethnology" (Crapanzano

1977: 69). Thus, ethnography is something you may do, study, use, read or write. The various usages reflect ways in which different scholars have appropriated the term, often for perfectly sound conceptual reasons. We would not wish to suggest that the word be employed in one sense only, even if it were possible to effectively dictate that this should be so. However, it is important to know that the differences, often subtle, exist.

Here we start from the proposition that definitions are necessary evils, even if they are subsequently rejected or modified. One service a series of this kind can render is to indicate where there is dissension and consensus in the use of particular terms and definitions. We will also sometimes make specific recommendations and set down guidelines for standardization, retention and change where this is thought desirable in order to clarify our own professional practices and aid concise and unambiguous description, analysis and comparison. In this sense, we are inheriting a role – at least as far as the definition of cultural traits and social arrangements is concerned – previously performed by *Notes and Queries*. Whilst not wishing to hinder in any way the insights which stem from idiosyncratic neologisms and from creative shifts in meaning, there are grounds for believing that the discipline could benefit from an institution comparable to the International Rules of Nomenclature for Zoology, although in his own sensitive discussion of the problems of terms and definitions, Nadel (1953: 107–111) once advocated "an anthropological Council of Nicaea" (p. 48). Rather than stifling controversy, the role of such an institution would be to clear the ground and enable description, analysis and debate to take place in a manner most conducive to its success. Many terms and definitions are already shared by professional anthropologists, and the role of such a body of rules in such cases would simply be to confirm existing usages. The nearest we have to such a set of guidelines at present are various specialized dictionaries (e.g. Gould and Kolb 1964).[3]

Finally, in a series of this kind it is necessary to ensure that four key terms are used as consistently as possible. These are "theory", "methodology",

---

[3] The International Social Science Council, which is affiliated to the International Union of Anthropological and Ethnological Sciences (IUAES), has a standing committee on conceptual and terminological problems (COCTA). On the basis of a pilot project funded by UNESCO, this committee has formulated an approach for establishing the newness of new concepts and for naming them in a way that should be acceptable to specialists in the subject field concerned. This procedure provides definitions of concepts that are needed in a subject field and identifies the terms which have been used (or may be used) to designate them. By arranging concept definitions in a systematic (classified) order (i.e. not alphabetically), it is possible to determine whether or not a new suggested concept has already been included. By entering these records in a computerized database, it is possible to update those who cannot tap the material on-line. Individual terms may be found, as in a dictionary, through the index. A pilot project illustrating the approach has been launched by COCTA, taking the field of "ethnicity" as its focus (Riggs 1982: 2).

"method" and "technique". These terms, and "methodology" in particular (see Holzner 1964: 425–426), are regularly used to mean rather different things to different people. Without claiming to provide the "correct" definitions, and while both recognizing that any definitions are (up to a point) arbitrary and that the content of each category profoundly affects the others, we propose the following:

(a) "theory": a supposition or body of suppositions designed to explain phenomena or data.
(b) "methodology": the systematic study of the principles guiding anthropological investigation and the ways in which theory finds its application; an articulated, theoretically informed approach to the production of data, e.g. "Marxist methodology", "ethnomethodology".
(c) "method": a general mode of yielding data, e.g. interviewing.
(d) "technique": a specific means of making particular methods effective, e.g. questionnaires, shorthand, kinship notations.

"Methods" and "techniques" together constitute "research procedures".

1.4 The sociology of anthropological research practices

It has been emphasized that this book primarily has a practical aim: to inform those interested of the methods and techniques which have proved useful in the production of ethnographic data, and which may therefore continue to be of some use in the future. But ethnographic research is always part of some wider social and political enterprise, while the quality and character of ethnography depends very much on the circumstances of its production. Some (e.g. Fabian 1979) have argued that any claims to objectivity are entirely dependent upon recognizing this. For this reason, we believe it important to encourage ethnographers to be reflective about their methods and the intellectual and institutional contexts in which they employ them. Although it is not our central purpose to discuss the sociology of anthropological practices, if ethnographers are to be anything more than mindless machines for the gathering and processing of data, it is most desirable that they should have some knowledge of the relationship between theory, methodology and research procedures (Chapter 2) and of the history of the development of research practices (Chapter 3). They should also have some understanding of the structure and organization of the various processes and activities in which they engage, of different styles of research practice (Chapter 4), and of the subjective experience of fieldwork (Chapter 5).

There is now a growing literature on the sociology of the discipline and its research practices. All this is relatively new, deriving in part from a sociology of knowledge with its roots in the writings of Marx and Mannheim, in part from the new philosophy and social analysis of science inspired by the work of, in particular, T. S. Kuhn, and in part from a humanistic social science tradition which gives prominence to the interaction of investigator and investigated, and is sensitive to the moral and political issues which the conduct of research highlights. At its more empirical end, these concerns have led to sociological studies of particular research projects (e.g. Hammond 1964; Platt 1976; Bell and Newby 1977), although much discussion has been at a very general level (e.g. Tiryakian 1971; Phillips 1973). As far as anthropology is concerned, a scholarly interest in the investigation of its own backyard became especially visible in the 1960s, with the growing awareness of the relationship between anthropology and colonialism, and of the relationship between the conduct of ethnography and the implementation of policy (Maquet 1964; Asad 1973) (Chapter 6). Only recently has this developed into a more general critical interest in the sociology of the discipline and its institutions, and into an examination of the status of fieldwork as professional boundary-maintenance and ritual (Firth 1972). Only more recently still have people begun to look seriously at *how* ethnography is produced, how experiences relate to data, how meanings get imposed on experience, how data is organized and transformed, how texts are written, and how (in the broadest possible sense) the research act is socially constructed. This is so despite the existence of a comparatively large literature on the fieldwork experience (§5.1). There is currently a trend towards studies of the social and semantic organization of anthropological knowledge (e.g. White 1966; Scholte 1980; Crick 1982), in relation to the rules, institutions, strategies, cultural organization and career structures of academic life (Bailey 1977), together with the sociology of anthropological literature and publication.

It has been suggested that anthropologists engage in two different kinds of mutually interacting but nevertheless disjunctive activity: doing fieldwork and writing books (Geertz 1973: 19; Crick 1982: 17). It is in part recognition of this which has occasioned scepticism about the positivist view of the research process. Rather than ethnographic "facts" being "collected", realities "perceived", data "presented" and "written-up", there has been a discernable shift to a view of the research process as one in which field "experiences" are radically "transformed" into data, through encounters between researcher and researched (Crapanzano 1977: 72), characterized as relations of transaction, information exchange and mutual exploitation (Wax 1952: 35; Lundberg 1968: 47; Hatfield 1973). This transformation, which may at the same time be seen as a process of translation from one cultural context to another, and as one of "construction" (or "reconstruction"), is highly symbolic, intersubjective and personal (Tonkin 1971; Scholte 1974: 439–442). Precisely "how" this takes place remains a crucial problem for investigation.

The ethnographic end-product – the text – generally purports to deal in objective facts: it is more accurately the result of a process of "objectivization". In this process, not only are data simplified and manipulated, but some would say that the original ethnographic confrontation is destroyed or, at best, alienated (Crapanzano 1977: 72). The kinds of influences which are brought to bear in the construction of a text are numerous, but at root the resulting text is seen as the end-product of a complex interaction between "self" and "other" (Dwyer 1977). The ethnographer's "explanations" are in part authored by informants, through a combination of plagiarizing their emic world and a "creative exchange" (Lewis 1973: 16; Clifford 1980). There can be little doubt that the extent to which the utterances and actions of our informants are manipulated through the processes of producing data and subjecting it to analysis are immense. Moreover, it has recently been argued (Ennew 1976: 44) that data are merely insertions in our own theoretical constructions, while we have also been reminded that persuasive texts are not necessarily those that are "accurate" (Campbell and Levine 1970: 385). It would seem that "reading texts" is almost as dangerous as writing them.

It is indeed curious that we should accept anthropological texts as knowledge yet pay so little attention to analysing our methods and the process by which ethnography is written (Clifford 1980: 519), compared with what we are prepared to pay to produce our data. It is perhaps more alarming still, once we have reflected on the dramatic extent of the transformations which take place in the process of producing ethnography. How important, then, that we should encourage a critical attitude to what we do and how we do it, and take active steps to dispel the pernicious notion that ethnography is a matter of "mere" description (Den Hollander 1967; Clifford 1980).

1.5 Bibliographical and general literature

Specifically bibliographical publications on research procedures and practices in social and cultural anthropology are relatively few; of these most are rather dated or limited in some other respect. Hymes (1959), for example, is biased towards linguistic methods. Unpublished bibliographies, on the other hand, are probably legion but highly inaccessible, with very local and limited departmental circulations. However, most general works on methods include bibliographies and these are sometimes quite lengthy. Most of them take the form of multi-authored collections of essays (Adams and Preiss 1960; Epstein 1967; Jongmans and Gutkind 1967; Freilich 1970; Naroll and Cohen 1970); only a small number consist of integrated texts by single writers (e.g. Williams 1967; Johnson 1978; Pelto and Pelto 1978; and Agar 1980). Of these, Agar (1980) contains a 223-item bibliography, Gutkind and Sankoff (1967: 214–271) an annotated fieldwork bibliography of 348 items, while Pelto and Pelto (1978) has 293 items. Although not specifically anthropological, Junker (1960: 159–207) contains an extended bibliographical essay and classified list of publications on research procedures up to 1960. By comparison, the present volume of this series alone contains a bibliography of 1130 entries. But although extensive, it does not necessarily supersede any of the bibliographies mentioned, which may be strong in certain specialist areas. To conclude this summary of the anthropological literature, mention should be made of a number of more general articles (e.g. Bennett 1949) and books containing chapters on methods (e.g. Herskovits 1948; Nadel 1953; Piddington 1957; Beattie 1964; and Lewis 1976), as well as books nominally about research methods in disciplines other than sociology, but which contain much of value for anthropologists. Among these should be included Goldstein (1964) on folklore.

By contrast, the equivalent sociological literature is very considerable, widely consulted by practitioners in the subject, and of more than passing relevance to ethnographic fieldworkers. The more useful, general and up-to-date works of importance include textbooks of a standard and introductory kind and works by single authors, covering a broad range of issues (Madge 1957; Selltiz *et al.* 1963; Cicourel 1964; Galtung 1967; Sjoberg and Nett 1968; Stacey 1969; Denzin 1970; Phillips 1971; Johnson 1975; Mann 1976; Mayntz *et al.* 1976; Nachmias and Nachmias 1976; Orenstein and Phillips 1978; Ackroyd and Hughes 1981). There are also readers (Festinger and Katz 1953; Denzin 1970; Filstead 1970; Forcese and Richer 1970; Bulmer 1977; Bynner and Stribley 1979; Burgess 1982), and collections of essays (e.g. Becker 1971). Wakeford (1979) contains a useful guide to materials used in teaching sociological methods courses in British universities and polytechnics. One journal which has maintained a consistently high level of interest in ethnographic research methods is *Human Organization*. For a dose of mental hygiene, and one which is both amusing and instructive in its healthy scepticism, see Ford (1975). In the critical evaluation and employment of methods, scepticism (rather than cynicism) is a prime virtue.

R. F. Ellen

2 Theory, methodology and the research process

2.1 Specific features of social science

One of the striking differences between natural and social sciences is the amount of attention which their practitioners devote to methodology. Physicists, for example, have a conspicuous lack of concern with methodology: books and articles on methodological aspects of physics are very few and remain unread by most practising physicists; a physicist obtains a degree – and thus a license to carry out independent research – without being required to pass an examination in a course on scientific methodology. In contrast, social sciences abound in methodological treatises,[1] courses in methodology form a considerable part of the curriculum, a social scientist could hardly qualify without passing examinations in methodology, his or her own research will hardly be of any interest to fellow practitioners if it is not methodologically significant, and the practitioner's status within the professional community is directly proportional to contributions to the development of the subject's methodology: a single methodological treatise is bound to bring more fame and prestige than several published results of empirically oriented research. It is a logical consequence of this importance attributed to methodology that, unlike in physical and natural science, much research in social science is designed not to yield new knowledge but to develop and test

[1] Some of these are cited below, although the literature is staggeringly large and diverse. Among those that are not cited, but which are in various respects prominent, illustrative or of particular interest to anthropologists are: Brown (1963), Blalock and Blalock (1968), Manners and Kaplan (1968), Harré and Secord (1972), Gellner (1973), Ryan (1970, 1973), Willer and Willer (1973), Keat and Urry (1975). See also §1.6.

ETHNOGRAPHIC RESEARCH
ISBN 0 12 237180 1

new research procedures. Excessive effort is diverted in this way from substantive to methodological problems so that, in Kaplan's words, "we are forever perfecting how to do something without ever getting around to doing it even imperfectly" (Kaplan 1964: 24).

This may be so, but the "myth of methodology", as Kaplan calls it, "the notion that the most serious difficulties which confront behavioral science are 'methodological'" (Kaplan 1964: 25), has good reasons for existing. It reflects the relationship between the observer and the observed phenomena in the two types of science: in physical and natural science there is a clear separation of the observer from the observed phenomena, whereas in social science there is either no such separation at all or it is (at least to a certain extent) blurred. The preoccupation with methodology in social science derives basically from the effort to deal – in one way or another – with the influence of the subject on the object and with the consequences of this influence for the process of knowing and the knowledge yielded. The main methodological problems of social science have thus typically been epistemo-logical problems concerned with the nature of knowledge and understanding and their relationship (Phillips 1971, 1973).

Though there exists this difference in the basic attitude towards method-ology in natural and social sciences, in both of them theory and method are directly linked and closely intertwined. Research procedures in natural sciences vary with the investigated phenomena and are related to ideas about the nature of those phenomena, in other words, to the current theory con-cerning their constitution. The same relationship obtains in social sciences. There too, research procedures are related to the basic assumptions about the nature and constitutive features of social phenomena, i.e. to the current theory of man and society.

2.2 Specific features of social anthropology

Because all social sciences investigate basically the same phenomena and share the same ideas about their constitution, the boundaries drawn between them are often blurred and at times questionable. Although no distinction can be made between social anthropology and other social sciences either in terms of the substantive problems to which they address themselves or in terms of the theory to which they subscribe, both anthropologists themselves (e.g. Johnson 1978: 9) as well as philosophers of science (e.g. Feyerabend 1975: 249 ff.), consider the unique method of yielding data though long-term "participant observation" as distinguishing anthropology from other social sciences.

The genesis of this approach is generally traced back to Malinowski's pioneering fieldwork among the Trobriand Islanders, and the importance he himself ascribed to it is probably best signified by the fact that he led his students "to believe that social anthropology began in the Trobriand Islands in 1914" (Leach 1957: 24). However simplistic this view may be, the role of fieldwork in shaping modern social anthropology has generally been recognized (Chapter 3): Malinowski is seen both as the pioneer of modern anthropological fieldwork carried out through participant observation, and as one of the founding fathers of modern social anthropology. The publication of his private diary (Malinowski 1967) has to a considerable extent shattered the image of a sympathetic observer of native life and an enthusiastic participant in native activities, fully immersed and involved in native life and culture. But in fairness to him, it has to be said that descriptions by other anthropologists of their feelings in the field (Evans-Pritchard 1940: 13; Bowen 1956: 207; Mead 1977; §5.2) should horrify the advocate of the anthropologist's full immersion in the native culture to no lesser degree; others have escaped the wrath which the publication of Malinowski's diaries aroused probably only because they have been far less explicit about their feelings than he was. Moreover, and more significantly, his personal feelings towards the Trobrianders can be seen as affecting the validity of his data only when participation is seen as the main data yielding technique. As it will be argued later, it did not have this value for Malinowski and the generation of anthropologists following him. The criticism based on the revelations in his diary and levelled against the conduct of his fieldwork appears thus to be misplaced, as it derives from the value put on participation within the context of a different set of assumptions about the nature and constitution of social reality rather than upon those which guided Malinowski's research.

However justified the evaluation by our contemporary standards of the style of fieldwork carried out almost 70 years ago may be, the fact is that Malinowski's role in the history of social anthropology derives from having done something which had not been done before. It is not that he invented participant observation, as is often believed. There were many others before him who did social research through participant observation, though they were not professional social scientists (cf. Wax 1971: 21 ff. for a useful survey). It is also not that he was the first anthropologist who produced his ethnography through his long-term residence in the native community and through communicating with its members in their own language, as is again often believed (Chapter 3). What makes Malinowski's place in the history of social anthropology unique is that he was the first to use participant observation to generate specific anthropological knowledge. At the root of this achievement lies the fact that he joined together phenomena which until then had had an independent existence: the anthropological theory aimed at explaining the available ethnography and the production of this ethnography through participant observation.

That much is agreed upon. What seems to be less agreed upon is the direction of causal links between the theory and the method of production of ethnography subject to explanation in terms of this theory. This is not surprising given that the relationship between observations, experimental results, or simply "facts", on the one hand, and theory on the other (or between observational terms and theoretical terms), is an unresolved problem in the contemporary history and philosophy of science and subject of much controversy (cf. Scheffler 1972; Taylor 1972; Putnam 1975, 1978; Worrall 1978).

It is quite clear that the data which Malinowski collected during his fieldwork were qualitatively different from those on which earlier anthropologists had relied for constructing their theories. This led some to argue that Malinowski's theoretical thinking must have been profoundly influenced by the kind of data he had collected through his long-term close observation of the life of the Trobriand Islanders and through his communication with them in their own language. Such data were obviously in need of explanations for which the current anthropological theories of the evolutionists and diffusionists were inadequate (Jarvie 1964: 38–39). When discussing this effect of the new method of the production of ethnography on anthropological theory, Gluckman (1967: xii) has likened it to developments in other disciplines:

> Clearly, if new ranges of data are suddenly made available to a subject, its whole theory and perspective are bound to undergo a radical change. A new technique of observation may virtually create a new discipline, as Leewenhoek's improvement of the microscope, and later the creation of radio-telescopes did. I consider Malinowski's field researches had this effect on anthropology, partly because of his long residence in the Trobriand Islands, partly because he worked through the Trobriand language, and partly because his temperament led him to a deep involvement with the people he was studying.

This view probably reflects more a belief in the inductive method of anthropology than an actual relationship between the method of yielding ethnographic data and the theoretical framework within which they are explained. Contrary to this view, and more convincingly, it has been argued that Malinowski's philosophical views on the nature of scientific explanation and the general methodology of scientific research had been formulated long before he embarked on his fieldwork and that, in consequence, the particular method of field research which he adopted had been determined by his theoretical concepts of man, society and culture (Paluch 1981).

The specific style of fieldwork which Malinowski advocated and urged his students to undertake had become the hallmark of British social anthropology in spite of the fact that most of Malinowski's students soon abandoned his notion of a custom functioning in relation to biological and cultural needs, and embraced Radcliffe-Brown's notion of the function of an institution within a social structure and the needs of societies. This shift of theoretical emphasis did not affect the methodology of the intensive study of small communities through participant observation: both Malinowski's functionalism and Radcliffe-Brown's structural-functionalism are underpinned by synchronic analysis of the interrelations of institutions within the existing socio-cultural whole.

A considerable move from structural-functional explanations has been noticeable in anthropology since at least the 1960s. Given the close interrelationship between the fieldwork methods of participant observation and the structural-functional style of analysis and explanation, it seems at first sight paradoxical that the recent proliferation of approaches to theory has not generated a corresponding proliferation of methods, as Strathern (1981: 30) has noted. Later on it will be argued that this view is deceptive and that one can talk about the continuation of the production of ethnography through the method of participant observation only if this method is understood as referring to

> a characteristic blend or combination of methods and techniques that ... involves some amount of genuinely social interaction in the field with the subjects of the study, some direct observations of relevant events, some formal and a great deal of informal interviewing, some systematic counting, some collection of documents and artifacts, and open-endedness in the direction the study takes (McCall and Simmons 1969: 1).[2]

If one pays closer attention to the way in which ethnography is actually produced by the researcher through interactions in the field, clear changes are noticeable which reflect the changing theoretical orientation of the discipline.

The reason for the continuation of fieldwork through participant observation, broadly understood, derives to a great extent from the fact that such fieldwork is seen as a distinguishing and defining feature of social anthropology: however else anthropology could be defined, to do anthropology meant to study a specific community through long-term participant observa-

---

[2] Johnson (1978) provides a useful discussion of research procedures adopted in various theoretical orientations within this broadly understood notion of anthropological participant observation. More generally, on the "de-composition" of the concept of participant-observation, on its various interpretations and ramifications, together with other matters discussed in this chapter, see §4.3, §8.2.

tion. Correspondingly, doing fieldwork in this way became to be seen as a "unique and necessary experience, amounting to a *rite de passage* by which the novice is transformed into the rounded anthropologist and initiated into the ranks of the profession" (Epstein 1967: vii). This value of fieldwork is clearly signified in the pejorative sense in which the expression "armchair anthropologist" has been used, and it is echoed any time an anthropologist's lack of fieldwork experience is invoked in the evaluation of his theories (Lévi-Strauss occasionally has been subjected to this treatment). This view of the formative role of fieldwork experience resulted in construing the production of ethnography as an activity to a great extent independent of its subsequent analysis (§4.1, §4.2).

This in turn had specific consequences for the recent development of social anthropology. Whereas the emergence of functionalism and structural-functionalism had the hallmark of a revolution and has been presented as such (Jarvie 1964; Gluckman 1965: 24; Kuper 1973), the continuation of fieldwork through participant observation enabled the recent proliferation of theoretical approaches to evolve gradually.

A specific theory about the constitution of the investigated object does not only shape the method of investigation; it also defines research problems and directs the researcher's observation of specific aspects of the object deemed theoretically significant. As Kaplan (1964: 133) points out:

> we do not observe 'everything that is there to be seen'. An observation is *made*; it is the product of an active choice, not of a passive exposure. Observing is a goal-directed behavior; an observational report is significant on the basis of a presumed relation to the goal . . . Data are always *data for* some hypothesis or other; if, as the etymology suggests, they are what is given, the observer must have hypotheses to be eligible to receive them.

Anthropological fieldwork is no exception in this respect: here too no observation can be made unless the observer has a point of view which guides his selection and interest. But even if the anthropologist's fieldwork is guided by a theoretical interest in particular problems, much of it still remains "undirected" (Jarvie 1964: 215) and open-ended, at least to a certain extent. What is observed and recorded is not only determined by a particular interest but also by what goes on around. As the significance for a particular problem of the many events which take place can never be determined *a priori*, they get recorded, if only as background information or the context into which to fit the specific problems to be investigated. It is the experience of probably all fieldworkers that they collect more data than they subsequently draw upon for analysis (as well as not having collected data which would have been ideally needed). Typically, participant observation yields a "surplus" of data, or what Kaplan calls "cryptic data", i.e. "those which, in a given state of science, are hard to make sense of in the light of the theories current at the time" (Kaplan 1964: 134). It is this surplus, or these cryptic data, which has enabled many anthropologists to interpret and explain their ethnography within a different theoretical framework from that which guided its production in the field (or at least within a *modified* framework if "different" is too strong a word) (Gulliver 1971; Harris 1972). Principally, all that is required to accomplish the change in style and mode of analysis and explanation is a change in the decision on what, at the level of ethnography, should be construed as the problem to be explained, and what the context is in which the problem should be placed.

To talk about the continuation of fieldwork through participant observation in spite of the changing theoretical orientation of the discipline, as we have done so far, is possible only when participant observation is understood in the above-mentioned sense of a combination of more specific methods and techniques. When closer attention is paid to these specific methods and techniques, distinct changes in the style of fieldwork are noticeable.

First of all, during the development of participant observation, the techniques of "statistical documentation by concrete evidence" for which Malinowski called, have become more sophisticated and, correspondingly, the techniques of sampling, quantification of variables and the application of statistical calculations to quantified variables have become subject to methodological scrutiny. Secondly, new technological gadgets like tape-recorders, cine-cameras and video-tapes have become available to modern fieldworkers. Their use in the field and its consequences for the anthropologist's interaction with the subjects of research, as well as for the nature of data, have again been extensively scrutinized (§7.9).

Though these methods and techniques are important and though their use affects the character and validity of the data yielded, it has to be recognized that it is a specific theoretical interest which induces the fieldworker to adopt and develop certain kinds of technique. The adoption of specific techniques is thus merely one aspect of an overall methodological stance determined by theoretical ideas concerning the constitution of the phenomena investigated.

2.3 Participant observation and positivism

As already mentioned, participant observation is a unique blend of specific methods and techniques developed within the context of structural-functional anthropology. In its turn, structural-functionalism owes much of its theoretical orientation to Durkheim. It was his assertion that social facts are external to individuals and exert pressures on them that lies at the root of the whole theoretical development of British social anthropology. Durkheim's (1897) dictum that "social life should be explained not by the notions of those who participate in it, but by more profound causes, which are unperceived by consciousness" has been strictly followed. When he asserted in his *Rules of Sociological Method* that "sociology does not need to choose between the great hypotheses which divide metaphysicians" (Durkheim 1964: 141), he did not mention that this is so mainly because it had already made the choice. What that choice had been was probably best epitomized later by the title of one of Radcliffe-Brown's books: *The Natural Science of Society*. Aiming at being scientific, social anthropology should not be, according to Radcliffe-Brown (1952: 192):

> concerned with the particular, the unique, but only with the general, with kinds, with events which recur. The actual relations of Tom, Dick and Harry or the behaviour of Jack and Jill may go down in our field note-books and may provide illustrations for a general description. But what we need for scientific purposes is an account of the form of the structure,

that is "the network of actually existing relations".

In Radcliffe-Brown's conception of social anthropology, the activities of individuals are used mainly as incidental descriptions, not as the object of study and explanation. The analysis is aimed at elucidating how society is structured, and for this purpose specific actions of individuals are mainly irrelevant. It is this particular aspect of his approach which became the main object of criticism by the proponents of situational analysis or the extended-case method (§4.9, §8.5). This criticism pointed out that the actual cases of interaction encountered in the field had not been used as the very stuff of analysis, but merely as apt illustrations for the structural schemata devised by the anthropologist. The extended-case method has concentrated analysis on the interactions themselves, with the assumption that the regularity of social relationships, i.e. structure, should be both elicited from such concrete cases and demonstrated on them (e.g. Turner 1957). Concentrating on concrete situations which can be encountered "on the ground" and making them the units of analysis brought a considerable methodological improvement. Theoretically, situational analysis does not, however, represent a departure from the positivistic paradigm of social anthropology and its "wish to establish the natural science of society which would possess the same sort of logical structure and pursue the same achievements as the sciences of nature" (Giddens 1976: 13).

But anthropology did not emulate natural science only in its logical structure and definition of achievements. If the science of society was to be built along the lines of natural science, then its subject matter also had to be conceived as having the same existential status as that of natural sciences and consequently its data had to be conceived as akin to the data of the scientist. Its subject matter had to be real, factual and it had to consist of empirical phenomena which exist somewhere "out there" in the world. It also had to be available to the anthropologist basically in the same way as the subject matter of natural science is available to the scientist: through direct observation which was seen as giving to the researcher something like sense-data, i.e. information about the social world gathered through sense experience. Non-observable relationships were conceived of as non-problematically following the proper arrangement of observable phenomena. These phenomena, when properly combined, fell into place and revealed "the form of social life" (Radcliffe-Brown 1952: 4). This "jigsaw-puzzle conception of social structure" (Lévi-Strauss 1960: 52) made social structure itself, through the empirical availability of its components through observation, a sort of empirical reality.

It was thus two mutually interrelated factors – the current ideas about the constitution of social life and the aim of anthropology of being scientific – which led to the establishment of fieldwork as the means of providing the anthropologist with the data needed for scholarly endeavour.

It was the aim of giving anthropology its scientific status which led to the requirement that anthropologists, like scientists, gather their own data instead of relying on information supplied by laymen without professional training. This was advocated as early as 1913 by Rivers (1913: 6) in his distinction between "survey work" and "intensive work":

> The essence of intensive work . . . is limitation in extent combined with intensity and thoroughness. A typical piece of intensive work is one in which the worker lives for a year or more among a community of perhaps four or five hundred people and studies every detail of their life and culture; in which he comes to know every member of the community personally; in which he is not content with generalized information, but studies every feature of life and custom in concrete detail and by means of the vernacular language. It is only by such work that one can realize the immense extent of the knowledge which is now awaiting the inquirer . . . It is only by such work that it is possible to discover the incomplete and even misleading character of much of the vast mass of survey work which forms the scientific material of anthropology.

Here we have a clear formulation of the method of conducting fieldwork which subsequently became known as participant observation. To illuminate the way in which this has been affected by the theoretical development of anthropology, it seems most profitable to concentrate on the relationship between the two activities which gave the method its name: participation and observation. As one obviously can participate without systematically observing for the purpose of gathering data for future analysis, and as one can observe without necessarily participating in any meaningful sense, the analytical dissection of participant observation into these two concurrent activities seems to be fully justified.

It is obvious that every kind of anthropological research, even one conducted through highly formal techniques like research survey procedures, structured interviews, formal experimental designs, etc., requires a certain, if minimal, amount of participation on the part of the researcher. Exchange of greetings, introductions, explanation of what is involved in the research and what tasks the subjects are expected to perform require some interaction between researcher and subjects, and hence a certain participation in their lives. Participation as understood in the phrase "participant observation" obviously means more than that. But what precisely?

To Malinowski participation did not mean more than camping right in the native village. His remarks about his participation in the life of the Trobrianders are scarce and both his monographs and his diary do not indicate that he was involved with them in any socially meaningful sense. He says:

> Soon after I had established myself in Omarakana (Trobriand Islands), I began to take part, *in a way*, in the village life, to look forward to the important or festive events, to take personal interest in the gossip and the developments of the village occurrences, to wake up every morning to a day, presenting itself to me more or less as it does to the native (Malinowski 1922: 7; emphasis added).

Further to the subject of his participation, he mentions only his capacity for enjoying the natives' company and sharing some of their games and amusements (ibid.: 8). For other researchers participation involves more than that. According to Florence Kluckhohn (1940: 331), it involves "conscious and systematic sharing, insofar as circumstances permit, in the life-activities and, on occasion, in the interests and affects of the group of persons". To Wax (1971: 6, 7, 15), participation means associating with the people one studies in all their activities. According to her, it means involvement "in the various kinds of human social relationships" (ibid.: 16) and "establishing some reciprocal relationships" with the people one studies (ibid.: 20). To McCall and Simmons (1969: 3), participation means "repeated, genuine social interaction on the scene with the subjects themselves".

Participation in the lives of the subjects during fieldwork thus seems to be a matter of degree, ranging from simply living in the community in which research is being conducted to emulating the natives as fully as possible. The aims of participation equally vary from case to case. It is suggested that most of them derive from demands and traditions different from the pragmatic demands of data gathering, which are themselves ultimately determined by the closely inter-related theoretical notions current in the discipline and its aim of achieving scientific status.

First, participation in the sense of living with and as the people one studies, is part of the romantic notion of "fieldwork culture" (Freilich 1977: 4 ff.), a part of the mystery of fieldwork as an initiation ritual in which the student of anthropology dies and a professional anthropologist is born (§1.1, §1.4).

Secondly, some anthropologists participate in the lives of the people they study primarily from a desire for a socially meaningful involvement in their affairs, rather than from the pragmatic demands of data gathering. Their participation stems from their own social philosophy or their ideological commitment to the cause of those they perceive and define as oppressed, exploited or discriminated against.

No more needs to be said here on the demand on participation deriving from the traditional "fieldwork culture" of anthropology or from the conception of the anthropologist as the champion of the underdog and moral protagonist of native societies. Methodologically more important are the aims of participation deriving from the pragmatic demands of data gathering.

From this point of view, the important aim of participation as part of anthropological fieldwork is to make it possible for the anthropologist to carry out research at all. A certain amount of participation in the activities of the people studied (its degree depending on the circumstances of the work and the problems researched) enables the anthropologist not only to observe the actual behaviour of subjects but to apply effectively all other possible research techniques: to conduct both informal and structured interviews, to collect statistical and census data, to carry out psychological tests, to photograph, film or tape-record rituals and ceremonies, or whatever other techniques s/he chooses to apply. From the point of view of making such complex research possible, the value of participation in typical anthropological fieldwork in an alien and most probably pre-literate and pre-industrial society differs from its value in typical sociological research in Western industrial society. In pre-industrial society the role "researcher" is meaningless to the people studied. In this situation participation aims at having the role of "a friend", "harmless foreigner", "our European", or something similar, ascribed to the anthropologist, who is anyway forced into a certain amount of participation because, as Wax points out, "respondents or potential respondents are judging him by what he *is* and what he *does*, and not by what he says he is or says he will do" (Wax 1971: 87).

When doing research in a Western industrial society, researchers are often presented with an opposite problem. They will be understood if perceived as researchers, but to have this role ascribed to them they have to be seen to be doing research. Filling in questionnaires is usually readily understood as being research, participation is not; depending on the circumstances, it either indicates that the researcher is spying on the subjects of his/her research for some ulterior reasons or is having a nice time at the taxpayer's expense. Wax (1971) describes how she was forced to use formal interviews when studying confined Japanese Americans during the Second World War to establish her role as a researcher.

The conception of participant observation as the standard manner in which anthropological fieldwork is conducted derives directly from the nature of typical anthropological research in a pre-industrial society. The anthropologist is, on the one hand, forced to conduct research through participant observation because in the culture of the people being studied there does not exist a role "researcher" into which s/he may slip; and, on the other hand, s/he is able to conduct research in this manner because the situation is so defined that the researcher is perceived by those studied as superior. It depends more on whether s/he wants to participate in the lives of those studied, and in the course of participation to observe their conduct, than on whether they allow the researcher to participate in their lives and observe them. Once anthropologists embark on research where their status is not automatically perceived as superior to those they study, the insistence on participation becomes problematic. Powdermaker (1966) found that she was unable to participate in, or even observe, much of the life she studied in Holywood and that her informants would talk to her only in formally arranged appointments.

It is clear from the preceding discussion that the real aim of the anthropologist's participation in the lives of subjects during the course of fieldwork is not to yield any research data; it is at best a means to carrying out the proper research activities which yield such data. Whatever other research techniques the anthropologist might employ during fieldwork, the main data yielding procedure is observation. The role of the researcher involved in this kind of fieldwork could best be described as that of a participant observer.

The importance ascribed to observation as the main data yielding procedure derives directly from anthropology's ideas about the constitution of its subject matter which, like the subject matter of natural science, should be directly observable, as well as from its insistence on empirical scholarship characteristic of science. This notion can be traced back to Malinowski's requirement of "the description of the imponderabilia of actual life"

(Malinowski 1922: 18) as one of the three types of evidence the anthropologist should gather in the field:

> In working out the rules and regularities of native custom, and in obtaining a precise formula for them from the collection of data and native statements, we find that this very precision is foreign to real life, which never adheres rigidly to any rules. It must be supplemented by the observation of the manner in which a given custom is carried out, of the behaviour of the natives in obeying the rules so exactly formulated by the ethnographer, of the very exceptions which in sociological phenomena almost always occur (Malinowski 1922: 17).

The demand for direct observation by the researcher, instead of relying on informants' reports, derives from the notion of analytical objectivity in anthropology as science. This has been clearly formulated by Van Velsen (1967: 134):

> Informants' statements . . . should be treated as the historian treats his sources: they are, that is to say, value judgements and should, therefore, be considered as falling within the category of data referred to . . . as observed behaviour. In other words, such statements should not be used as if they were objective, analytical observations by outsiders. The sociological evaluation of actions and other behaviour is the anthropologist's job, and the sociological evaluation of the same actions, etc., may well be very different from their social evaluation by local informants. After all, one cannot expect untrained informants, be they Bemba headmen or white-collar workers in London, to present the anthropologist with sociological analyses of behaviour observed in their respective communities. To do so would be to assume, as many laymen do, that to be a member of a community is to understand it sociologically.

The very process of observation implies the sensory perception of things that can be seen, heard or felt. And there is only one kind of phenomena in the whole realm of social life which is observable in this sense: specific actions of individuals (be they physical acts or speech acts). Given the limited scope of things which can truly be observed, the insistence on direct observations and refusal to be content with informants' statements derives, on the one hand, from a concern with scientific rigour, and has, on the other hand, led to an exclusive concern with interactions (social structure) and an intentional neglect of notions (culture). This logical circle in which positivistic anthropology has enclosed itself has been aptly expressed by Radcliffe-Brown in his insistence that it is possible to have a science of society, but impossible to have a science of culture.

It has been argued that the anthropologist's direct observation has been envisaged in positivist anthropology as the primary method of data gathering, and that carrying out this observation through the anthropologist's participa-

tion in the social life of the people observed has been dictated by the pragmatic necessities of making this observation at all possible. From the point of
view of the theoretical notions of positivist anthropology, as well as from the
point of view of its aim of being scientific, participation is a highly problematic and, from the methodological point of view, rather unnecessary part of
fieldwork. Far from being, in itself, methodologically desirable, the anthropologist's participation violates the separation of observer and observed
phenomena (characteristic of observation in natural sciences), and thus goes
contrary to the scientific canons of anthropology. It makes problematic the
basic belief that the object of the research exists in an external world in the
same sense as the object of any science does and that, like any other object of
scientific inquiry, it has knowable characteristics which must not be disturbed
in the process of observation. The anthropologist's presence affects the
observed situation so that observation of it cannot be assumed to be completely free of distortion, thus rendering problematic the basic aim of
reporting "what is really happening there".

One of the conscious efforts at eliminating this disturbing effect of the
researcher's participation can be seen in the insistence on the length of a stay
in the field (§7.8): the longer the stay, so it is argued, the less disturbing the
effect of the ethnographer. The longer the stay, the more likely s/he is to
resemble the researcher who observes a situation through a one-way mirror
without being seen and thus without disturbing what goes on. Witness to the
awareness that participation is a highly problematic part of the way in which
the anthropologist gathers basic data is the fact that discussion of participant
observation has been for the most part concerned with the technical problems
it presents. The strictly methodological problems of participation which are
usually the subject of discussion, are also concerned with the effects participation has on the observed subjects. The typical issue of discussion is the role the
participant should assume so that it imposes minimal limits on observation
and minimally disturbs the observed situation.

2.4 Participant observation and the interpretative paradigm

We have so far discussed three aspects of positivistic social science: its ideas
about the constitution and existential status of its subject matter (or its theory
of social phenomena), its ideas about itself (or its theory of social science as a
science) and its ideas about the way its practitioners gather their data (or its
research methods). We started the discussion by asserting that its ideas about
the constitution of its subject matter and its ideas about its status as a specific

branch of specialized knowledge are closely interconnected. The discussion tried to show that its research methods are equally closely connected with the former two sets of ideas forming together a paradigm in the sense of "the entire constellation of beliefs, values, techniques and so on shared by the members of a given community" (Kuhn 1970: 175).

The interdependence of theory, research methods and the knowledge of social phenomena which the research yields has been generally recognized. Talking about sociology, Phillips (1973: 78) has formulated the problem in the following way:

> A fundamental problem in sociology is that what we know about social be-
> haviour (and, indeed, most social phenomena) is dependent on our methods for
> studying it, while our methods for studying it depend on what we know about
> social behaviour. So in order to know more about social behaviour and inter-
> action, we need better methods; and to obtain better methods, we need to know
> more about behaviour and interaction. This constitutes a kind of vicious circle
> which we must break out if the social sciences are to move beyond their present
> stage of development.

Breaking this vicious circle depends on deciding whether our priority lies with theory or with methods of research. This question has been asked by philosophers of science, and Peter Winch (1958) answers it by arguing that many of the important theoretical issues in social science can only be settled by *a priori* conceptual analysis rather than by empirical research. This position is in line with that of other philosophers of science who argued that progress in science does not derive from new research techniques or from accumulation of new data but from looking at existing data in a new way (Polanyi 1958; Kuhn 1970; Feyerabend 1975).

This kind of conceptual re-thinking or philosophical reflection on social science has led to an explicit critique of the positivistic paradigm. Such a critique represents a radical departure in all three spheres of ideas. In the sphere of ideas about the constitution of social phenomena there is a clear move from the theory of social facts as things to the theory of social facts as constructions. This theory holds that facts exist only within a frame of reference, that "there is no such thing as 'pure experience', no such thing as 'facts' that are recorded directly 'from nature'. Theoretical presuppositions are always involved" (Phillips 1973: 115) and, in consequence, "a fact" is always the product of some interpretation. In the sphere of ideas about the status of social science, there is a distinct move away from the notion of the methodological unity of the natural and social sciences towards the realization that the social sciences require different methods of inquiry from those used in natural science investigations due to the subjective quality of social

phenomena. And in the sphere of ideas about the way the researcher gathers his data, there is a distinct move from the notion of observation as the primary method of data gathering. These three sets of ideas are again closely inter-related and we are thus justified in talking of a new paradigm replacing, or at least challenging, the paradigm of positivistic social science.

The new paradigm, in sociology found in symbolic interactionism (Blumer 1969; Stryker 1980), ethnomethodology and phenomenological sociology (Cicourel 1964; Garfinkel 1967; Glaser and Strauss 1967; Douglas 1971; Filmer *et al.* 1972), is usually talked of as interpretative social science, this label deriving from the realization that a social scientist is not simply observing things, but interpreting meaning. This notion, in its turn, stems directly from the idea that the social world is not a real objective world external to man in the same sense as any other objectively existing reality (natural world) but is a world constituted by meaning. It does not exist independently of the social meanings that its members use to account for it and, hence, to constitute it. Social facts are thus not things which can be simply observed.[3]

In discussing observation, it has been mentioned that there is only one kind of phenomenon in the whole realm of social life which is observable: specific actions of individuals. Even these specific actions are not observable in their entirety. What are observable are physical movements; it is their meaning which transforms physical movements into the actions of individuals. Obviously, the same physical movements may have different meanings (cf. Anscombe 1957: 40 ff.); what limits the scope of possible meanings are preconceived criteria which the actors have and which they apply to their actions. Applying criteria is not, of course, observation; actions are therefore experienced by the actors simultaneously through senses and through thought processes (cf. Gorman 1977: 6). The researcher cannot apply to the observed actions his own meaning which may be different from that of the actors, for that would ultimately lead to the distortion of the observed reality, or, more precisely, would result not in studying the existing social reality, but in creating it. Unless the researcher wants to distort the meaning of the observed actions, the actions have to be available in the same way as they are to the actors. And as they are not available to the actors by simple sense experience,

---

[3] Cohen (1978) admirably illustrates such difficulties in relation to the ethnographic identification of "community". He concludes that the senses in which the concept is locally employed, and therefore sociologically relevant, can only be discovered through the generation of a "theory" grounded in the particulars of specific fieldwork experience and practice, rather than through attempts to apply any pre-conceived comparativist definition. See Bensman and Vidich (1965) on the importance of unsystematic middle-range theory in field research and Glaser and Strauss (1967) on "grounded" theory in general.

they cannot be so available to the observer. Like the actors, s/he too has to experience them simultaneously through senses and through thought processes. This means that observation cannot be the only or even the main process through which data are gathered.

Since people do not behave in isolated actions but in interactions with and towards others, they must make their actions meaningful to others. This means that the criteria for ascribing meaning and the ways of interpreting actions must be known and shared by them: otherwise social life would not exist. As long as members of society are able to comprehend their actions, i.e. if they non-problematically and automatically ascribe meaning to them, the researcher should also be able to comprehend them in the same way in which they do: in the course of their practical accomplishments.

A logical corollary of the theory of social world as constructed through its members' interactions and as intrinsically meaningful is thus a theory of its cognitive availability through participation in the construction of its meaning, which implies a research procedure in which the notion of participation in the subject's activities replaces the notion of their simple observation as the main data yielding technique. It is a research procedure in which the researcher does not participate in the lives of subjects in order to observe them, but rather observes while participating fully in their lives. Such a procedure defines the role of the researcher not as that of participant observer but of observing participant, and it consciously eliminates the distinction between the observer and the observed phenomena and thus radically departs from the scientific attitude of the positivistic paradigm.

The roles of the participant observer and of the observing participant are not simply mirror images of one another. While for the participant observer the observation is the *main* data gathering method, which is usually complemented in fieldwork by the use of other research techniques, for the observing participant active participation in the social life studied is virtually the *only* data gathering method. If this research attitude is vigorously adopted, the researcher should refrain from asking even simple questions if they are ones which a subject would not ordinarily ask. Asking questions prompted by the researcher's current theory means forcing the subjects to adopt an attitude which is not ordinarily part of their praxis (Bourdieu 1977) and thus shapes the social reality being studied.

The epistemological problems of and reasons for the researcher's participation as the main data gathering method have been extensively discussed before (Cicourel 1964; Bruyn 1966; Blumer 1969; Spradley 1980). Suffice it to say here that the crucial importance of the researcher's active participation in the lives of subjects derives from making it possible to learn the meaning of

actions: through living with the people being studied, and through the necessity of behaving towards them and communicating with them, s/he comes to
share the same meanings with them in the process of active participation in
their social life. The social scientist's research means, in this sense, socialization into the culture being studied, culture being understood here in
Goodenough's sense as "whatever it is one has to know or believe in order to
operate in a manner acceptable to its members and to do it in any role that
they accept for any one of themselves" (Goodenough 1966: 36).

In this respect, active and conscious participation in the social life of the
people one studies acquires meaning in addition to being the chief method of
discovering the actors' cultural meaning, their emic rules and their logic. In
positivistic social science, the problem of the validation of the researcher's
account of the studied culture and society was resolved by the scientific
community through the application of the criteria they had themselves
devised for this purpose. Whether the actors themselves would subscribe to
the account was not the issue: they are not professional social scientists
trained to understand their society sociologically. The account would not
have been falsified if the actors rejected it. In interpretative social science, the
validity of the researcher's account is not tested against the corpus of scientific
knowledge. It is tested against the everyday experience of the community of
people. However, approval of the subjects with the published analysis of their
culture or social relations (an argument often resorted to by anthropologists)
is not the proof of the accuracy of the account: in interpretative as in
positivistic social science it may only indicate that the actors are capable of
playing the social science game or that they have acquired a false consciousness – in this case an anthropological or sociological one. The subject as
a check on the accuracy of the researcher's knowledge or comprehension can
only be used during the actual process of data collecting. Researchers must
demonstrate to the actors that they can talk as they talk, see as they see and do
as they do. This implies a social science methodology which, in yet another
way, radically departs from the scientific attitude of positivist social science:
it confounds data gathering (research) with analysis. It is a methodology
where the notion of success replaces truth as criterion of validity and where
the participation of the researcher becomes the main means of verifying his
account. If able to interact successfully with and towards subjects, i.e. if able
to pass for a member, the anthropologist's understanding of their culture is
right. And it is, of course, the group which defines the terms of acceptance and
rejection of new members.

2.5 The anthropologist's problem

More and more anthropologists are currently accepting the theory of the social world as formulated within the paradigm of interpretative social science. Not only the recent interest in ethnoscience and cognitive anthropology, but primarily the preoccupation with meaning and the growing strength of what is usually called symbolic or semantic anthropology clearly point in that direction. But having accepted the theory of a social world constituted by meaning, the anthropologists doing fieldwork in an alien society and culture are faced with an added difficulty in their research which need not be faced, or least not faced to that degree, by researchers who study a group or community which is part of their own society.

It is patently obvious that an anthropologist studying some alien, pre-industrial people is less equipped to cope with the methodological demands on research embedded in the paradigm of interpretative social science than is a researcher who works in a complex, and preferably natal, society; for s/he is much more severely restricted in any effort to become its member. Successful membership requires the availability of a role which the observing participant can assume in the course of research. Complex societies offer a variety of jobs, occupations or statuses into which a researcher may be admitted and thus pass for a member of the group or community studied (Gillin 1949). In classic anthropological research, where a Western anthropologist studies non-Western peoples, it is more difficult to pass for a member: s/he lacks the necessary genealogical qualifications for entering into many roles, speaks the language inadequately or with an accent thus proclaiming a status as foreigner, and is effectively prevented from becoming a member by the non-socially defined criteria which enter into the definition of membership and which remain basically non-manipulable (like the colour of the skin). Given these obvious difficulties, it is not surprising that the recent change in the theoretical orientation of anthropology is accompanied by a search for a research method genuinely adopted to the anthropological study of meaning, and that anthropologists reflect on the way of conducting fieldwork more than ever before. It seems that there are basically two ways of overcoming the difficulties.

The first one is obvious: to do research in a group or community of which it is not impossible for a researcher to become a member. It might be expected that with the growing commitment to the study of meaning, we shall see Western anthropologists embarking more and more on research in Western society, whereas research into non-Western societies will be carried out more by anthropologists who are themselves members of these societies or can more easily pass for members than their Western colleagues.

The second possibility consists in realizing what precisely are the disadvantages which the anthropologist who wants to become an observing participant in the life of an alien, pre-industrial society faces in comparison with his or her counterpart participating as a member of a specific group or community in their own society, and in consciously exploiting these disadvantages to heuristic advantage. The obvious predicament of the anthropologist is performative inadequacy, and the main problem to which anthropologists currently address themselves is how their inevitable incompetence in the native culture can be turned into their research tool.

A solution to this problem lies in realizing that it is the social relationship of the researcher and subject which has the status of the research method, and in recognizing the reflexive nature of this relationship (Rabinow, 1977). In other words, it is the interaction between anthropologist and host which can be used as the method of yielding data and developed into the main research tool.

As the anthropologist does not know at the beginning of fieldwork the normative rules and constraints of the culture studied, s/he is bound to disregard them in encounters with informants. This should inform more directly about what is and what is not fundamental to their social life than observations of their everyday activities which follow the normative constraints but do not make them explicit.

The anthropologist can take advantage of a performative inadequacy not by trying to accomplish the imposible task of becoming a member, but by systematically exploiting the fact that s/he is not a member and acting on the basis of his or her own cultural rules to find out to what extent they differ from those of the actors. In the long run such a systematic violation of native rules is obviously inadvisable if for no other reason than because it might ultimately lead to the interruption of any interaction on the part of the natives and thus jeopardize the researcher's whole endeavour. But at the beginning of fieldwork it is inevitable and can be taken advantage of; for nowhere is the process of constructing social reality as apparent as during the beginning stages of fieldwork, when the anthropologist is being accepted and defined by the members of the group being studied. Though s/he starts immediately establishing relations with them, the exact contents of these relations are not known beforehand, either to the anthropologist or to his hosts. They have to be created by some sort of mutual consent, through successful and failed encounters. Since both the anthropologist and the people studied live, at least during the beginning stages of fieldwork, in different worlds, there is no mutuality of meaning between their actions and their norms, they do not share or even know each other's "cognitive maps". In this process of mutual discovery, both the researcher and his subjects are inquirers: they are both engaged in the work of finding out the meaning of events, activities and situations they are engaged in and confronted with. Exploiting this situation to methodological advantage means realizing that there is less difference between the researcher and those studied than we would like to believe, and that "those being studied are also avid students of human relations; they too have their social theories and conduct their investigations" (Gouldner 1970: 496). When the anthropologist discusses with the actors "what is going on" in the search for the meaning of the encounters in which they are jointly engaged and the situations they are jointly confronted with, s/he is engaging with them in negotiating meaning. Through this process, a competence at meaning construction equal to theirs is gradually acquired. Although never a member of the society studied, s/he must acquire a competence at meaning construction equal to that of a native if s/he is to understand in any intelligible way. To do so, s/he must participate in the lives of subjects in the sense of actively interacting with them, for only through interaction can we gain any insight into the meaning construction in the culture studied.

The level of achievement in participation, i.e. the extent to which an ethnographer really becomes a member of that culture, has no direct bearing on the investigation. What is of significance is competence of introspection and ability to reflect on experience. So far, the anthropologist's own experience has been conspicuously absent from anthropological monographs whose main subject is what the natives do (or, more accurately, what the anthropologist understands them doing), and how they perceive, organize and manipulate their world (or, again, more accurately, how the anthropologist has understood them perceiving, organizing and manipulating their world). And yet, the anthropologists' experience has always played a major role in their analyses. Every anthropologist who has done fieldwork as a participant observer and subsequently analysed data in the process of writing up has realized that these are not only field-notes, but also experience which has never been written down and whose importance only becomes obvious later (§8.1). This was the main reason why it was possible for a number of anthropologists to explain their data within a different theoretical framework than the one which guided their fieldwork: they were able to fill in the gaps in their data with their memories and recollections of events and occurences not recorded at the time. It is precisely this unwritten experience which is acquired in an increased amount during the process of the researcher's data gathering as an observing participant. In this respect, to a certain degree, the anthropologist's experience as a participant observer of some alien society and culture approximates the experience of an observing participant, although for the reasons mentioned above it differs, certainly in degree if not in kind, from the experience of the researcher working at home (§5.5). An inevitable performative inadequacy can become methodologically significant if treated as basic data; not experience with the alien culture after it has been learned, but experience gained in the process of learning it. This inevitably requires an increased awareness of the process of participation and logically leads to the need to analyse more fully than ever before the researcher's own experience (Kaplan 1964: 141 ff.; Wilson 1970; B. S. Phillips 1971; Phillipson 1972; Phillips 1973: 164; Okely 1975).

The fact that the anthropologist's performative inadequacy is methodologically significant, and that it is the main issue of the current discussion of the problems of fieldwork, indicates more than anything else anthropologists' awareness that their fieldwork has to change to reflect recent developments in the anthropological theory.

Ladislav Holy

3 A history of field methods

3.1 Introduction

Fieldwork holds a position of crucial importance in modern anthropology. Through fieldwork the discipline is supplied with new ethnographic information, on the basis of this material theories are developed and ideas are tested; by "doing" fieldwork new anthropologists undergo a *rite-de-passage*, living and participating in alien cultures and thereby are admitted into the discipline (Chapter 4). But fieldwork has not always held such a prominent position. There was a time when anthropologists only knew other cultures through the secondary accounts of travellers, missionaries, administrators and others. The process by which anthropologists themselves became involved in the study of other cultures did not occur suddenly; nor was it the invention of a single person or of one particular anthropological tradition, in spite of certain well established claims to the contrary. The development of fieldwork, and of the technique associated with such research, was a gradual process. It went hand in hand with other developments both within and outside anthropology: a more critical approach to ethnographic sources, changing theoretical interests, the increasing professionalization of academic anthropology and the easy access to remote areas of the world due to the expansion of European colonial control and improved methods of communication.

The aim of this chapter is to provide a short history of the development of fieldwork traditions in British and American anthropology. In concentrating on these two traditions I do not mean to imply that they alone were responsible for the development of fieldwork methodologies, nor that anthropologists in France, Germany, the Netherlands, Scandinavia or Russia, to name but a few of the other major traditions, did not develop distinctive

ETHNOGRAPHIC RESEARCH
ISBN 0 12 237180 1

methods and contribute to ethnographic traditions in Britain and America. Limitations on space have not allowed me to widen the discussion.[1] Nor are methodologies examined in any real detail. Instead, the broad outlines of the traditions are discussed in general terms, within the contexts of the development of the discipline itself.[2]

3.2 Institutions and questionnaires

Until recently anthropology has been largely Eurocentric, concerned with exotic and distant cultures. This concern is ancient in European thought, certainly older and wider than the existence of a specialized subject such as anthropology. Rowe (1965) has argued that the Renaissance rediscovery of European classical antiquity prepared Europeans to accept the diversity of alien cultures encountered from the sixteenth century onwards during the age of reconnaissance and conquest in the Americas, Africa and Asia. The initial period of European expansion and the contemporaneous development of printing meant that many accounts of other cultures were published. These were utilized in a number of ways: to extend literary themes and to enhance philosophical discourse, but rarely as sources inherently interesting and requiring further information and analysis.

Most of the information on other cultures collected between the sixteenth and nineteenth centuries was obtained in an unsystematic fashion; the exotic and peculiar was valued and often elaborated upon, while most material was recorded after only a cursory acquaintance with other cultures. The systematic collection of specimens, careful recording of observations and synthesizing of sources first occurred in the natural sciences, particularly botany, from the middle of the eighteenth century onwards. These new methods are most apparent in the work of European voyagers to the Pacific in the late eighteenth

[1] One might briefly mention here the Germanic tradition of fieldwork with an emphasis on the collection of not only information but also museum specimens and the use of the material to construct regional hypotheses (Heine-Geldern 1964); the French ethnographic work with a strong emphasis on fieldwork, particularly under the influence of the Durkheimians (see the manuals of Mauss (1947) and Griaule (1957)); the Dutch tradition with its emphasis on the academic training of colonial officials not only in anthropology but also in the language and classical literatures of the areas of research (Ellen 1976). On the formation of other national anthropologies see Gerholm and Hannerz (1982). For discussions of the development of field methods in sociology see, for example, Madge (1963) and Easthope (1974). Wax (1971: 21–41) looks briefly at the development of the fieldwork concept in both sociology and anthropology.
[2] Changes in fieldwork methods in recent years have been associated particularly with technological innovations. These have a long history but have not been discussed here; for a good outline see Rowe (1953), though a great deal more information could be added to his account.

century. Rich collections, particularly of the flora, were brought back to Europe and detailed accounts of Oceanic cultures were published. While many of these accounts deal with odd and peculiar customs, the reflective comments of the explorers as they compared their own culture and age with the new worlds they encountered hint at the critical and comparative anthropology to come.

Baudin's French naval expedition to Australia (1800–1803) followed the pattern of earlier Pacific scientific voyages. Before leaving France, Baudin requested the newly formed *Société des Observateurs de l'Homme*, the first "anthropological" society, to provide guidance and instruction to the expedition. Georges Cuvier advised on the collection of anatomical data and Joseph-Marie Degérando provided instructions on recording of the customs of life of other cultures (Stocking 1969; Moore 1969). Degérando's instructions included a critique of earlier work, a consideration of the difficulties and essential skills needed to collect information and instruction on the categories of information to be recorded, often in the form of questions (Degérando [1800] 1969). Though the expedition failed to return with either the type or quality of information Degérando had suggested (Moore 1969), the instructions influenced the compilation of later questionnaires which were to play an important part in nineteenth century ethnographic inquiry (Urry 1972).

Anthropology, or more correctly ethnology, was established as a recognized field of study in the 1840s both in America and Europe.[3] "Science", as an organized activity, was just being established and common interests and concerns in the advancement of scientific research were recognized. Older scientific societies were reformed, new specialized institutions were founded and conferences were arranged to bring together the leading scholars in various fields. People interested in ethnology were actively involved in these developments and a number of learned societies were founded (Société Ethnologique de Paris 1839–1848, Ethnological Society of London 1843–1871, American Ethnological Society 1842–1870). The subject was included in the meetings of the wider scientific community (for instance at the gatherings of the British Association for the Advancement of Science).

These ethnological institutions not only encouraged debate about ethnological issues but also actively promoted the collection and publication of

---

[3] Anthropology in both Britain and America had included more than just cultural or social anthropology: human biology and archaeology have played an important role in the subject. This was particularly so in the nineteenth century when the unity of the study of man was more apparent than it is today and when terms such as ethnology, ethnography and anthropology possessed somewhat different connotations. On the development of the original term "ethnology" see Stocking (1973).

new information about other cultures. As the major concern of early ethnologists and later anthropologists was to compare the customs and institutions of disparate cultures, either to trace their history and diffusion or to establish the laws of cultural evolution, it was essential that information from as wide a variety of cultures as possible be recorded. The need to secure more information was seen as urgent, for it was believed that many cultures and "races" were on the verge of extinction. The view that the task of anthropology was to salvage information on other cultures (cf. Gruber 1970), a view still prevalent today, influenced not only how "facts" were collected but also the types of information recorded. The basic strategy to secure information was to send out comprehensive lists of queries in the form of questionnaires.

The use of questionnaires to elicit information on other cultures antedates the foundation of the ethnological institutions (Fowler 1975). The foundation of the institutions, however, provided a firm base for the construction and distribution of questionnaires. Many of these were based upon earlier queries but the most important questionnaire issued in the nineteenth century was the [Royal] Anthropological Institute's *Notes and Queries on Anthropology*. This work, and its predecessors issued by the Ethnological Society, was produced with the assistance of the British Association and appeared in six editions between 1874 and 1951. The changes in the various editions clearly reflect alterations, refinements and improvements in the methods of ethnographic inquiry, as simple questions gave way to detailed instructions. *Notes and Queries* was also to have a considerable impact on the standard of ethnographic inquiry up to 1914 (see Urry (1972) for details).

Individual anthropologists with particular interests also issued questionnaires in the nineteenth century, the most notable of which was L. H. Morgan's *Circular* relating to kinship terminologies, issued through the Smithsonian Institution in Washington (Morgan 1862; see Spoehr (1981) for replies from the Pacific). This questionnaire was to result not only in the compilation of Morgan's massive *Systems of Consanguinity and Affinity* (1871) but also encouraged a number of his correspondents to take up ethnographic research. Another important individual questionnaire was issued by Sir James G. Frazer in a number of editions betwen 1887 and 1916 (Frazer 1887, 1907 etc.).

Such questionnaires produced varied responses (for examples of replies to Frazer see Stretton 1893; Frazer *et al.* 1894) and their problems and limitations were recognized quite early. Many of the questions proved difficult to put to speakers of poorly understood languages while collectors themselves often experienced difficulty comprehending the import of the queries, lacking any real knowledge of the interests of the experts. But certain individuals who received the queries and who were ideally situated to collect further information responded to the enquiries and produced important accounts of other cultures. A good example is the work of A. W. Howitt in Australia from 1872 until his death in 1908 (Mulvaney 1971; Walker 1971). Howitt, along with his early co-worker, the missionary Lorimer Fison, was stimulated by Morgan's *Circular* of 1862. Fison and Howitt produced detailed accounts of Aborigines for Morgan (Fison and Howitt 1880) and maintained an active correspondence with Morgan until his death in 1881 (Stern 1930). Howitt then developed a working relationship with British anthropologists, first E. B. Tylor and later Frazer. Within Australia, Howitt issued his own questionnaires to gather information (see examples reprinted in Walker 1971: 323–329), as did a number of other Australian collectors interested in Aborigines (see Taplin 1879; Curr 1886). Howitt had considerable personal knowledge of Aborigines, and conducted his own investigations, which included getting Aborigines to re-enact rituals they had abandoned in the face of European colonization (Mulvaney 1970; §4.4.3, §8.10.4).

Information on Australian Aborigines was in great demand among European anthropologists in the late nineteenth century; in their evolutionary schemes Aborigines provided essential evidence for the primeval state of society and religion (Mulvaney 1981). As Europeans colonized Australia it was believed Aborigines would soon become extinct, hence the sense of urgency in the collection of ethnographic "facts" which motivated many collectors and indeed which led to competition between many of them (cf., for example, the conflict between Howitt and R. H. Mathews (Elkin 1975, 1976)). A division of labour thus was established in Australia, and elsewhere in the world, between collectors and experts. Its most sophisticated form was established in the relationship between W. Baldwin Spencer, assisted by his co-worker F. J. Gillen, and Frazer after 1896. Spencer, Professor of Biology at Melbourne University, began collecting information on Aborigines on a central Australian scientific expedition in 1894. On later expeditions, devoted entirely to ethnographic research, he was assisted by Gillen, a magistrate resident in central Australia with a detailed knowledge of Aboriginal life. Spencer wrote up the research and published it with Frazer's help, and his writings became critical sources in the development of numerous theories in anthropology, sociology and psychology in the early years of the twentieth century.

Spencer's relationship with Frazer clearly illustrates the limitations of a division of labour between collectors and experts. Frazer argued for the collection of pure "facts" for later synthesis by scholars:

It is our business to prepare . . . by collecting, sifting, and arranging the records in order that when, in the fulness of time, the mastermind shall arise and survey them, he may be able to detect at once that unity in multiplicity, that universal in the particulars, which has escaped us (Frazer 1905: 6).

Little thought was paid to how the "facts" should be collected, and there was little appreciation of the difficulties involved in eliciting information. Other cultures were viewed as objects to be recorded by scientific method, as if ethnographic facts could be easily assembled as natural history specimens (§4.4.1). Spencer wrote to Frazer of his advice to Gillen:

I send him up endless questions and things to find out, and by mutual agreement he reads no one else's work so as to keep him quite unprejudiced in the way of theories (Spencer in Marett and Penniman 1932: 10).

Such an attitude was not conducive to the development of finer ethnographic techniques nor to critical inquiry.

By the late nineteenth century the number of Europeans living in close proximity to exotic cultures was increasing with the spread of European trade and colonization. Accounts of explorers and missionaries were now supplemented by those of colonists, traders and particularly administrators, the last group often including many highly educated people. The interest in anthropology increased among such people who often became "official correspondents" for anthropological institutions which published their accounts. The journal *Man* was founded by the [Royal] Anthropological Institute in 1900 in part to provide an outlet for the writings of such people (Myres 1951). The issuing of questionnaires and the encouragement of local collectors thus resulted by 1900 in a rich and increasing flow of ethnographic material on other cultures. Articles which claimed to cover all aspects of cultures gave way to brief accounts of selected aspects of cultures and eventually to detailed and extensive monographs on single cultures which often followed a standardized pattern with details on material culture, social life, religion, etc. As the data accumulated it became increasingly obvious that "primitive" cultures were far more complex than had previously been thought and that ethnographic descriptions were valuable in their own right. People began to relate the information contained in such ethnographic accounts not to some general theory of the evolution of all mankind, but to the specific development of cultures within particular areas of the world. At the same time scholars began to realize that the quality of much of the ethnographic description was poor, that different types of "facts" were needed, and perhaps the experts should become collectors in their own right.

3.3 The early American experience

In the United States anthropology developed with a particular interest in the origin, development and inter-relationships of the indigenous cultures of the Americas. Unlike their European contemporaries, American anthropologists had extensive and varied Indian cultures to observe within easy reach of their studies, while the westward expansion to the frontier and the building of railways facilitated further research. The collection of ethnographic materials, the use of informants and local collectors and the synthesizing of the material was, therefore, often very different from the methods of European anthropologists (see Rohner and Rohner (1969: xxiii–xxvii) on early American methods). One of the most marked differences was the emphasis on linguistic material, both in the form of word-lists and grammatical data and the collation of native texts. The first major collection of texts and ethnographic material in this form was made by Henry R. Schoolcraft who published his material from the 1840s onwards (Hallowell 1960), and this tradition of rich ethnographic reporting, supplemented by texts, was to continue in America through the rise of "professional" anthropologists in the late nineteenth century to reach its zenith in the work of Franz Boas and his students in the early decades of the twentieth century.

During most of the nineteenth century anthropology in Europe, particularly in Britain, was a concern of amateurs and learned societies but in America anthropology was supported by the government and became established quite early in museums and universities. One of the most important supporters of anthropological research in the United States was the Smithsonian Institution founded in 1846 and in that year Schoolcraft presented the Institution with a plan for ethnological investigation (Schoolcraft 1886). The Smithsonian supported archaeological and ethnological investigation and distributed questionnaires (Fowler 1975: 22), including Morgan's *Circular*. The Smithsonian's intensive concern with anthropology occurred, however, after the American government had established the Bureau of [American] Ethnology in 1879 under its auspices to gather, collate and publish information on Indian affairs. For the next 20 years, under its director John Wesley Powell, the Bureau was to dominate American anthropology (Hinsley 1981).

Powell had considerable personal experience of Indian culture and had carried out research among Indians of the South West before the Bureau was founded (Fowler and Fowler 1971). Under his leadership the Bureau actively encouraged research, local correspondents supplied information and in turn were given questionnaires and guides to assist their work (Fowler 1975: 22).

The Bureau also employed directly people to conduct studies and surveys (Hinsley 1976) and this work was influenced by Powell's experience with geographical and geological surveys (in fact Powell was not only Director of the Bureau but for a time also ran the US Geological Survey). Many field-workers worked under Powell including James Mooney, J. O. Dorsey and the enigmatic Frank H. Cushing who spent four years with the Zuñi Indians after 1879, learning the language and being initiated into certain Zuñi rituals (Green 1979; Mark 1980). Ethnographic details covering a wide range of subjects were collected by the Bureau, but there was a particular stress on linguistic material, including texts of which examples were given in the first *Report* of the Bureau to guide future researchers (Illustration of the method . . . 1879–1880). The results of this work were published in the massive Annual Reports of the Bureau and these lavishly printed and illustrated accounts set new standards in ethnographic reporting which were admired in anthro-pological circles throughout the world. In spite of this apparent professional-ism in both fieldwork and publication, the fieldworkers of the Bureau were subordinated to those in Washington who published and synthesized their results to produce grand theories (Hinsley 1976).

British anthropologists were particularly impressed by the Bureau's work and in 1884 the British Association, meeting in Canada, established a Committee to promote research among the Canadian Indians with particular reference to the Northwest Coast tribes. The Committee issued a research guide (Circular of inquiry . . . 1887) drawn up either by the Chairman, E. B. Tylor, or by the newly appointed director of the project, the veteran linguist Horatio Hale (Gruber 1967). The aim of the research was to carry out detailed surveys of the physical anthropology, cultures and languages of the region. Hale recruited a German anthropologist, Franz Boas, who had already carried out field research among Eskimos (1883–1884) and in the Northwest Coast (1886) (cf. Stocking (1968) on his early career). The German ethnographic tradition, with which Boas was associated in his early years, stressed first-hand field observation; A. Bastian, Boas' associate, had travelled widely in the pursuit of information and there was also the con-temporary research of F. von Luschan and J. Kubary in Oceania, K. von den Steinen and T. Koch Grünberg in South America and the later work of L. Frobenius in Africa. It has been argued that Hale influenced Boas' research strategy by stressing the importance of collecting linguistic material, par-ticularly in the form of texts (Gruber 1967: 32), but this approach was already commonplace in American anthropology.[4] Undoubtedly Boas gained

[4] Boas' interest in texts in ethnographic contexts was probably derived from three sources: first, existing American interests in Indian literature; secondly, German concerns with European

considerably from Hale's direction but he was also developing his own approach to ethnographic investigation. The Committee of the British Association and Hale wanted a general survey of the cultures completed, but after carrying out research Boas expressed a desire to carry out more intensive research in individual cultures, an attitude which brought him into conflict with Hale (Boas in Rohner 1969: 106–108; Rohner 1966: 175–176, 177; Gruber 1967: 30–31). In his later research, initially as part of the Jesup Expedition (1897–1900) which he directed, Boas concentrated on the Kwakiutl and tended to dismiss his earlier research as of little value (cf. Codere 1959, 1966: xiii–xiv).

In his mature phase Boas was quite clear about the value of particular kinds of ethnographic evidence. His ultimate aim was to produce ethnographic material which reflected the "mind" of the people studied (Stocking 1974). He criticized many ethnographic accounts as being full of description which could not be verified as they depended upon the subjective opinions of collectors and were thus superficial and unscientific. Instead Boas argued for accounts which showed what "the people . . . speak about, what they think and what they do" recorded by the ethnographer in "their own words" (1906: 642; see also Rohner and Rohner 1969: xxiii). The only way to achieve this was through the collection of artefacts and the extensive recording of texts in the native language (§4.4.4). Only when such material had been collected, documented, classified and printed could anthropology have adequate materials on which to found an objective field of study. It has been argued that Boas' approach to the collection of ethnographic material reflected his "natural science" approach to material (Smith 1959; see also Lesser 1981). Stocking (1977), however, has cogently argued that Boas' approach reflected his humanistic as much as his scientific training. Boas, Stocking suggests, attempted to establish a body of material equivalent to that which European humanistic scholars had at their disposal; artefacts which reflected the art and industry of cultures, literary texts which illustrated the life and history of a people and grammatical material which expressed their "genius" (see also Boas to W. Holmes in Stocking 1974: 122). Whatever the origins and motives of Boas' work, his aims were quite clear: raw data was needed before theory.

Though Boas' aims in ethnographic collection were quite clear, his exact methods were never explicitly stated (Smith 1959: 53; Codere 1966: xiii); by "method" Boas usually referred to analysis rather than field techniques (Boas

folklore; and thirdly (with direct reference to his Eskimo research), the collections of Danish scholars, especially H. Rink (see Rink 1866–1871), who knew Boas and advised him on matters of Eskimo linguistics.

1920). It is possible, however, to reconstruct the outline of these techniques. Though Boas certainly participated in the cultures he studied, at least during the early years of his fieldwork, his research emphasis was on the collection of data through the intensive use of particular informants. Indians were encouraged to record information on their own cultures in the native tongue; this use of Indians as equal partners in field research had a long tradition in American anthropology. Boas, in fact, taught his chief informant on Kwakiutl, George Hunt, how to read and write. Hunt was to be a major source of Boas' ethnographic data, working in the field and mailing the texts to Boas (Codere 1966: xxviii–xxxi; Rohner 1966). This approach to ethnographic inquiry resulted in massive compilations of material, accounts, texts and esoteric details of cultures, dense and difficult to handle, but no general account of the cultures or a description of everyday life. There was a lack of integration in the accounts, as Kroeber noted (1935: 543). Indeed, Boas never completed his ethnography of the Kwakiutl and Codere had to complete and publish it long after his death.

3.4 The foundation of intensive fieldwork in Britain

The funding of research among Northwest Coast Indians, as well as the subsidies toward the production and distribution of the various editions of *Notes and Queries*, were only some of the ways in which the British Association encouraged anthropological research in the late nineteenth century. At its annual meetings the section devoted to anthropology at the British Association would establish Committees to investigate particular problems sometimes endowing them with limited funds and encouraging local societies at home and abroad to assist in the collection of information (Brabrook 1893). Various Anthropometric committees (1877–1884, 1893–1894, 1903–1908) collected information on physical form and the data were analysed statistically under the direction of Francis Galton (see Forrest 1974). In 1892 an ambitious programme for an Ethnographic Survey of the United Kingdom was begun combining physical anthropology, ethnography, archaeology and folklore studies (see Haddon 1898: 434–489 for details) and plans were made for a similar survey of India. During the same period the British Association, in conjunction with the Anthropological Institute, attempted to interest the government in a Bureau of Ethnology (later named an Imperial Bureau of Ethnology or a Bureau of Anthropology) on the lines of the American Bureau. The aim was to provide the funds and assistance for the collection and collation of ethnographic information from across the Empire with a vague idea that the information might somehow be of use in colonial administration. First proposed in 1896 (*Rep. British Association* 1896) the suggestion was coolly received by the government and, though approaches were again made in 1908 (Bureau of Anthropology 1908) and immediately before and after the First World War, the plans came to nothing (Myres 1929; Feuchtwang 1973: 81–82). Indirectly, however, these efforts, along with other approaches which stressed the value of anthropology to colonial government, were a major factor in the establishment of the teaching of anthropology in British universities in the first years of the twentieth century.

As teaching became established in universities, a new profession of trained anthropologists was formed and the role of the Anthropological Institute and the British Association in the promotion of ethnographic research declined. The new teachers of anthropology nearly all possessed academic experience but most had a background in the natural sciences rather than the arts, a fact which was to have a considerable impact on the development of ethnographic techniques. Alfred C. Haddon was undoubtedly the most important person to enter anthropology with this background in the late nineteenth century (Urry, 1982).

Haddon was a zoologist with a particular interest in marine biology which involved him in field research and in 1888 he carried out research in Torres Strait (Quiggin 1942: 77–90). Though his research was primarily biological, Haddon took with him to Torres Strait anthropological questionnaires to record native customs (*Notes and Queries* and Frazer 1887; cf. Haddon 1890: 300). His subsequent publications on the customs of the Islanders brought him into the mainstream of late nineteenth century anthropology which, under the influence of Imperialism, folklore studies and the writings of men like Frazer, was experiencing a new lease of life and wide popular acclaim. Haddon's experience in Torres Strait, and the success of his ethnographic writings, drew him away from zoology towards anthropology and Haddon brought with him the natural scientists' methods of research.

Marine biological research in the late nineteenth century was dominated by the success of the *Challenger* Expedition (1872–1876) and many research projects and scientific expeditions, including those by Haddon, were influenced by its work. As Professor of Zoology at Dublin Haddon conducted marine research in Irish waters from 1885 onwards (Quiggin 1942: 64–67; Went 1971–1972); he visited remote areas of Ireland in the course of this research and collected anthropological material. In 1893 Haddon organized the Ethnographic Survey of Ireland as part of the Ethnographic Survey of the United Kingdom sponsored by the British Association, and many papers on Irish anthropology were published including an ethnographic study of Aran Island (Haddon and Browne 1893). Haddon, however, wished to return to

Oceania and in 1895 he was instrumental in the establishment of the British Association Committee to investigate the biology of Oceanic islands (*Rep. British Association* 1895). In 1896 Haddon managed to get anthropology included in the list of topics to be investigated, but as British Association funding was not forthcoming, Haddon raised research funds from other sources. In 1897 the Committee reported (*Rep. British Association* 1897: 352) that an expedition was planned to the Torres Strait and that the focus of research was now anthropological. In 1898 the Cambridge Anthropological Expedition to the Torres Straits left for the field (see Wichmann 1912: 683– 685; Quiggin 1942: 95–105 for details).

The Torres Straits Expedition (1898–1899) is important in the develop- ment of ethnographic techniques not so much because of what it achieved but rather because of the people it brought into anthropology and the approaches they subsequently developed from their experiences in the field. Indeed, many of the research projects initiated by the Expedition, for instance psychological testing, were never to form a major part of anthropological research in the decades to follow. Haddon recruited an odd assortment of experts for the Expedition, most of whom had little or no experience of anthropology. W. H. R. Rivers, assisted by W. McDougall, was to carry out psychological experiments, C. S. Myers was to study music, S. H. Ray linguistics, A. Wilken to take photographs and C. G. Seligman[n] to study medicine (see Quiggin (1942: 95–99) on their recruitment). Seligman and Rivers were later to make significant contributions to anthropology, but Rivers undoubtedly was to have the most profound effect on both theory and method (Slobodin 1978).

While carrying out psychological tests in Torres Strait Rivers collected details on the kin relationships of his subjects in the form of genealogies. Rivers realized that such information provided a basis for an understanding of the social life of the Islanders and the "genealogical method", as he termed his approach, was a means by which anthropologists could study "abstract problems . . . by means of concrete facts" (Rivers 1900: 82). The problem of "method" was critical for Rivers and his generation of anthropologists; only through clearly stated methodologies and systematic terminology could anthropology really be established as a true science. Using the genealogical method Rivers set out to test his approach, first among the Toda in India (1901–1902; see Rooksby 1971; Slobodin 1978: 98–114; Mandelbaum 1980) and later, though with different objectives, in Melanesia (at various times between 1907 and 1914). This stress on science, methodology and terminology can be most clearly seen in the production of an entirely new edition of *Notes and Queries*, begun in 1907 but not published until 1912. Rivers made substantial contributions to the new edition and the quality of the work reflected the considerable thought which was given to problems of method after 1900 (Urry 1972: 50–52). The new edition also suggested that people carrying out anthropological research needed some kind of instruction in such methods.

By 1903 Haddon was arguing that "trained observers" were required to carry out anthropological research and that every scientific expedition should include a skilled anthropologist (1903: 22).[5] Expeditions and surveys, whether or not they were specifically ethnographic, were indeed a major source of new ethnographic information from the time of the British Association's investigation of the Northwest Coast Indians onwards. The Torres Straits Expedition was part of this trend and was succeeded by other expeditions to the Malay peninsula, central Africa, Western Australia and by a number of forays into Melanesia. Such work, conducted by either individuals or groups, concentrated on regional studies and was termed "survey work". The results of the research contributed greatly to an understanding of specific regions (cf. Seligman's surveys of Melanesia (published 1910; cf. Firth 1975) and the Sudan based on fieldwork carried out mainly before 1914 but published in 1932) and undoubtedly encouraged the development of studies of culture history and diffusion common at this period (cf. Rivers (1914) for an example based upon his own "survey" work in Melanesia). But, just as Boas had found in Canada in the 1890s, British anthropologists came to realize that detailed studies of *individual* cultures provided richer accounts of other cultures than most "survey work" which was often superficial. Of particular importance in this regard were the works of missionaries and administrators who, having lived in single communities for extended periods, produced detailed ethnographic accounts (see especially Callaway (1870) on the Zulu, Codrington (1891) on Melanesia, Junod (1898) on the Baronga, Man (1885) on the Andamans, etc.). This emphasis on regional and individual ethnographies also reflected the changing interests of British anthropologists, including an increasing distrust of the comparative method (Wallis 1912).

In 1904 Haddon suggested that a new approach to field research should involve:

> exhaustive studies of limited groups of people, tracing all the ramifications of their genealogies in the comprehensive method adopted by Dr Rivers for the Torres Straits Islanders and for the Todas (1905: 478).

Rivers' work among the Toda (1901–1902, published 1906) had pioneered this approach and his example was followed by C. G. and B. Z. Seligman

[5] It is possible that Haddon was influenced in his views by his visits to America in 1901 and by his observations of American research methods (see Haddon 1902).

among the Vedda (1907–1908, published 1911) and A. R. [Radcliffe-] Brown in the Andamans (1906–1908, published 1922). By 1908 Haddon could argue in a testimonial for two Finnish ethnographers, G. Landtman and R. Karsten, that: "Our watchword must now be 'the intensive study of limited areas'" (Haddon in Stocking 1979: 10; on Haddon's use of this term see also Haddon (1910: 154) and Myers (1940: 230)). The need for intensive fieldwork was echoed by Rivers in his report to the Carnegie Institution in 1913. He appealed for intensive research in single communities "in which the worker lives for a year or more among a community of perhaps four or five hundred people and studies every detail of their life and culture" (Rivers 1913: 7; see also Urry 1972: 50).

Though the professionalization of British anthropology had begun later than in America, by the end of the first decade of the twentieth century ethnographers, trained in anthropology, were being sent to do intensive field research. Barbara Freire-Marreco (later Mrs Aitkin) had worked in North America before 1910, G. Landtman was sent to Papua, R. Karsten to South America and Diamond Jenness to Melanesia in 1911; A. M. Hocart and J. Layard were taken to Melanesia by Rivers in 1914; Bronislaw Malinowski was sent to Papua and Maria Czaplicka to Siberia, also in 1914. Landtman, Karsten and Malinowski were all students of Edward Westermarck in London, though they received additional advice from Seligman and Haddon; Layard and Hocart were students of Rivers; Jenness, Freire-Marreco and Czaplicka were students of R. R. Marett at Oxford. Mrs A. Hoernlé, who had studied under Haddon and Rivers, also began fieldwork in South Africa in 1912 (Krige 1960). It would appear, therefore, that all the leading anthropologists in Britian who trained students in anthropology had agreed by 1910 that fieldwork was an essential part of a career in the subject and that intensive fieldwork by individuals for extended periods was the best strategy for future research.

3.5 Malinowski, Radcliffe-Brown and their legacy

The intervention of the First World War halted further moves towards the development of intensive field studies; most of the leading British anthropologists engaged in war-work and few new anthropologists were trained or sent to the field. Those who had left for the field flourished in the war-time conditions, particularly Malinowski who extended his research in Melanesia well beyond the time he had originally intended (Laracy 1976). It is clear from his diary that during his fieldwork Malinowski was very concerned with methodological issues and critical of his own research. In the field, long before he wrote his first major ethnographic account of the Trobriand Islanders, Malinowski planned to preface his work with comments on methodology which would distance his work from earlier ethnographic descriptions (Malinowski 1967: 155, 215, 229–230, 290).

In the opening chapter of this first book Malinowski grouped his "principles of method" under three headings. First, he argued the ethnographer must possess "real scientific aims" and a knowlege of "modern ethnography"; secondly, he must live among the people themselves; and finally, the ethnographer must apply "a number of special methods of collecting, manipulating and fixing his evidence" (1922a: 6). The goals of ethnographic research Malinowski placed under three headings to which he gave titles. First, there was the outline of native customs which he called the method of "statistical documentation by concrete evidence" (1922a: 17, 24) and which included the collection, through direct questioning, of genealogies, details on technology, the village census, etc. The second goal was derived from his second principle of method, from the ethnographer living with the natives, and which he termed the collection of the "imponderabilia of actual life and of typical behaviour" (1922a: 20, 24). Such methods revealed the minute details of everyday life as lived by the people and observed by the ethnographer. The third method required the ethnographer to become competent in the language of the people for it involved the recording of everyday speech, magical formulae and myths to be presented as "*a corpus inscriptionum*, as documents of native mentality" (1922a: 24). These three methods were needed to achieve the ultimate goal of ethnography: "to grasp the native's point of view, his relation to life, to realise *his* vision of *his* world" (Malinowski 1922a: 25).

Leach has claimed that Malinowski "created a theory of ethnographic fieldwork" (1957: 118) and Gluckman (1961) that his ideas caused a revolution in anthropology, but all Malinowski's principles and goals of fieldwork had been recognized before 1914 (Urry 1972: 53). Malinowski was the first clearly to articulate these methods while at the same time demonstrating in the writing of ethnography how valuable they were in improving ethnographic accounts. The value of his work was perhaps more in terms of this demonstration than in terms of his claims to methodological rigour; it was the *style* of his ethnography which convinced people of the value of his methods (see also Payne 1981). As Lowie put it, Malinowski:

> Endowed with an unusual literary sense, . . . thus succeeded in creating a 'flesh and blood' picture of Melanesians (1937: 231).

Earlier ethnographic accounts, Malinowski's own study of Mailu included (1915), tended to be rather dry, each chapter attempting to deal with separate aspects of culture and the whole lacking any integrated approach. Malinowski's descriptions of actual people and behaviour was in part derived from what became known as participant observation, but it was also a necessary part of his functional analysis of the interdependence of institutions in cultures. Malinowski recognized this quite clearly when he stated that functionalism was a "theory which, begun in field-work, leads back to field-work again" (1932: xxix).

In line with this functional approach, Malinowski recommended that information be collected in the form of charts or tables which could then be cross-referenced to establish inter-relationships between cultural institutions (1922a: 14–15). This was the method Malinowski taught his students and indeed was the approach he took in his theoretical analysis. Fieldwork method and functional theory were thus intimately interwoven. Fortes (1957a: 160) has pointed out that, for Malinowski, direct field observation was the only reliable source of ethnographic information (see also Kaberry 1957). His own writings were firmly grounded in his Trobriand material; only occasionally did he refer to earlier ethnographic accounts and, unlike Haddon and Seligman, he did not maintain an extensive correspondence with colonial officials, missionaries or traders who had extensive knowledge of other cultures. His students likewise relied heavily upon their own fieldwork; they tended to be dismissive of the work of earlier ethnographers and most made little effort to carry out historical research into earlier sources on the people they studied.

Malinowski was a gifted linguist and he exploited this ability in his field-work. Unlike their contemporary American colleagues, however, British anthropologists in the period before 1914 were not greatly interested in the study of language as a central and integral part of ethnographic inquiry. They preferred to utilize the skills of specialist linguists and to work in pidgin or with bilingual informants. Malinowski's interest in language was thus a personal trait but he only became keenly aware of the importance of language after his initial Trobriand fieldwork, and during his second period of field-work he concentrated on linguistic work (Malinowski 1920: 73–74; see J. R. Firth 1957: 97). It was during this period that Malinowski developed his third goal of fieldwork, the recording of native texts to reveal aspects of "native mentality".[6] But while Malinowski saw this linguistic work as an

---

[6] This goal sounds as if it were derived from Boas, but a more likely influence on Malinowski was R. Thurnwald who appealed for such an approach after fieldwork in Melanesia (see Thurnwald 1912). Firth (1981: 127) has noted Malinowski's respect for Thurnwald. Firth's other claim (1981: 123) that Malinowski's fieldwork techniques, particularly his use of the vernacular, were derived from Westermarck's influence, is less convincing.

integral part of his field methods, and indeed published important papers on language and Trobriand text material (Malinowski 1920, 1923, 1935(2)), very few of his students followed his example. This is not to deny that they all stressed the importance of learning the native language and recording facts in the vernacular, but few published in this vein. Language was not seen as the basic means to collect concrete information, least of all to establish an understanding of the mentality of people studied, but more as an adjunct to the collection of the details of everyday life. Language was to be "used" rather than analysed; it was merely "speech in action", as Richards put it (1939a: 302–303).

The basic approach to fieldwork outlined by Malinowski in 1922 was accepted by all his students, but the stress on life as lived was the aspect of his field methods that most of the students emphasized. They stressed the need for participant observation and how descriptions of everyday action needed to complement and enrich details of custom (Richards 1939a: 305–307). Indeed there was something more real and important about such descriptions and they were given a prominent position in research designs (Fortes 1957a: 159). Malinowski himself did little to discourage such interpretations of his research methods, as this was the feature of his method which gave his writings their vitality. Malinowski, through his teaching and his writings, encouraged his students to produce detailed ethnographic monographs in which their material was contextualized into a coherent account in quite a different way from the ethnographic writings of Haddon, Rivers or Seligman (see, for example, the monographs of Firth (1936) and Evans-Pritchard (1937)).

Malinowski, however, had failed to develop one of the key methods of pre-1914 fieldwork: the use of the genealogical method to collect information with its emphasis on the study of kinship. Rivers' methods were continued by his Cambridge students, particularly Radcliffe-Brown, Layard, Hocart, T. T. Barnard, W. E. Armstrong and later, under Haddon's aegis, A. B. Deacon (see Langham 1981, Chapter IV). Although his own fieldwork belonged more to the tradition of pre-1914 survey work, Radcliffe-Brown was most influential in the continuation of kinship studies backed-up by the intensive fieldwork of his students which he actively encouraged (see Radcliffe-Brown 1931: 276–277, 1958 [1923]: 34–35, etc.). Radcliffe-Brown's numerous teaching posts in various parts of the world spread his influence over many continents and he helped establish a new tradition of intensive fieldwork in South Africa (see also Schapera 1934), Australia, North America and elsewhere. While Radcliffe-Brown's only specific methodological contribution was a short note on the recording of kinship systems (Radcliffe-Brown 1930; but see

Needham 1971: xxiv–xxv), his major achievements lay in his insistence on the need for a clear, standardized system of terminology in the study of society, the combination of both ethnographic observation and interpretation by trained fieldworkers (Radcliffe-Brown 1958 [1931]: 67–71) and his persistent attempts to define the proper task of anthropology as the comparative study of society, or, as he termed it, "social anthropology".

When Malinowski had reviewed the tasks of anthropology in 1922 (under the title ethnology; Malinowski 1922b), it was unclear if anthropology was really a branch of history, part of the emerging subjects of psychology and sociology, or a discipline in its own right. By the late 1930s most of the younger British anthropologists, including nearly all of Malinowski's students, had come to accept Radcliffe-Brown's definition of the subject as "social anthropology", though not all agreed with his claims to what this entailed. This shift in the vision of how anthropology should be defined, however, had been accompanied by other changes, most notably in the ethnographic focus of anthropology and the growing claims for the practical application of anthropology in colonial administration.

Haddon, Rivers, Seligman and other anthropologists of their generation had conceived research as salvage work and all stressed the urgency of the task as cultures were either changing beyond recognition or were becoming extinct (Haddon 1897; Rivers 1922, etc.). This emphasis on salvage work was particularly apparent in the research in Oceania which became the major area of study after 1900. Malinowski saw his own research in the same light, in spite of his claims to record actual behaviour. In 1922 he described Trobriand culture as "Neolithic", preserved into the present (Malinowski 1922b: 216), and in 1930 he noted of his Trobriand research:

> I was still able with but little effort to *re-live* and *reconstruct* a type of human life moulded by the implements of a stone age (Malinowski 1930: 406; my emphasis).

Malinowski, though, was to change this view as he became interested in the study of social change and culture contact, as he became concerned with the problems of practical anthropology and as the attention of his students shifted away from Oceania to Africa.

The shift from Oceanic to African research was to have a profound effect on British anthropology. By the late 1920s it was obvious that African societies were not threatened by imminent extinction but were changing in response to European influences. These changes provided a challenge to all concerned with African affairs: colonial officials, missionaries, businessmen and anthropologists. With his essentially pragmatic vision of anthropology,

Malinowski was attracted to the study of culture contact with the promise that his vision could provide practical solutions to the problems faced by both Africans and Europeans.

Malinowski and his students recognized that the study of African cultures and the contact situation needed additional field methods to those originally proposed by Malinowski. Lucy Mair brought together a number of papers mainly written by Malinowski's students (Mair, M. Hunter [Wilson], I. Schapera, A. Richards and M. Fortes) concerned with approaches to the study of culture contact which were published first in the journal *Africa* in the early 1930s and later in a separate volume (*Methods of Study of Culture Contact in Africa* 1938). Most of the papers are anecdotal rather than dealing specifically with details of actual methodologies to be followed while conducting field research. In general terms, however, they suggested that anthropologists needed to devise sampling techniques to deal with the large complex societies encountered in Africa and specialized skills in economics, politics and the study of such subjects as nutrition, education, etc. The results of such fieldwork are perhaps a better indication of the methods adopted in such work. For example, the work of Richards (1939b) on Bemba agriculture and diet, Fortes (1938) on Tallensi education and Wilson (1936) on culture contact among the Pondo, to name a few.

Many of Malinowski's students entered anthropology from other disciplines and some came from countries with their own distinctive ethnographic traditions, particularly New Zealand (Firth, R. F. Fortune, F. Keesing) and South Africa (Schapera, Fortes, G. and M. Wilson and later M. Gluckman).[7] While some were undoubtedly influenced by their connections with these ethnographic traditions, others show few signs of influence. The same could be said of their earlier training in other disciplines, though Firth certainly utilized his background in economics in his anthropological writing. In the field of methodological innovations, however, the influence of both other ethnographic traditions and training in other disciplines is more difficult to discern. Schapera's interest in history was apparent from the outset, but the most noticeable influence can be seen in the writings of S. F. Nadel. Nadel, originally trained in psychology and philosophy (cf. Fortes 1957b), was interested in applying psychological methods to anthropological problems

---

[7] The continuing tradition of intensive fieldwork by colonial officers, many of whom had received training in anthropology at university, and the work of government anthropologists during the inter-war period also should be noted. The research and publications of people like J. H. Hutton and J. P. Mills in Nagaland, W. W. Skeat and other Malay Civil Servants, R. S. Rattray, C. K. Meek, N. Thomas, M. Green and S. Leith-Ross in West Africa and F. E. Williams and E. W. P. Chinnery in New Guinea, to name but a few, provide good examples of the work of such officials.

(Bartlett 1937; Nadel 1937a) and he developed new techniques to measure intelligence in the field (Nadel 1937b, 1939a). At the same time he produced one of a few purely methodological papers on field methods, specifically on interviewing techniques, ever published by Malinowski's students (1939b).

In a sense, the complexity of issues with which anthropologists were faced in the study of social change required the acquisition of specialized skills for field investigation. It might have been better to have abandoned the principle of individual fieldwork in favour of group studies, but by the 1930s the ideal of individual fieldwork in a single culture had become the accepted norm of anthropological investigation. Fieldwork itself had become a matter of dogma among "professional" anthropologists: it was how an individual gained entry to the profession and indeed it was the method of investigation, rather than any distinctive theory, which separated anthropology from other disciplines such as sociology, politics or economics.

The shift in focus in British anthropology away from the study of small-scale "traditional" societies towards more complex societies undergoing change not only occurred in Africa but also, during the late 1930s, involved other areas, including China (see Freedman 1963) and Malaya where anthropologists were faced with problems of studying peasant societies (Firth 1946: 307–317). But these changes in ethnographic and theoretical interests did not drastically alter field methods, but merely the emphases of inquiry. Research tended to concentrate on subjects such as kinship and on the economic, political and legal aspects of social life. Radcliffe-Brown's claims that social anthropology was a generalizing science and that its task was to establish scientific laws of society was to have a more profound effect on the conduct of fieldwork. A spirit of positivism in anthropological inquiry was further encouraged and the belief that "objective" data could be collected through the utilization of rigorous techniques encouraged the development of new field methods. These changes were most apparent in the research concerns of the anthropologists at the Rhodes-Livingstone Institute founded in 1937 (Brown 1973), and in the work of its second director, Max Gluckman.

Gluckman encouraged the development of new techniques including the recording of detailed case studies of particular social situations (Gluckman 1961) and the adoption of techniques from other social sciences, such as economics, demography and sociology. Some of his students were interested in large-scale surveys or rural and urban communities and in the utilization of statistical techniques (Barnes 1949; Colson 1954). A number of papers concerned with methodology were published in the Institute's *Journal* (Marwick 1947; Silberman 1947; Barnes 1947; Mitchell 1949, etc.), and Gluckman and his students later up-dated some of these papers and, with new articles, produced one of the few works in British anthropology devoted specifically to methods of research (Epstein 1967). Otherwise, until quite recently, British anthropologists rarely discussed specific aspects of ethnographic inquiry; it merely became an accepted fact that to "become" an anthropologist one had to do a period of intensive fieldwork for one or two years (§1.1, §4.2).

3.6 American anthropology and changing perspectives on field methods

American anthropology after 1900 was increasingly dominated by Boas and his students. Unlike the situation in Britain, American anthropologists did not view the revolution that occurred in their subject in terms of a transformation in field methods but instead in terms of methods of analysis – a victory for Boas' critiques of evolutionary explanations. In terms of ethnographic inquiry there had been no major change, there had never been a major division of labour between collectors and experts in American anthropology and first hand ethnographic research into American Indian cultures had always been the norm. In part, this was a reflection of the essentially pragmatic nature of nineteenth century American life and of the fact that anthropology was viewed as a science rather than an adjunct to a literary tradition as it had been in nineteenth century Britain. But if methods of ethnographic inquiry had not altered drastically in form, they had changed in terms of the quality of material collected and this in turn was a reflection of the better training anthropologists received, particularly in linguistics.

Kroeber (1943: 14) pointed out the only two major courses Boas taught throughout his career at Columbia were "Statistical theory" and "American Indian languages", both cornerstones in his training of anthropologists. Statistics were mainly for physical anthropological inquiries and, of Boas' students, only Kroeber attempted to apply the techniques to cultural data and then only in regional studies, not in fieldwork situations (Golbeck 1980). Linguistic skills, however, had to be mastered as they were essential for fieldwork, the key to understanding other cultures, recording texts and building up the data base for the discipline. Though not all Boas' students were equally competent in linguistics, all possessed a basic proficiency and agreed on the importance of linguistic analysis in fieldwork. In 1913 Edward Sapir, undoubtedly the most able of Boas' students in linguistic matters, wrote to W. D. Wallis:

> It is highly useful, I think, for one making sociological studies among primitive peoples, to know enough about linguistic matters to take down Indian words and

even texts with reasonable accuracy. . . . I have always been struck by a certain
externality about all such studies that are not based on linguistic knowledge. I
always have an uneasy feeling that misunderstandings bristle in such writings. . . .
And then, to look at it somewhat more broadly, what we are after in studying
primitive peoples is, to a large extent, to get their scheme of classification. This
scheme must be more or less reflected in their own language (Sapir quoted in
Darnell 1976: 107–108).

Most ethnographic research in America conducted from 1900 to 1940
was carried out individually, often for expanded periods in short visits.
Researchers had to contend with meagre grants and could only get research
leave during their teaching vacations. To maximize their time in the field,
researchers concentrated on particular problems, sometimes suggested by
their superiors, to fill gaps in the ethnographic record. Given these conditions,
and the fact that the Indian societies being studied were undergoing massive
cultural change, anthropologists often worked with a few articulate individuals, recording in texts the memory culture of their informants rather than
participating in everyday life. But there was a long-term commitment to the
people being studied and information collected in the field was supplemented
by detailed analyses of neighbouring cultures, later accompanied by historical research (for a good account of such field research see Lowie's (1959)
autobiography).

Nearly all of Boas' students worked with North American Indian cultures,
producing detailed accounts of the languages, religions, mythology and social
structures usually accompanied by extensive texts. Most of the research was
published not as separate monographs but as part of a series of papers; the
Smithsonian anthropological publications continued and new anthropological
series were established (American Museum of Natural History Papers,
University of California, Publications in American Archaeology and Ethnology, etc.).

Although many of Boas' students became dissatisfied with his theoretical
interests, all agreed that his strategy for the collection of ethnographic
material was correct. This does not mean, however, they believed they could
not be improved. Radin disagreed with Boas' natural history approach both
in analysis and in his consideration of what constituted ethnographic "facts".
Radin accused Boas of thinking that ethnographic "facts" were of the same
order as physical "facts" and noted: "*Cultural facts do not speak for
themselves, but physical facts do*" (Radin 1965 [1933]: 9; his emphasis). But
Radin continued to follow Boas in arguing that texts, recorded in the native
language, were the only way to collect ethnographic information (1965
[1933]: 106, 108–109) in order to produce "records . . . in the same category

as the archives of history" (1965 [1933]: 117). For Radin, and for most of Boas' students, the "primary function" of the ethnographer was to "present facts in the words of the informant" and to keep "record and comment distinct" (Radin 1965 [1933]: 119, ft. 18). On this score he criticized British anthropologists such as Malinowski and Radcliffe-Brown, as well as the American Margaret Mead, of having:

> entangled their own data in such a maze of discussions, impressions, reinterpretations, and implications that it is even more difficult to get to the original record than it is in the case of the worse offenders among their predecessors [whom they often disparaged] (Radin 1965 [1933]: 119).

Radin included Margaret Mead in his strictures because, although one of Boas' students, by the early 1930s she had begun to move away from his position and appealed for fieldwork methods similar to those proposed by Malinowski. Unlike most of Boas' students, her first fieldwork experience was outside North America in Samoa, and in her subsequent work in Melanesia she was heavily influenced by British fieldwork techniques through working with Fortune and Gregory Bateson. Her appeal for more comprehensive fieldwork techniques (1933) stressed the need for participant observation and the recording of everyday life, although her stress on psychology was different from most British anthropologists. It is obvious that her numerous fieldwork expeditions to different locations, her changing methodological focus and her many publications were viewed with suspicion by Boas' senior students; Radin described her as "essentially a journalist in the best sense of the term", and doubted that either Fortune or herself could achieve the results they claimed from their brief periods of fieldwork (Radin 1965 [1933]: 170, 178; see also the comments of Lowie 1933: 296, 1935: xx).

Obviously stung by these comments Mead was later to launch a thinly veiled attack on the Boasian language-text-method of fieldwork (Mead 1939). Focusing specifically on the problem of language competence in fieldwork, Mead argued that anthropologists did not need to know how to speak the language of the people they studied, but only how to use the language in order to comprehend everyday discourse, establish rapport and ask basic questions. She argued that the stress on the need to know the language well (which she caricatured as claims to linguistic virtuosity), the emphasis on text collection rather than on everyday language, and the concentration on individual informants rather than on observation of the flow of everyday life, were all misguided. Lowie (1940) countered most of Mead's arguments pointing out there was very little difference between the field methods of his generation ("horse and buggy ethnographers") and the claims of the new generation ("stream-lined ethnographers") except rhetoric.

Lowie, however, repeated his doubts concerning the value of short-term field research, and the claims made by Mead and others for such "new" methodologies (§7.5.4).

Mead's criticisms of the role of linguistics in anthropology reflected not only changes in anthropology but also the development of linguistics as a separate discipline in the 1930s, particularly under Sapir and L. Bloomfield (Hymes and Fought 1975). Rather than as an adjunct to the study of culture, the investigation of languages became a field of study in itself; linguistic data and texts were used to construct new methodologies and theoretical positions rather than as keys for understanding other cultures (Hoijer 1973: 671). As linguistic methodologies and theories became increasingly complex, older anthropologists felt left out of linguistic discussions (Hymes 1977: 54–55); the competence of many younger anthropologists in linguistics declined and with it the emphasis on texts.[8] The whole basis of the Boasian field methodology was obviously changing.

Stocking (1976) has suggested that American anthropology after 1930 was characterized by what he terms "centrifugal forces of specialization". Some of these were derived from suggestions made by Boas and followed up by his students, but others came from outside the Boasian tradition. Psychology and psychoanalysis influenced theoretical orientations and, eventually, field methods. Radin, with an interest in psychology, suggested that the individual ought to be the focus of attention and that the collection of life-history texts was essential (see his famous account of Crashing Thunder (1926) which was to influence anthropologists in both America and Europe; Langness 1965). But Radin wished to stress the individual not so much for the study of psychology but more to emphasize the crucial role of individuals in history (1965 [1933]: 184–185). The rise of culture and personality studies raised new issues of how to collect information on individuals in cultures. Sapir, who increasingly stressed the need to study culture through the examination of individuals, failed to develop any methodological guides as to how this could be achieved (Preston 1966); the same could be said of Ruth Benedict. It was left to Mead, specifically in her fieldwork with Bateson in Bali and New Guinea, to attempt such new methods. The careful use of minute observation, photographs and films to observe everyday life and personal interaction (Bateson and Mead 1942), however, was not generally adopted and it was

---

[8] This is not to suggest that interest in linguistic matters ceased. The methodological and theoretical developments in linguistics of this period and later were ultimately to have a profound influence on American anthropological approaches to fieldwork and theory in the work of those interested in "ethnoscience" (see Goodenough 1956).

not widely repeated until recent developments in new techniques and technology. However, psychological approaches gained increasing popularity in American anthropology, and methodologies were adopted from psychology in fields such as the study of child rearing (see Kluckhohn 1939), cognition, mental health, etc. (see Klineberg 1980).

Mead was not the only American anthropologist to be influenced by British fieldwork techniques. A number of Americans were trained by Malinowski and Radcliffe-Brown and the influence of the latter increased after he came to Chicago in 1931. Radcliffe-Brown encouraged his students to carry out intensive field research, but his theoretical interests meant that they were expected to pay particular attention to the study of society, especially kinship. A number of Radcliffe-Brown's students carried out research into Indian social organization (see the papers in Eggan (1937)), but such interest in the study of society was matched by other trends in American anthropology after 1930, particularly those inspired by sociology.

By the early twentieth century, sociology in America was a more established discipline than in Britain. One branch of American sociology specialized in direct observation of communities, of which the most important example was the study of Middletown by R. S. and H. M. Lynd (1929), who were themselves influenced by anthropological methods. Anthropologists became involved in similar research in American communities during the 1930s. W. L. Warner, an American anthropologist who had carried out fieldwork among Australian Aborigines under Radcliffe-Brown's direction, returned to the United States and, using anthropological techniques, began a large-scale study of an American community (Newburyport, Mass.; studied 1930–1934, published from 1941 onwards). Other American anthropologists carried out research on specific aspects of American society using anthropological field techniques (see the account of Powdermaker (1966) on her work in Mississippi as an example). Studies of peasant communities in Latin America (see Redfield (1930) for one of the earliest; see also Hansen (1976) for Redfield's later methods) and further afield in Ireland (Arensberg and Kimball 1940) were also begun. In studying such complex large-scale societies, American anthropologists encountered many of the problems faced by their contemporary British colleagues in Africa, problems of taking surveys, constructing questionnaires and using methodologies from other disciplines. There was a growing recognition that not only peasant communities, but all the societies anthropologists studied, were part of an increasingly complex modern world and this is clearly stated in the appeal for acculturation studies made in 1936 (Redfield et al. 1936).

In spite of all these new sociological approaches, however, American anthropology still continued to study cultural problems. But the vision of anthropology had altered; in Britain social anthropology triumphed over ethnology and in America cultural anthropology replaced ethnology (Stocking 1976: 23). American anthropology, though, was richer and more varied than its British counterpart in terms of its interests, methodological approaches and theoretical concerns. The ethnographic concerns of American scholars also expanded after 1945. Whereas most American anthropological research since 1900 had concentrated on American Indians, with a few people working with the government in the Philippines (Eggan 1974) and further abroad (like Herskovits who worked for part of his time in Africa), the changing position of America in world affairs after 1945 brought new opportunities for foreign fieldwork. This, combined with a considerable expansion in the universities, meant that most anthropologists, as part of their pre-doctoral research, conducted a period of intensive fieldwork in a single culture, often remote and exotic. In a sense, therefore, there was a gradual convergence between British and American anthropologists in the strategies adopted for fieldwork and similarities in the techniques employed. The writing of ethnography also followed the British practice; no longer were "facts" presented in collections of papers, but single ethnographies combining new information with methodological and theoretical points were produced to contribute to ongoing academic debate.

3.7 Conclusion

Anthropological interest in the collection of ethnographic information at first had developed from an increasing critical awareness of the need for reliable accounts of other cultures to establish a true "scientific" study of man. The methods of direct observation developed in the other scientific disciplines during the nineteenth century, particularly in the natural sciences, were to play an important role in the adoption of anthropological field methods. In both America and in Britain in the late nineteenth century it was men trained in such scientific methods, and who took up anthropology at a later stage in their careers, who pioneered changes in ethnographic study. But the response of these men in the contexts of British and American anthropology was quite different. The American tradition, particularly under Boas, attempted to establish a firm data base, epitomized in the text, from which to develop theories. The British were interested in concrete facts, explicit in accounts of customs and society, which could either be arranged "taxonomically", into their "anatomical" parts or related to each other to explain the "physiology" of society. This tradition, strongly influenced by the natural sciences, runs directly from Haddon through Rivers to Malinowski and Radcliffe-Brown.

These new approaches to ethnographic literature raised new issues concerning methodology, though more explicitly in Britain than in America. In Britain the claims that anthropology not only studied a distinctive body of data but also that it possessed a sophisticated methodology to collect these data, was an important factor in the establishment of anthropology as an academic discipline. This was less necessary in America where, by the late nineteenth century, anthropology was already established in universities, museums and government agencies. But in spite of claims to scientific methodology, particularly in the British tradition, there are surprisingly few details about the actual methods anthropologists used in the field, beyond a few first principles and illustrative anecdotes. There was a wide belief among British anthropologists that fieldwork could not be taught to new recruits, but could only be experienced by individuals in the field. In the American tradition texts provided what was regarded as an objective body of data, whereas the British tradition was more a matter of subjective experience. It is a strange paradox in the history of the development of field methods that the scientific study of other cultures should have been built upon such a foundation.

James Urry

4 Approaches to ethnographic research

4.1 Introduction

Fieldwork has come, through a fairly long period of the historical evolution of the discipline, its own growing and changing self-image and the progressive development of an institutional base for the subject in the universities – departments, chairs, degrees, professional associations and outlets for publications – to play a dominant part in the actual activity of *doing* anthropology (Condominas 1973: 2; see also §1.4). And, indeed, not only in doing it. In many cases it is not simply a case of ethnographic literature based directly upon fieldwork providing the empirical data base of the discipline, but also of the strongly held (although not always fully articulated) feeling that the act of having done fieldwork is a *sine qua non* for admission to full professional standing and to the recognition by one's peers of the validity of a claim to be an anthropologist. (The first question that is so frequently asked by one anthropologist meeting another who he does not know, is "where did you do your fieldwork?" – a question incidently institutionalized in the criteria for membership of at least one major professional body, the Association of Social Anthropologists.)

But actually lying behind this "taken-for-grantedness", this rather naïve belief that having shared "the experience" of fieldwork somehow binds anthropologists together into an almost mystical siblinghood, are to be found major divergences in opinion as to what really constitutes fieldwork in practice – its precise intentions, the ways it is to be carried out, the philosophical or ideological presuppositions which colour it (Jarvie 1964; Salamone 1979; see also Chapter 2). This problem is the subject of this

ETHNOGRAPHIC RESEARCH
ISBN 0 12 237180 1

chapter. It has to be admitted at the very beginning that it is a subtle (although very significant) problem – not only because it is rarely raised but equally because, even when the problem is recognized, there is a central difficulty in dealing with it – notably that very few fieldworkers anywhere (whether in their ethnographic monographs or in their theoretical writings) actually make explicit the precise "model", "orientation" or "style" of fieldwork they are employing. And there are many such "styles": the conception of field-work as the collection of texts, for instance; the model of "encyclopaedism" (i.e. attempting to collect all possible data on everything); fieldwork as the "translation" of one culture into the terms and idiom of another; and so on. A little probing indeed proves very quickly that even the most commonly used terms and concepts, such as "participant-observation" or the "village study", conceal more than they reveal about the detailed and often contradictory conceptions of fieldwork practice.

The intention of this chapter is to describe the leading examples of such "styles" of ethnographic research, to assess them in a fairly critical manner and to relate them to their intellectual and even philosophical backgrounds in an attempt to contextualize them as representing what are quite often very different conceptions of the discipline of anthropology itself (see also Edgerton and Langness 1974).

4.2 The fieldwork concept[1]

The "fieldwork concept" has not always existed in anthropology, and even since its introduction it has been subject to varying emphases and inter-pretations (§3.1). In some traditions, such as that emanating from the Durkheimians in France, and perhaps best represented by figures such as Marcel Mauss, fieldwork even took second place to analysis based upon numerous sources, not only first-hand reports of travellers and foreign residents, but also from the literatures of antiquity. The primary concern of this group of anthropologists and their descendants down to the present day was conceptual elaboration and synthesis, not data collecting (Clammer 1983). But in the British and American traditions, in those of much of northern Europe and for many French anthropologists, fieldwork has come to play an epistemologically central role in their whole endeavour. Fieldwork from this point of view is *the* source of knowledge; philosophical and theoretical elaboration comes later; fieldwork is the only (or the most important) characteristic of social anthropology, separating it from the other

[1] See also Junker (1960: Ch. 1), Powdermaker (1968), Freilich (1970a) and Burgess (1982).

social science disciplines, and having done it is the primary badge of membership in the guild (§1.1).

But this "fieldwork concept" is the product of history in two senses – it is
on the one hand the result of the internal elaboration of the discipline from
within, through contact with the other evolving human-oriented disciplines
and through its progressive separation from the natural sciences (especially
zoology) and the literary subjects (especially classics). But on the other hand it
was a product of the changing social context of anthropology, especially in
relation to the role of colonialism, which not only helped shape the discipline,
but which also created the opportunities and openings for an increasing
interest in field data and field experience (Asad 1973). This, in turn, raised a
critical issue about the fieldwork concept which has never really been fully
resolved: the grave danger of fieldwork creating a subject–object relationship
between the fieldworker and those observed. This relationship is often based
upon unequal power and resources, is temporary and exploitative for the
observer, but may have permanent and unforseen results for the observed;
and is asymmetrical, in that the fieldworker reports and analyses, but the
observed may never have access to his data or have the opportunity to react to
or criticize them. The fieldwork concept itself then, while usually justified in
terms of basic data collection, greater insight into the culture studied and so
on, can actually be a very problematic one when viewed historically, politically
or morally, and this needs to be said at the very beginning, especially as the
idea of fieldwork is now so entrenched and commonplace in anthropology.

Seen in its positive aspect, however, the fieldwork concept – that is the idea
of the anthropologist going in person, either individually (itself a matter of
some methodological and even ideological interest) or, more rarely, in a team
to the actual location in which the data is to be collected, and then actually
collecting that data over a period of time (often a fairly long one of between
one and two years or even longer) by way of first-hand observation, participation, collection of census material, interviewing and questionnaire administration – is an admirable advance. It represents a break with the older
methods of relying either on second-hand and often totally unverifiable
information collected by a rather random selection of individuals not trained
at all in systematic data acquisition methods (e.g. colonial administrators) or
those (most especially missionaries) who often did possess immense insight
into the particular culture, including an intimate knowledge of the language,
but whose objectives in being there were quite different from those of
anthropologists. The other method favoured prior to the emergence of
fieldwork was the sending around of circulars or questionnaires by home-
based scholars, who then based their own analyses on the rather erratic

responses that they received back (§3.2). The fieldwork approach furthermore had the salutary effect of breaking down the distinction between
"collectors" and "experts": the new breed of anthropologist had to be both.

The fieldwork concept then, as it emerged in the early years of this century,
had certain key characteristics. It was seen largely as an individual activity; it
was based on the intensive study of individual cultures; it involved detailed
first-hand involvement by the anthropologist in the life of the culture being
studied; and it was comprehensive in intent, being designed to explore every
facet of the life and culture of the people concerned, including not only
kinship and beliefs, but also political and economic structures and processes,
and very often material culture and ecology as well. As the century progressed
certain "refinements" in this method gradually appeared. On the one hand
there were increasingly complex methods of data collection, the anthropologist frequently dispensing with interpreters and learning the language;
the use of survey and statistical techniques; the development of the "life-
cycle" concept for studying an individual or a whole culture; and, finally, the
use of technical aids: cameras and even video for the making of a visual
record; tape-recorders for the preservation of texts, i.e. songs, speeches,
recitation and so on (§7.9). On the other hand, there was a tendency for the
ethnographer to narrow his or her sights, concentrating for example on
kinship, or on the political system, and abandoning in practice the older
holistic approach (to which lip service is no doubt still paid). The result was
the disappearance of some important features of culture from most ethnographic accounts, the most prominent casualty being material culture.
Refinement of method has thus, paradoxically, gone hand in hand with a
narrowing of the scope of the typical modern ethnography.

At the same time, there has been a growing interest in reflecting on the
methods of fieldwork from a more systematic point of view as well as an
upsurge in literature describing the experience of fieldwork from a personal
perspective (§1.5, §5.2). To a great extent this has clarified what is actually
meant by the term "fieldwork", although it has also added and formalized a
large number of procedures that may be included under this general term. But
certain general points can be made. Fieldwork, although it often purports to
be about a whole society or culture, is actually usually only carried out within
a single community – typically a village – within that society, and it is
assumed (often on rather shaky grounds) that the particular community is
somehow "representative" of the wider society. In practice a particular
community is often selected for a variety of reasons – accident, contact in the
community, advice from a student's supervisor, its proximity to a convenient
university or town (see Freilich (1970) for quite explicit examples of this

kind) — and rarely on the basis of a systematic survey of the full range of communities available to the investigator or on the basis of any rigorously argued claim to true representativeness (§7.7). Once the choice of the community has been made, a range of events typically follow: selecting or being placed in a residence, establishing a rationale for being there at all, gaining the trust of the locals, establishing a social network, language learning, choice of special informants, basic surveying (census, households) of the community and the eventual setting up of more systematic techniques of data gathering — interviews, land-tenure surveys, life-cycle analysis, case studies, network analysis, participant observation of events, rituals, and characteristic social processes.

4.3 Units of study

Two key ideas in this conception of fieldwork are very frequently encountered: the idea of the (village) community as the unit of study (Arensberg 1954), and the idea of participant observation as the typical (and most anthropologically distinctive) mode of information collecting. Both these ideas need briefly to have something said about them. It is becoming more and more widely accepted that the "village community" is a myth (e.g. Dumont 1966). This is not, of course, to say that villages do not exist: of course they do, but rarely if ever as isolated, self-contained entities. In fact, they exist as part of a wider society with links (political, economic and social) with other villages and with the larger locality, the district, even the whole nation-state. With the process of modernization this is becoming truer by the minute. Most anthropologists, either privately or even in print, will admit this, but yet persist in treating the village community as their typical unit of study (it is "convenient", "relatively bounded", etc.). A similar error creeps in with the familiar concept of the "ethnographic present", i.e. the treating of this community as if it were frozen in time, thus obviating the necessity of worrying about history, change and the diachronic aspects of any culture, especially those being forced on the community by outside agencies. This is, in part, the legacy of functionalism, in part theoretical timidity, in part (in some cases and parts of the world) of upsetting local or government authorities, and in part the desire to see the unit of analysis as "pure", as untrammelled and "uncontaminated" as far as possible by the "outside" world. Essential as fieldwork may be to the gathering of in-depth information (it is a much more effective tool than the questionnaire, for example, in this respect), this is also its crucial limitation: it can only work within a very

limited sphere of space and time. A close and critical reading of many
ethnographic monographs reveals how true this is, and the generalizations so
typically found in this genre are often simply extrapolations from a very
particular and often narrow data base.

The other idea which so strongly pervades the fieldwork concept is that of
participant-observation (§8.2): the immersion of the fieldworker in the
everyday life and activities of the people of the culture being studied. This is
such an article of faith with most anthropologists that it is rarely questioned,
but question it here we must. Again, no one would doubt that participant-
observation is *one* approach amongst many others (although many anthropologists talk and write as if it were the *sole* technique), and that living
amongst the people studied is essential. But how far is the technique applied in
practice? Very many anthropologists who claim to use the method actually
live apart from the people studied, often in a house or apartment with much
better facilities than those of the informants, and surround themselves with
equipment (justified as being "necessary" to the fieldwork) and comforts
(often characteristic of their own culture). This often extends even to the
consumption of their own rather than the native food. Very few anthropologists, in a one or two year period of fieldwork, ever actually fulfil or even
attempt to fulfil Malinowski's demand that the fieldworker know the local
language intimately, even though this is a primary requirement of genuine
participation. The recording (on film, tape or in a network) of peoples'
behaviour as they go about their daily business, renders spontaneous participant observation difficult. Finally, there are severe limitations on how far
such observation is actually possible. In some cases it is not possible because
the anthropologist is excluded from secret activities, or from activities which
do not accord with his or her ascribed social or gender status in the
community; because participation is technically unfeasible (as in a local
dance, say or complex ritual), because it is morally difficult for the observer to
participate (say in head-hunting!), or because, beyond a certain point, it
becomes ridiculous to try to participate appropriately. However much the
observer's "identification" is with the people, they know and the ethnographer knows that he or she is not a member of the culture, only a
temporary and transient member. Finally, observation may be constrained
through fear of what will be reported. And such fears are often justified. Not a
few anthropologists have adopted questionable tactics in order to discover
information that is held to be confidential by the community on the grounds
that it is "essential" for an "objective" view of native activities (Chapter 6).

When one comes across claims about participant observation one should
therefore proceed with care and enquire much further into what was actually

done, how well qualified the observer was and what is meant by such dubious terms as "understanding" and "empathy" when used in the context of this technique. This is particularly true when one realizes that many anthropologists in the field actually represent themselves as something else: as teacher or writer for example (Freilich 1970: 3). How the varieties of these techniques actually manifest themselves in the field must therefore be the next question to which we should pass.

4.4 Styles in the practice of fieldwork[2]

Despite the increasing body of pedagogical information available for formal instruction on how to do fieldwork, it must be admitted that the activity has always been, and is likely to continue to be, a matter of personal *style*. Such styles have four basic sources: the individual and idiosyncratic characteristics of specific fieldworkers; ideological and philosophical presuppositions (for example, "action" work or "pure" research, Marxism or functionalism); the general conception of method (e.g. a predilection for case studies, or quantitative methods); and the nature of the problems to which the research is addressed: is it the economy, for example, or the process of socialization, or perhaps a re-examination of a society worked in before by the same or another anthropologist? Taken together these four sources generate a range of fairly typical styles of ethnographic work. The following sections try to identify and explore the leading examples of such styles and, in doing so, to illustrate the range of possibilities that exist and co-exist within or under the general rubric of "fieldwork". These "styles", as we have already suggested, are not usually made explicit by the individual fieldworkers themselves, but nevertheless there are several fairly consistently found orientations. It is these which we will try to identify.

4.4.1 Ethnographic encyclopaedism

This is perhaps the oldest style which we can identify, and one which is still respected in many circles. In one sense, the term epitomizes an essentially nineteenth century natural history approach to ethnography: the drawing up of inventories of so many customs in response to lists of queries of the kind enshrined in *Notes and Queries*. It has relatively little to say about the inter-relationship of facts, or their interpretation. In this form it is typified in the work of some of the great expeditions (e.g. Veth 1881–1897; §3.4), the

---

[2] On styles within a more explicitly sociological tradition, see Douglas (1976).

celebrated missionary and colonial anthropologists (e.g. Junod 1898; Skeat and Blagden 1906; §3.7, n. 7), and in the continental European ethnological traditions. In a more contemporary sense it is reflected in the view that an ethnographic monograph *should* be encyclopaedic in character. That is to say it should present, as far as possible, a totally comprehensive view of the culture or community in question, and should range over the entire social field: kinship, economics, politics, belief systems, ecology and perhaps material culture and the performing and decorative arts as well. Under these terms of reference, a study which is not encyclopaedic, in intention if not in execution, does not qualify as an ethnography, but is something else – a thematic study of some sort, for instance. But in practice such an approach has been hard to sustain for a variety of reasons: the sheer difficulty of studying *every* aspect of culture; a bias on the part of the ethnographer towards some particular theme – say the political system – which is wittingly or unwittingly emphasized at the expense of other aspects; the belief amongst many anthropologists that the raw data from the field needs to be theorized to make it respectable, which leads, for example to elaborate structural analyses behind which the empirical evidence tends to disappear; or the effects of specialization within the discipline and of personal limitations. Few contemporary anthropologists would claim to be the extraordinary polymaths that encyclopaedic fieldwork would require them to be, and a person who is entirely competent to deal with, say, the economic system, may be entirely incompetent at dealing with ethnobotany. Additionally, there is a feeling amongst many, and not least amongst graduate students who will make up the future personnel of the profession, that encyclopaedism is simply *old-fashioned*, and that the way forward is in fact through the definition of theoretical problems, for the solution of which fieldwork of some sort is necessary. This, of course, is the reason why large numbers of anthropologists persist in working on the same and already well documented societies (e.g. the Minangkabau), where they are not really doing ethnographic work at all in the old sense of the word, but rather working on a theoretical theme, such as the position of women in petty commodity production, or whatever. But of course this is to be expected, as the intention of the fieldworker will to a great extent determine the style of fieldwork.

4.4.2 Theory and description

Encyclopaedism as a style, while not unknown today (for example it is still widely practised by many Indian anthropologists), has tended to flounder on the two rocks of the sheer difficulty of doing it well on the one hand and, on the other, the growing interest in theoretical questions which has (rightly or wrongly) tended to reduce descriptive writing and research to a secondary and under-labourer role. Evidence of this is to be found in the lists of publishers of anthropological books: it is very hard these days to publish a purely descriptive monograph, rather one must address oneself to a "question" for one's work to have much impact. The same is true when applying for a research grant. Indeed, in making these remarks, we have already identified a second prominent approach of fieldwork: that of *fieldwork as designed to illuminate a theoretical issue*. This approach, which today probably dominates the profession, is of course not new. It was indeed typical of much of the work of Malinowski and just about all of the work of Radcliffe-Brown, although there were naturally major differences in style here (Radcliffe-Brown's structural-functionalism and his view of anthropology as comparative micro-sociology, for example, determined to a great extent his approach to fieldwork and to the analysis of fieldwork data).

Lying behind this divergence between the encyclopaedic approach and the approach which sees data primarily in the light of theoretical preoccupations, however, lies an important issue about the role of *description* in ethnographic research, and we need to say something about this here. "Description" is sometimes thought of as being the opposite of "theory", and is furthermore thought of as being presuppositionless, "pure" and detached from any taint of ethnocentricity – at least ideally. On these grounds, the criticism is levelled at ethnography that it not only lacks theoretical rigour but, more damningly, also fails to live up to its own canons of descriptive purity, if they ever did. In fact the style of thought to be found in the so-called descriptive ethnographies was actually (although presumably quite unconsciously and unintentionally) close to the methods of phenomenology, a philosophical school which, while it has had little direct impact on anthropology, has had a profound influence on the development of the theory and practice of ethnomethodology in sociology.

In terms of the phenomenological method, the intention was to reject *a priori* constructions and systems and to concentrate on the description of *experience*, including not only what the subjects *do* (their physical action), but also their intentions and projects, the way they classify, and the ways in which they endow their world with senses and meanings. As a method then, phenomenology had a great deal in common not only with latter-day ethnomethodology, but also the development of the anthropological approach to ethnoscience, i.e. the study of indigenous sytems of classification, and of the emergence and nature of indigenous *theories* about the world, plants, animals or whatever. In essence the phenomenological method is about making explicit what is implicit (Turner 1974; §2.4).

It is really in the light of this seeming digression that we should see the project of descriptive ethnography: as a means of circumventing *a priori* approaches to society and culture, as a way of retaining the world of phenomena which gets obscured behind so many apparent methodological "advances", as well as a source of data. Many of these requirements remain for anthropology, although naturally there have been numerous improvements in style. The old division of ethnographies into discrete chapters, each dealing with a supposedly distinct aspect of the culture in question – the economy, the political system, the belief system – has happily disappeared, not only because such divisions are now seen to be largely arbitrary, but also because of changing literary standards and theoretical issues. Nevertheless, all ethnographies contain a heavy dose of description (as they must, not only for data, but because the genre demands it) and an ordered sequence of parts, if not in terms of the old systematic approach, then according to some newer idea: the theory of the life-cycle, the domestic cycle, the agricultural year, and so on. But whatever the format and the literary devices employed, a basic point remains: that the link between description and theory building has always been close in the history of ethnography, and an attempt to divorce them is to create or perpetuate a dangerous mythology.

If the problems of "pure description", which seems to be the philosophical basis of a lot of the encyclopaedic style of ethnographic research, are fairly major, what is to be done? The answer lies in the development of numerous styles of fieldwork which fall somewhere in between the poles of pure description and pure theory; and in the corresponding development of techniques and methodologies to express these styles. We can attempt to draw up a rough catalogue of the main varieties.

4.4.3 Simulation (§8.10.4)

Originally, the idea of simulation was to get the participants in a culture to do one of two things: to re-enact activities which formerly they had performed but had since abandoned (this is how Howitt (§3.2) interpreted the idea in his work with Australian Aborigines); or to perform an activity out of context for the purpose of an ethnographic description, if no such activity was likely to be performed spontaneously during the ethnographer's visit – for example a wedding ceremony. For either to be successful several requirements must be met: the participants must be able to remember accurately what to do and be good actors, while it must be socially and/or psychologically possible to perform the actions (e.g. a ritual sequence, a special dance) in circumstances in which they would not normally be performed. Only if these conditions are met can a successful and valid simulation take place, and this is hard to guarantee or verify. For these reasons, most ethnographers avoid simulations except in the last resort – if absolutely unable to record the event as a spontaneous manifestation, for example, or if faced with a situation in which the culture is rapidly being forgotten or destroyed.

Latterly, the idea of simulation has come to take on an additional and rather different meaning, when it is applied to the idea of computer simulation. In this case, the observer may build a mathematical model of such social processes as marriage exchanges, economic transactions or demographic patterns and then experiment on his/her model, since s/he cannot, of course, experiment on the actual people s/he is studying. This form of simulation can be very fruitful if the data are sufficiently good and sufficiently quantifiable to sustain it, since it can identify relationships between numerous hitherto obscurely related variables. The main problems with this technique are that frequently the data base is inadequate for such sophisticated model building; that false causal conclusions are drawn from the results of such data; and the fact that such models are always provisional and may be overthrown by the discovery of new data or by the discovery that the wrong assumptions had been made about the human and cultural behaviour underlying the mathematical data (Pelto and Pelto 1978: 307–308). The idea that such simulations, or their less developed predecessors (Leach 1968), actually "objectify" human behaviour and thus render it free from subjective factors or errors of analysis is, of course, false. The value of simulations of this mathematical kind is essentially as a heuristic device and an analytical tool rather than as an end in themselves.

4.4.4 The collection of texts (§4.4.7)

The recording, collecting, compilation and analysis of texts in the native language was one of the earliest styles of fieldwork and ethnographic analysis. It was very much a characteristic of the mature work of Franz Boas (§3.3), and reflected a number of beliefs: that data comes before theory, that textual work avoids the subjectivity and unverifiable quality of much allegedly descriptive ethnographic work, and that the vernacular texts themselves provide the most accurate and objective picture of the minds and preoccupations of the people studied. This is, in principle, undoubtedly true, but the textual method as developed by Boas had certain drawbacks.

First, the recorder of the texts (or the translator, if they are written down by literate informants or recorded by mechanical means) needs a highly sophisticated grasp of the native language, its colloquialisms and idioms. Secondly, the texts, by their very nature, are usually recorded from a very small number of informants and the validity of the method thus depends heavily on the truthfulness, accuracy and memory of these informants. Thirdly, the text is only one kind of information: an account of something rather than the event itself, unintelligible unless its context is fully understood. It is thus no substitute for fieldwork of a more extensive kind. Furthermore, a great deal of ethnographic information simply cannot be collected in the form of texts: actions, behaviour, events in themselves, processes, say, of economic exchange. Texts, rather, are talking *about* these things, and are no substitutes for observation of the actual events. Fourthly, the text itself in a sense may be an abstraction: a structured and therefore distorted condensation of complex events. The now large methodological literature on oral traditions makes this abundantly clear: texts can be distortions, selective, and subject to the informants elaborations and lapses of memory (Vansina 1969). The text itself, as a written document, may thus be even further removed from the event it records. Boas taught his Kwakiutl informant, Hunt, to read and write, and the process of literacy must also itself change the way in which events are seen, recorded and transmitted. Fifthly, the textual method can introduce a most unhealthy division between the informant and the analyst in cases where the former records the text and then sends them out of the field to the anthropologist to analyse. This technique, which leads to a separation between the data and its living context, was used by Boas and it is by no means dead: this is almost the only method that Lévi-Strauss has employed in his own huge synthesis and compilations, and it is perhaps not accidental that Boas is one of his sources – which, in this case, places the analyst at *two* removes from the data (Lévi-Strauss 1969). Finally, the textual method can lead to another variety of ethnographic encyclopaedism: vast compilations of material, but without form or structure or obvious links to the actual dynamics of the culture or the realities of everyday life.

Of course, if handled sympathetically, the textual method can yield very valuable results. It does lead to the generation of a body of material that is "fixed"; it does not disappear, as does so much ethnographic data, with the passing moment. It can be worked over more than once; it can provide data for other anthropologists with comparative, historical or structural interests to analyse and work with; it has led to important methodological advances in the related field of oral tradition; it produces a body of important information *per se*; and when it *is* integrated with observation, participation and a deep understanding of the everyday life of the culture in question, it provides an excellent and fruitful mode of analysis of events. This is why it still finds its adherents (e.g. Crane 1982). The textual method, in any case, was never historically confined to Boas: Malinowski used and recommended it, for example, but as one method amongst others (notably, the collection of genealogies and other objective material, and participant observation).

Some of the "spin-offs" from the textual method have in fact been very fruitful. One has been the recognition of the fact that literary sources are important in their own right (a fact obscured by the fetish for fieldwork-as-participant-observation) and may indeed be *necessary* to the study of some cultures – especially the complex literate ones (cf. Gillin 1949; Freedman 1963). Another has been the recognition of the limitations of informants divorced from first-hand experience by the ethnographer. A third has been the growing emphasis on both the ability of the anthropologist to know and use field languages well and on language and linguistic behaviour as central areas of ethnographic concern and investigation *in themselves* (Ardener 1971). The necessity for the recorder of texts to be a good linguist has thus almost entirely undermined the older tradition of working through inter-preters (except in the early stages of fieldwork), a move which has immeasur-ably increased the quality and insight of good ethnographic reportage and made access to such subtle but central areas of culture as classification increasingly possible.

Finally, the textual method gave rise to another method which has acquired some prominence of its own and is somewhat separate from that of its parent – notably, the *life-history* (§8.7). The life history (as originally conceived by Radin in the United States) is essentially the compilation of a text by an informant recording, in his own words or with the help of the ethnographer, the history of his own life, the events in his personal history, the cultural and economic influences bearing on him, etc. (for a brief example, see Freeman 1978). The attraction of such a technique is that it can provide a very personal and vivid account of a culture from within – through the eyes of one of its own members – and, as such, not only provides a document of great intrinsic ethnographic interest, but also provides material for the development of the life-cycle concept (the concept of describing a culture by following the personal development of a "typical" member of that culture through life). As such, this extension of the textual method has had considerable impact, especially on the anthropological study of socialization (Mayer 1972).

We have touched here on an important point – that changes in fieldwork styles, or the evolution of what is supposed to be meant by the concept of fieldwork, often take place because of one of a number of factors. The first is cross-fertilization from (at first sight) seemingly unrelated areas within the discipline (thus the textual method influenced linguistic attitudes, and in turn linguistics has had an enormous, although often indirect, influence on fieldwork in general). The second is changes occasioned by technological innovations, such as computation. A third is progressive changes in the dominant paradigm of what constitutes anthropology. An example of this is the movement in Britain (less marked in the United States) away from the older conception of anthropology as a holistic science of man (which incorporated biological, linguistic and archaeological approaches as well as cultural and sociological ones) towards a much more sociological view: towards, in fact, a very Radcliffe-Brownian view of anthropology as micro-sociology. This latter idea, which is what is usually designated by the name "social anthropology" in the United Kingdom, has had a great influence on the development of fieldwork styles, which have, correspondingly, moved closer to those of sociology. In many cases, two of these three factors have come together, in particular the "sociologizing" of anthropology and technological innovation to produce a style of fieldwork which may form our next general category.

4.4.5 Survey methods (§8.8)

Within this category there are several important sub-divisions. One of these is the camp that favours survey methods and their associated quantitative techniques because this kind of numeracy seems to offer a route along which empirical and theoretical advances are likely to be possible, or because no distinction between anthropological and sociological techniques is thought to be necessary. A second camp consists of those who have taken on the study of large and complex societies (typically in Africa) and find that the traditional style of individual face-to-face fieldwork simply is not adequate for the scale and problem of the societies that they confront. A common stylistic development here has been the mixing of traditional fieldwork with some sort of quantitative survey (e.g. Pons 1960). There is also a third camp which overlaps with this second one but deserves separate treatment because of its growing importance, notably urban anthropology.

For the members of each of these camps, certain methods, typically associated with a more sociological style of work, have come to prominence; in particular the making of large-scale social surveys (of city wards for example, or of migrant flows), the use of secondary statistical data (e.g. census returns), the use of questionnaires as a research tool, and the statistical analysis of the returns from these questionnaires using computer processing techniques. Such innovations, necessary as they may be, have radically changed fieldwork styles. In many such cases, no longer does the ethnographer labour alone in a remote community with few aids other than a notebook. Now he may operate in a large city, quite possibly as a member of a team, perhaps an interdisciplinary one, and much of the actual interviewing, formerly the *forte* of the individual anthropologist, may now be done by paid research assistants. Despite this, there are those who would maintain that quantitative methods are still subordinate to the "ethnographic work proper", which is still of the traditional style – quantification being seen as a means to an end which statistics cannot capture (Mitchell 1967: 21).

4.4.6 Humanistic and subjectivist approaches

In some respects the survey approach represents one end of a stylistic spectrum: the end tending towards an identification with sociology, with technique and with a growing interest in quantitative methodology as an area of substantive concern in itself. But of course all spectra have two ends, and some would see the opposite end as represented in a set of styles which might loosely be characterized as "subjective soaking". To some extent, however, it must be admitted that this polarization is artificial: as we have just noted, many of the quantifiers do not see numerical methods as an *alternative* to traditional, participant-observation style fieldwork, so much as techniques which can enhance it.

Put simply, the idea of *subjective soaking* is that the fieldworker from the outset abandons the idea of absolute objectivity or scientific neutrality and attempts rather to merge him/herself into the culture being studied. As a style, it has many affinities with the phenomenological method and with the more commonly accepted view of participant observation, from which it differs, however, in its eschewing of "objective" methods as an adjunct to the participation and in its elaboration of certain ideas perhaps inherent in the idea of participant observation – for example, that of "empathy". Indeed, the key concept here is that of *participation*, in a sense very close to that used by Lévy-Bruhl in his studies of the so-called "primitive mentality", which really comes very close to a mystical identity of fieldworker and subject. One of the terms most frequently encountered here is that of empathy: the attempt, that is, to somehow think and feel oneself into the minds and emotions of one's subjects (e.g. Marwick 1967: 231, 241).

The problem here lies largely in the meaning of the associated term "identification". Clearly any sensitive researcher needs to be sympathetic towards informants (although, despite his theoretical and methodological claims about this, even Malinowski in many respects failed to achieve this: Malinowski 1967). S/he likewise needs to attempt to grasp their world view as they see it, to understand their life-projects in their own terms, to penetrate their rationality. But (except in the most exceptional cases) the anthropologist does not *become* one of them. S/he spends a relatively brief period with them, may not have a perfect grasp of their language, and is always regarded by the subjects as an outsider (however sympathetic) or even as someone to be exploited for payment, for a job or contacts in the outside world. The difficulty of empathizing with someone known intimately in one's own culture is great: the possibility of doing so with the members of an alien society are very much greater. Part of the problem is that one never knows for sure if one has empathized at all or empathized correctly: verification is often impossible, and much resulting anthropological understanding is hence illusory. (A recent *cause celébre* being the debate surrounding the works of Carlos Casteneda and the reliability of his reports: Casteneda 1972.)

"Subjective soaking", then, should be regarded as a part of all participant-observation style fieldwork, and is valuable as long as its major limitations are recognized and provided that claims to knowledge are not made where they cannot be tested or documented. The issues surrounding the subjectivist style have nevertheless given rise to some interesting developments. They have, for example, spilled over into debates in philosophical anthropology about the nature of rationality in other cultures and the possibility of achieving understanding of alien belief systems (Wilson 1970).

4.4.7 Fieldwork as translation, ethnography as text

Subjectivist approaches have also given rise to the idea of fieldwork as *translation*, regarding the other culture as an opaque text in an unknown language. This approach is perhaps best illustrated in the Oxford tradition emanating from the pioneer work of Evans-Pritchard. The objective of the fieldwork and the subsequent analysis is thus to "read" the text, and the problems of fieldwork are seen as being like those of a literary translator: in referring, say, to "witchcraft", "money" or "soul", how does one know that one has chosen the *right* term or concept to render an alien notion? The link with the subjectivist method here is through the concept of *hermeneutics*.

In a more recent sense, the "text" which is to be examined is not simply the culture in the abstract but the actual notebooks and memories of the ethnographer. Thus the text is seen as the product of an interactive process between described and describer, analysed and analyst (Clifford 1980; §1.4).

4.4.8 Formal emic approaches

The traditions of fieldwork associated with Malinowski and Evans-Pritchard, have both been especially concerned with understanding the native mind. There is here a direct connection with the notion of ethnographic translation and the subjectivist approaches discussed above. However, it was in response to the perceived lack of rigour in these methods that, in the decades following 1950, there developed in the United States more formal procedures for eliciting information concerning indigenous categories and concepts and the relations between them. These procedures, based on analogies with linguistics, formed the basis for a new methodological vision of anthropology which we now know as *ethnoscience*. Practitioners of ethnoscience have recorded "cultural grammars" in an attempt to provide accurate descriptions of what a person would have to do to perform effectively in a particular culture. Culture is, in this view, seen as a tool-kit for getting things done; and fieldwork conducted in this tradition relies extensively on the use of formal "question frames" and other standardized procedures (§8.4). For some, this has been no less than a "quiet revolution", resulting in a "new ethnography" and a new conception of ethnography. The result has been a degree of confusion, not least because some of its proponents have seemingly redefined ethnography itself as a set of highly formal techniques for extracting cognitive data (e.g. Spradley 1980; Frake 1964; for a critique, see Manning and Fabrega 1976).

4.4.9 The extended case method (§8.5)

As we can see, there are often "styles within styles", and occasionally attempts are made to synthesize several of these varieties into a single entity. An example of this approach would be that of the *case-study*, often applied in the study of legal anthropology. Here there are several problems, both conceptual and methodological, relating to what law is, if it is to be defined, how legal systems (if they exist at all) operate in non-centralized societies, whether the absence of codified law means that there is no concept of justice, and how disputes are adjudicated. The style of working through case studies, i.e. individual instances of the behavioural realm under analysis, from which rules and generalized insights can be derived, is seen by some fieldworkers as an approach which requires participant observation, allows focus on certain relatively bounded units of behaviour (not "the village" here, but "the case") and should provide a series of building blocks out of which theory can later be constructed (Epstein 1967: 205–230).

4.4.10 Action anthropology

There is another major stylistic area which must be touched upon – that of action anthropology[3] in its various manifestations. Again we are faced with a number of sub-varieties – *applied anthropology* (applied, that is, these days typically in the context of development projects, and in the past typically to problems of colonial administration); with *partisan anthropology* (the involvement of the anthropologist quite consciously in political and social action by and for the community studied), with the much maligned *missionary anthropology*, and finally with *salvage anthropology* (§4.4.3).

Stylistically these are all related, varied as they are in their specific intentions. Fieldwork undertaken in the context of action-anthropology is often very different from that undertaken for "pure" purposes – it is often short-term, oriented to the solution of a practical problem and renders quite impossible long-term participant observation or subjective soaking. Nevertheless, with increasing interest in development issues and with the shrinkage of the academic job market, action anthropology or operational research (e.g. Spillius 1957) is likely to expand out of its presently very marginal role in the profession. (For a summary of the issues and problems in applied anthropology see Bastide 1973.) Partisan anthropology is really a politicized variety of applied anthropology, and its precise nature will be heavily influenced by the particular political ideology involved, as well as by the specific problems to which it is directed. As a style, it implies close involvement with the community in which the anthropologist is involved, identification with and activity in their socio-politico-economic struggles, and a willingness to take considerable political, professional and even physical risks. Siding with a guerrilla movement, for example, can be dangerous, to oneself as well as to one's objectivity – which is the criticism usually directed at this style of work (Jaspan 1964). Again, it is not always clear what fieldwork would actually mean in practical terms in partisan anthropology – uncovering a relationship of exploitation perhaps, or mobilizing of grass roots protest movements.

Missionary anthropology tends to have a rather bad reputation amongst academic anthropologists, but this again can usually be traced to three

[3] For an account of *action research* in sociology, which in Britain has been closely associated with the work of the Tavistock Institute of Human Relations, see Jacques (1951) and Trist and Bamforth (1951).

sources; differences in intention (the missionary being concerned say, with the religious conversion of the locals), differences in ideology and differences in opinion as to the value of the local culture – the missionary often (at least in the past) seeing it as negative and an obstacle to his other intentions; the academic anthropologist as a thing of intrinsic interest. The resulting clash of attitudes has, however, obscured certain very important positive features of missionary anthropology: the great amounts of data collected by missionaries, their laying of the groundwork (for example, recording of languages) which has made it possible for the shorter-term academic anthropologist to do his work at all, and their tendency to stay for very long periods with the chosen culture (20 or 30 years being not exceptional) and thereby to achieve a degree of insight and intimacy with the culture that no academic anthropologist ever achieves. Some rare academic commentators have indeed noticed this (e.g. Burridge 1975), that missionary anthropology (like other forms of action anthropology) *can* be a uniquely insightful and valuable approach. Finally, *salvage anthropology* is concerned with the urgent recording of rapidly vanishing cultures before they disappear. It tends to be short-term, utilizing the textual method, simulation (§4.4.3), interviewing of elderly remaining informants and, these days, often involves the use of visual aids, especially ethnographic film. It is also sometimes described under the label of *urgent anthropology* (Mead 1965: 5).

4.4.11 National fieldwork styles

Finally, some mention must be made of national styles of ethnographic research. Two differences are immediately apparent between British and American styles for example. The first being the high degree of British casualness towards systematic *training* for fieldwork (of graduate students in particular) contrasted with the thoroughness and formalism of the American approach (contrast Epstein (1967) with Pelto and Pelto (1977) for this difference to be vividly highlighted; §7.1). The second is the British tradition of the isolated fieldworker, working entirely alone, without much in the way of backing facilities or resources, and in professional communication only with his thesis supervisor, contrasted with the common American style of team-fieldwork. Two well known examples of the latter were the comprehensive study of the Thai village of Bang Chan by Hanks and others (for documentation see Keyes 1977: 167) and the studies carried out in east central Java by Geertz and a team of colleagues in the 1950s (Geertz 1959: ix). Very few British examples of teamwork (§7.11) of this kind exist (since the Torres

Straits expedition), and even where a couple have worked in the same location, it is usually on totally different topics (for example, Freedman on the Chinese community in Singapore and his wife on the Malays: Freedman 1957 and Djamour 1959). Even here, however, there are certain convergences – the most conspicuous being the independent publication of individually authored monographs (one looks in vain for any collective writings or syntheses), and the disdain in which "expeditions" are held: they are considered to be useful for surveying the ethnographic possibilities, but not for "serious" fieldwork. The French (Lewis 1973: 6; Richards 1967), German, Scandinavian and other national traditions (on which see Diamond 1980; Gerholm and Hannerz 1982: part 2) in turn reveal numerous differing emphases, foci and nuances of style (§3.1, note 1).

4.5 Conclusion

Clearly a survey of styles can never be totally comprehensive. We could, for example, single-out "sub-styles" within the broad stream of participant observation, such as mass-observation (Madge 1957; Stanley 1981), or various orientations informed by particular theoretical developments (e.g. transactionalism, culture and personality) (Johnson 1978). Nevertheless, we can draw the discussion together by looking at certain issues which may help to contextualize the debate. We have seen that the fieldwork concept, at first sight apparently unproblematic, is actually really rather complex: in terms of the multiplicity of styles that the idea covers and in the theoretical and methodological problems that it raises. Furthermore, the situation is dynamic: changes in emphasis within anthropology, such as the increasing importance of urban studies, are having a major effect on the fieldwork concept. Similarly, changing political circumstances have had an impact. Many "first world" anthropologists now do fieldwork in their own cultures because of lack of access to "exotic"ones, others have redefined their identity (perhaps calling themselves sociologists or educationalists) in order to gain entry to field situations abroad, while most "third world" anthropologists, although they may be trained abroad, tend to do their fieldwork at home (in some cases in their own natal village) (§5.5).

These changes raise new problems. Thus, styles of doing fieldwork in one's own culture are often of necessity different from doing fieldwork in another, there being pros (such as knowledge of the language) and cons (such as the difficulty of seeing what is really significant in a "taken for granted" situation that the fieldworker has grown up in). This increasingly familiar state of affairs raises also the insider/outsider question in fieldwork, issues of objectivity and subjectivity, ideological questions and questions about the verification of field data. There are no simple answers to these, but rather than engage in methodological debate, we may point out, once again, that the style of work to be expected from an insider is likely to be rather different from that of an outsider, *even when they are using the same formal methods.* This illustrates perhaps more clearly than any other example, the significance of *style.* (On the outsider/insider question, see Brislin and Holwill 1979.)

The determination of which style to use is not simply a question of the personality of the fieldworker (important and underrated as this factor may be). It is also heavily influenced by the subject being studied. Nowhere is this more clear than in the field of urban anthropology, not only because of the unit of analysis (the town) and the corresponding complexity of social processes, but also because of the crisis of confidence that urban ethnography has precipitated in the continuing validity and assumptions of the traditional styles and methods of fieldwork. To some extent this crisis is unnecessary because it is based on the entirely false belief that anthropology is the science of primitive, rural and small-scale societies. The growing emphasis on the town is thus seen as a major retreat: an attempt to shore up the discipline against the progressive collapse of its traditional subject matter. Hopefully we can agree that this is a warped view (for discussion, see Fox 1977; Eames and Goode 1977). The real challenge of urban anthropology comes from the question it raises about the applicability of the usual styles of fieldwork in an urban setting. How does one do participant observation in a community of perhaps one million people? How do you study, say, an ethnic group when its members are scattered all over a huge urban area? How do you integrate the traditional qualitative with the newer quantitative approaches in this sort of context? (for a relatively early attempt to answer these questions, see Gillin 1949).

Numerous answers are beginning to emerge to these sorts of questions. Some anthropologists have predictably fallen back on studying enclaves, bars, street corners and marginal groups, or the kinds of relationships that are traditionally studied by anthropologists and which also exist in town – for example, kinship (for a good example of this narrow and mistaken approach, see Wolf 1968). Others have selected certain processes in the adaptation of tribal or rural cultures to city life and have concentrated on the development of surveys and other methods appropriate to processes such as migration (Mitchell 1967). Yet others have attempted to develop a system of analytical categories (e.g. "networks", "quasi-groups") to characterize social relations in urban settings (Mayer 1968). This process of stylistic elaboration is by no means at an end and innovations continue to appear (e.g. Clammer 1982), including the application of symbolic analyses to urban settings; the resurgence of interest in ethnicity; and linkages between the urban sector and questions of under-development, for example the "tertiary sector" or "informal economy" (Rimmer *et al.* 1978). Clearly urban anthropology is an area to be watched closely, for within it one can expect to see stylistic and methodological advances most fully highlighted; indeed they have to be if anthropology in urban settings is to achieve any degree of rigour, insight and professional respectability.

Three final points need to be made. The first of these is that the style of fieldwork adopted is a reflection of at least three factors: the personality and personal tastes of the individual fieldworker, the nature of the problem selected for study and the general philosophical models held by the fieldworker in the prevailing climate of the profession at that moment. In this latter category would come both ideological factors and ones pertaining to beliefs about the nature of social facts.

The second point is that despite the almost universal assent to the idea of fieldwork as the foundation of method in anthropology (and particularly in social anthropology), the concept has its critics, and some of their criticisms are very telling. Thus, while it might be agreed that fieldwork is a source of great insight, of data that can be collected in no other way, and is itself an enriching and even revolutionary experience in the personal and professional life of the individual fieldworker, it also has its severe limitations. Some of these – for example the problems with the concept of participant observation, and the serious conceptual limitations of the "village study" – we have already mentioned. There are also others – the main ones being the impossibility of building up a macro-picture from fieldwork data, the difficulty of generalizing from the narrow base of a small community study, the impossibility of dealing adequately with diachronic and historical forces working on and within the particular community, and above all, the frequent inability of fieldwork in the traditional sense to generate the correct kinds of data at all (for example, concerning the existence of class relationships). To put it the other way round, there are certain kinds of questions that one simply cannot ask of traditional fieldwork data, since the techniques inherent in that method have no way to answer or even discover them. When one is concerned, for example, with social change or with the working of politico-economic structures in the wider society on the smaller community, the limitations of fieldwork are glaringly revealed. Fieldwork, therefore, cannot be the only *method* in anthropology, but needs to be complemented by others, which also points out the foolishness of identifying anthropology *with* a particular method which is supposedly its defining characteristic and unique to itself (for a very good critique of fieldwork along some of these lines see Joshi 1979).

Finally, we must briefly return to the idea of *style*. It has been suggested here that style is not merely a matter of personal nuances, of taste. In reality, it also reflects theoretical intentions, ideological presuppositions, the specific definition of the object of study, and constraints imposed upon the fieldworker by the context (social, cultural, economic or ecological) in which he or she works. But that is not all: the ethnography itself, the written interpretation (rather than *record*) of fieldwork is a product of varying beliefs about (and ability to manipulate successfully) styles. The translation of fieldwork into ethnography is itself, in other words, an interesting and problematic area for investigation. As at least one writer (Marcus 1980) has pointed out, ethnographic writing is itself a genre with its own rules of evolution, conventions, implicit standards, hidden links with phenomenology and hermeneutics, modes of self-revelation, rhetoric and expectations about what ethnographies are like, what they should do and their rules of evidence and presentation of "facts", including their adherence to traditional motifs such as the model of holism. Anthropology at the present time, while possessing a loosely defined body of customs about how these things should be done, lacks any single organizing paradigm. The consequence is the proliferation of styles, both in fieldwork and in the presentation of the results of fieldwork in an ethnographic text. This diversity need not be seen in a negative way: it is one of the sources for the richness and variety of anthropological methods and projects. Both the strengths and the weaknesses, however, of the fieldwork methods need to be borne in mind. Not only is it not the only method available to anthropologists but, with changing techniques and theoretical interests (socio-biology, Marxist anthropology etc.), the fieldwork concept (and the concept of "the ethnography") will inevitably change. The historical resilience of anthropology suggests that this will only prove to be for the progressive enrichment of the discipline as a whole.

John Clammer

5 *The fieldwork experience*

5.1 Introduction

It has been said many times before that research in the social sciences differs radically from that in the natural sciences in that it involves an intimate and special interaction between investigator and object of study. Ethnographic research in particular does not permit the separation of occupational and extra-occupational spheres (Geertz 1982: 157).[1] Clearly, this presents a large number of problems which, in turn, explain the apparent obsession among social scientists with methodology (§2.1). These include the presentation of self, communication, the interpretation of others, the ways in which personal relationships affect data, how research affects the researcher, and the consequences of the interaction of all these. Thus, personality, psychological stress, latent ethnocentrism and other background experiences may become major practical matters (see Clarke 1973: 17). Even feeling may not simply be an added attraction or source of discomfort; it may be an epistemologically vital precursor to understanding (Geertz 1975: 47).

It has been recognized in the past that so many ethnographic accounts have suppressed the personal element (Beals 1970; Crapanzano 1970; Scholte 1980) that it has led to an increasing awareness of the importance of the subjective role of the anthropologist in the production of data over the last 20 years. Some of the historical forces responsible for this have already been discussed (§1.4), but we should now add to these the increasing personal involvement by anthropologists in their work, the "democratization" of anthropology, multiple studies of the same culture (sometimes the same

---

[1] For two further illuminating discussions of this fundamental issue, see Shils (1959) and Clarke (1973).

ETHNOGRAPHIC RESEARCH
ISBN 0 12 237180 1

village), and assertions of independence by native peoples (Nash and Wintrob 1972). On the other hand, we should not become so obsessively concerned with empathy that we overlook the possibility that objectivity may at times be reduced by overengagement (see, e.g. Miller 1952). Personal experience is obviously important, but we forget only at our peril that the fieldworker is also a representative of a discipline and of a cultural system (Dwyer 1979). Relationships entered into are social and cultural as well as personal. For such reasons it is necessary to review what is already on record concerning the various aspects of the subjective experience of fieldwork and on problems of personal interaction and adjustment.

5.2 Publications on fieldwork experience

5.2.1 Introduction

There is now an extensive literature on the personal experience of doing fieldwork, covering a wide range of field situations in most parts of the world (§5.2.2). The bulk of it may broadly be categorized as follows:

(a) Reports in the context of accounts of field methods (e.g. Hubert *et al.* 1968; Chagnon 1974) and in ethnographic monographs (e.g. Whyte 1955).
(b) Individual reports on particular field experiences in periodicals or edited thematic collections (e.g. the contributions to Spindler 1970).
(c) Autobiographies (e.g. Powdermaker 1966; Mead 1972).
(d) Reflective philosophical accounts (e.g. Read 1965; Lévi-Strauss 1955; Turnbull 1972).
(e) Diary accounts (e.g. Malinowski 1967; and, on Boas: Rohner and Rohner 1969) and letters (e.g. Mead 1977), presumably not intended for publication.
(f) Book-length popular accounts (e.g. Turnbull 1961), travelogues and fictionalized accounts (e.g. Bohannan 1964) and "ethnographic novels" (Coon 1932).

Apart from the fact that the boundaries between these categories are sometimes unclear, such a classification cuts across common themes. It is according to theme that the literature is best discussed if the reader is seeking advice or a comparison of experiences in relation to specific issues (§5.3, §5.4, §5.5).

First of all, however, something must be said about the individual *genre* listed above and their qualities.

Until the 1960s, published accounts of personal fieldwork experiences were generally limited to a few introductory remarks (e.g. Malinowski 1922: 1–25; Evans-Pritchard 1940: 7–15, 82–84; Hilda Kuper 1947: 1–10), or an optional methodological appendix (Whyte 1955: 279–358) in an ethnographic monograph.

One important exception was Lévi-Strauss's *Tristes Tropiques*, first published in 1955. A version in English, omitting a number of chapters, appeared in 1961 as *World on the Wane*; a comprehensive and satisfactory translation did not appear until 1973. The book is a curious combination of travelogue and theoretical musings in which fieldwork is portrayed as both an initiation ordeal and as a mystical experience. In substance, it is a dramatic and descriptive chronology of events, people and settings in a variety of places of his professional acquaintance, seen through the anthropologist's eye. It is a colourful and elegantly written piece, the author's occupational identity betrayed only through his choice of subject matter. The book changes towards the end, however, when the author narrates a resumé of an unfinished play he wrote over a six-day period while awaiting companions stranded by epidemic 80 km away. The story is recounted to illustrate "the mental disorder to which the traveller is exposed through abnormal living conditions over a prolonged period". The book concludes with two chapters, one contemplating the nature of societies and the comparative role of the anthropologist, the other reflecting upon aspects of his own existence as evidence of the personal wisdom he has acquired through being a reflective anthropologist. Whilst the book makes good reading, it is difficult to see its practical value to fieldworkers.

That the dramatic rise in the number of personal accounts of fieldwork experience took place in the 1960s was not entirely fortuitous. This was a period of intense introspection in anthropology. Challenges from without led to self-doubt, questioning and a new openness in the discussion of methods. These developments took their most extreme form in the, frankly, mystical writings of Castaneda (1968, 1971, 1973; see also Douglas 1973) concerning his experiences apprenticed to a Yaqui sorcerer. The accounts that we have are therefore often autobiographical in a general sense as well as reporting on the day-to-day personal events of fieldwork. Fieldwork is often seen as a kind of therapy, the ethnographer emerging from it with increased self-knowledge, specifically concerning limits and capabilities when coping with a completely new environment and new kinds of personal relations.

Not only does fieldwork involve the kind of education which is inevitable

in having to adapt to a new (and often exotic) social and physical environment (e.g. Norbeck 1970); it also generally involves the development of a new
social personality for the field situation, or a new self-image and social
projection, and very often profoundly affects the fieldworker's own values
and world view. Sometimes these effects may be extreme. Castaneda enters a
world so different that he comes to accept reality itself as nothing but a social
construct, with effects so devastating on the anthropologist that ethnography
becomes mysticism. For other writers the effects of fieldwork are less spectacular but subjectively no less important. For Joshi (1979: 78), for example,
fieldwork was a period of intellectual and emotional self-examination. Joshi's
account is an instructive example of how, because a fieldworker has to
"suspend his meaning system for the time being and to understand the
people's meaning system" (Joshi 1979: 73), part of the fieldwork experience
is the gradual recognition that the theories or ideas that the anthropologist
takes into the field are generally being modified through experience and new
information. For Raman Unni (1979: 70), the personal rewards were more
general: excitement and the ability to see cultural phenomena in their total
context.

Retrospectively, most anthropologists see fieldwork, and particularly their
first period of fieldwork, as a very important episode in their own psychological and intellectual development. As Gullick (1970: 124) points out, in
addition to having to cope with one's familiar emotions – boredom, elation,
embarrassment, contentment, anger, joy, anxiety, and so forth – in the field, it
is also necessary to be continually alert, learning new routines and cues which
highlight previously hidden aspects of a personality.

The discovery of these can be emotionally devastating, Read (1965: 8)
suggests. Before going into the field, he confesses (1965: 5–6), he was not the
gregarious kind and found it difficult to interact in "the polite and superficial
fashion generally expected of us", preventing him from having a large
number of friends. He had feared that this (and the added uncertainty of not
knowing what would be expected of him or how he should treat his
informants) would handicap his fieldwork, which would depend upon
establishing personal rapport. Despite these fears, he found that he made
deep and lasting friendships in the field, friendships which had profound
effects on his inner development, affecting him the more deeply, he said,
because there seemed to be such little common ground on which to base a
personal relationship. Moreover, he came to see, through fieldwork, that his
preference for his own company in his own culture, which he had regarded as
part of some personal inadequacy that prevented him from making the right
response and experiencing the usual enthusiasms of most of his acquain-

tances, was more a personal strength than a weakness. He realized, positively, that he was capable of being quite alone, without others of his own kind, for a considerable length of time. There may be some truth in the suggestion made by Gullick (1970: 124):

> that many anthropologists are attracted to the field less out of love for other cultures than out of the feeling of being inadequate in their own.

So what can we learn from reading autobiographical accounts of field-work? What makes one kind of account more useful than another? What kinds of "usefulness" should we have in mind?

What appears to have prompted the publication of many accounts is a consciousness that the rapport struck between an individual ethnographer and the subjects of research is likely to be unique, together with a concern about the effects of this on the results of research. This might lead us to expect that such reports will be so idiosyncratic as to have little instructive value for others. However, whilst each fieldwork experience is bound to be unique in many respects, it is possible to identify certain *kinds* of problems common to most situations in which ethnographers find themselves. The edited collec-tions of short articles on the experiences of a number of anthropologists, where the authors have had to summarize their experiences, are valuable because they permit comparison and certain generalizations to be made. Henry and Saberwal (1969), for example, include accounts by five young fieldworkers in the 1960s of their particular stresses in fieldwork. The final chapter, written by a psychiatrist, identifies the kinds of personal, social and political stresses most fieldworkers have to contend with. The editors sum-marize these kinds of stresses into three phases: a primary stress of involve-ment, of establishing rapport with strangers, of participation and reciprocity; a secondary stress of commitment, when intense involvement with "my people" varying from a warm emotional relationship to mutual hate (cf. Chagnon 1974; Turnbull 1972), may produce conflicts in the performance of professional duties (Kloos 1969); and a tertiary stress on finding that one's research design is inadequate, even irrelevant – which is almost inevitable given that the anthropologist rarely has adequate information about a society before embarking upon fieldwork.

The problem with these and other generalizations is that they are *so* general. If any ethnographer preparing to go to the field for the first time sits down and thinks about the kinds of problems s/he is likely to have to face, s/he is bound to list problems of entry and coping with new kinds of physical, emotional and social demands upon him/herself, bound to expect problems concomitant with the role of participant-observer, bound to anticipate changes in a research design prepared before going into the field. What intending fieldworkers really want to know about are not the abstract *kinds* of problems about which generalizations can be made, but rather how they are likely to be able to cope with the practical and emotional ones *they* are going to face. This they can never know beforehand, and reading through the literature summarizing the experience of others only seems to confirm the idea that, except in the most general sense, all fieldwork experiences are different. In order to get a feeling of the emotional impact and practical consequences of particular problems, frustrations, sources of elation, boredom, ethical conflicts and misunderstandings, prospective fieldworkers are perhaps better off reading more detailed accounts by a single author. Many of the summary accounts in edited volumes or in periodicals are too far from the immediacy of emotion, too predigested and analysed to convey a real sense of what it was like. The sections of a field journal which Hitchcock (1970) includes in his piece are one notable exception and are far more informative than the dry and somewhat turgid prose he uses in the text.

The best "what it was like" as opposed to "how to do it" presentations are, however, generally found in the longer accounts, of which the works of Berreman (1962), Bohannan (1964) and Chagnon (1974) are notable examples. Chagnon, in particular, is powerful prose; others verge towards purple. Such accounts have the space to explore particular relationships, emotional involvements and reactions, and the day-to-day aspects of living in a strange environment.

Laura Bohannan's "anthropological novel", *Return to Laughter* (1964), was first published pseudonymously to detach the mistakes, irrationalities, stupidities and embarrassments recounted in the novel from the scientific treatise based on her fieldwork in Nigeria. Despite any literary licence used in the story, the novel conveys better than many of the later non-fictionalized accounts what fieldwork may be like. It could only have been written by an experienced fieldworker and a clever writer. The kind of bland statements about experiences found in the summary accounts of real experience come to life and are conveyed more subtly through the narrative of the novel. The reader experiences the heroine's sense of isolation in a crowd, appreciates how routines and servants kept her sane, how she arrested the loss of her identity through constant note-taking and consciously playing the role of the aloof anthropologist, how misunderstandings and moral conflicts arose. In fact she recounts many of the kinds of joys and tribulations detachedly related in the literature by anthropologists working in a wide variety of contexts.

Read's *The High Valley* (1965) perhaps epitomizes the kind of mystical voyeurism of which Lévi-Strauss's book is the prototype. His intentions, stated in the preface, are promising: he sets out to correct the imbalance in the anthropological literature which he regards as "too antiseptic . . . so devoid of anything which brings it to life". Like Lévi-Strauss' book, Read's revolves around individuals and events, with a good deal of descriptive ethnography thrown in. The author is located somewhere in the background of events, doing the feeling and reacting but never interacting. We thus never really share in the experiences of the working anthropologist: we never see him take out his notebook or talk to people, despite the fact that the book focuses on the author's relationships with three key individuals. The book fails, in part, precisely because it is not written as a novel. It lacks the novel's sense of chronological and emotional progression and such techniques as first-person dialogue in parentheses to break the narrative, techniques on which the Bohannan book flows, arresting and carrying the reader's imagination.

Read describes his own reactions to traumatic rites of initiation, his aesthetic and emotional reactions to filth and lack of personal hygiene. At the same time, he accepts that, although he "felt for" those going through such ordeals as the initiation rituals, he could never really know what was going through their minds. While this is true, Bohannan is able to convey cultural differences in patterns of thought and their consequences on fieldwork through the speech and actions of the characters in her novel: thought differences are not taken for granted but dramatically revealed. Because Bohannan's book is written as a novel, there is a more or less implicit sense of the heroine's emotional progression and development, greater scope to explore the kinds of culture clashes which may be experienced by an individual who "enters" a foreign culture and the ways these conflicts may be resolved. To this extent it is instructive.

Although Read is also dramatic in places, the reader is left feeling that the book is more catharsis for the author than a "lesson by example" for the reader. It is too personal an account and tells us much more about the author than what it would be like for anyone else attempting fieldwork in New Guinea. Throughout the book, he lapses into somewhat abstract reveries, metaphysical reflections on the human condition, as when illness fore-shortened his fieldwork (1965: 22). It is a curious book, and I wonder why it was written. Between his own psychological musings are large chunks of ethnographic description, like the long section on the relationship between the sexes, which seem to have been incorporated somewhat fortuitously and would better have been included in a more conventional ethnographic monograph.

More private still is Malinowski's posthumously published *Diary in the Strict Sense of the Term* (1967; see also Forge 1967; Young 1979). His incessant fantasies about female spectres from his past, his longings for his mother and latterly the woman he was to marry, his obsessive concern with his health and medication, however, do show us something about fieldwork as much as about the man. Fieldwork is caricatured as a kind of limbo, a kind of self-exile and feeling of imprisonment away from "civilization". His lecherous thoughts are symptomatic of his yearning for the white civilization of which the objects of his fantasies were an integral part. The limbo of fieldwork was also a time for him to reflect upon his life in the "civilized" world before re-entering it. His reflections included "a strong upsurge of pro-British feelings" (1967: 209) and guilt about not doing his bit for the war effort, shortly followed by an equally intense hatred for England and the English; and constant resolutions on his own moral reform. Fieldwork itself seems to have been a part-time occupation, secondary to the pursuit of pleasure, leisure and exercise, fitted in between novels and *Punch*, taking tea or an enema, or confined to the period before breakfast when he "did the rounds" of the village he happened to be in. "Ethnographic problems", he once admits (1967: 272) "don't preoccupy me at all. At bottom I'm living outside Kiriwina".

The sense of isolation and alienation, of living in a place and yet not being a part of it, pervades the book. Malinowski does not appear to have related to the Trobrianders as people at all (Hsu 1979: 251). Such feelings are also reported in many of the more recently written accounts of fieldwork, but Malinowski's diary is naturally of special historical interest. Although particular individuals may be irritating, aversion to one's informants is probably less common today (although, cf. Turnbull 1974) – and probably tells us as much about colonial attitudes as Read's book tells us about American culture of the 1960s. "Homesickness", however, in one more or less acute form or another, plagues many anthropologists alone in a strange environment, and those with stronger emotional ties "back home" more than others.

Personal accounts of fieldwork are very mixed in content, orientation, intention and quality. Some are written by anthropologists who have clearly thoroughly enjoyed their fieldwork and are full of entertaining anecdotes and contain little about the authors' problems and how they coped with them (cf. Norbeck 1970; Honigmann 1970); others seem to play down the pleasures and emphasize the hardships as if they feel the need to impress upon themselves and the reader that what they were doing was legitimate work and not just a glorified vacation abroad or professional perquisite.

Much of the literature on methods also includes some information about personal experiences, since the two aspects of fieldwork are not readily separated. Specific techniques, theories and research designs, ways of using data and final reports often evolve directly out of, or are affected by, specific research experiences; and problems with research design or technique usually give rise to personal problems as well. Writers who concentrate on comparing fieldwork methods in different contexts, like Gutkind (1969) who studied unemployment in Lagos after working in rural areas, may also describe the effects on relations between informants and fieldworkers when working in societies of different size, composition and institutional structure. Srinivas *et al.* (1979) comprises accounts of "the problems and challenges in sociological investigation" by mostly Indian sociologists working in a variety of settings, mostly in India. While accounts focus on methods, there is also a good deal of information on experiences included in discussions about role in the field, the degree to which the participant-observer participates, problems of associating with the élite and getting the "establishment myth", problems of working in situations where there are sharply differentiated and opposed groups, the degree of involvement in local personal and public affairs and the methodological and moral problems to which this gives rise, and how different writers coped with their personal biases. Jongmans and Gutkind (1967) also include notes on the personal aspects of fieldwork experience.

Some personal accounts of fieldwork are more prescriptive than others (e.g. Spindler 1970), and the editors of many of the volumes on fieldwork see their role as one of drawing lessons from the experiences of their contributors. Freilich (1970), for example, includes a final chapter on the lessons learned from the articles in his book and advice for other fieldworkers. For Freilich, traditional anthropological fieldwork may broadly be divided into two phases, the first a phase of "passive research" involving learning to survive physically, psychologically and morally in a strange setting, the second an "active" phase when the fieldworker attempts to obtain data. In practice, of course, it is difficult to see how the two may be separated. The papers in the book vary in the amounts of personal experience they include (Perlman (1970), for example, is particularly valuable) but all contain useful comparative material on initial experiences and settling in and most divide into sections on "problem, theory and research design", "passive or adaptational research", "active research", "bowing out" and "summary and evaluation", following the editorial design. The Freilich analysis, based on comparative illustrations from his own fieldwork experiences among the Navaho Mohawks and Trinidadians and from the material in the volume, includes advice on such topics as how to handle the problems of being suspected (as were many of his contributors) of being spies. He suggests that the fieldworker should explain why connections with the local administration are necessary, bring one's spouse and children into the field, dress in local styles, learn to converse well in the native language and use local "magical words" whenever appropriate, eat and drink with the natives as often as possible and participate in minor illicit activities. More "advice" of this kind is to be found in Hanna (1965), who states that fieldwork is most successful where the satisfaction of one's hosts is maximized and one's study is legitimate and poses a minimum threat. This ideal is approached by describing one's investigation as non-political, one of several, informal and helpful, and oneself as, for example, professor, peer, friendly participant, sympathetic neutral, stranger and American.

Many of the things listed above are what most anthropologists do anyway in attempting to establish and maintain rapport. Such "advice" may be reassuring to the professional fieldworker, but it might be thought to be of questionable value to the novice – particularly when, as Freilich himself points out, each in the end adjusts in his or her own way, as is appropriate to the situation. Apparently, however, such points *do* need to be made to some intending fieldworkers. Shelton (1963) reports on the disastrous effects that three young anthropologists brought upon relationships and their research in Africa because they were unable to adapt to situations involving physical contact (refusing food and drink, refusal to shake hands appropriately, unwillingness to enter the house), leading the author to urge a greater self-examination by potential fieldworkers before embarking on fieldwork.

Nevertheless, it is necessary to ask whether the purpose of such accounts is to help us "see through" the ethnographic monographs and permit alternative interpretations of events and analyses, or whether their value is simply in reassuring an anthropologist who feels that having such emotions is wrong or a failing in professional abilities that it is really something that is shared with all colleagues (Freilich 1970). Both may be true and, moreover, there is nothing in the literature to suggest that the more one "experiences" fieldwork, either directly or indirectly through the literature on the experiences of others, the better equipped one is to tackle subsequent periods of fieldwork in different settings. Even though a researcher may be technically more efficient in subsequent fieldwork, every field situation *is* different and initial luck in meeting good informants, being in the right place at the right time and striking the right note in relationships may be just as important as skill in technique. Indeed, many successful episodes in the field do come about through good luck as much as through sophisticated planning, and many unsuccessful episodes are due as much to bad luck as to bad judgement. These cannot be anticipated in advance. Simply having done fieldwork is no guarantee of being able to handle new fieldwork any better. Maybury-Lewis (1965: 151) remarks how, despite having done fieldwork among the Sharente, he and his wife nevertheless had a strong feeling of not knowing what they were letting themselves in for (particularly since they now had a baby son) as they embarked for fieldwork among the Shavante.

A number of anthropologists have compared their several periods of fieldwork. The work of Powdermaker (1962, 1966, 1968) is a notable example. She conducted fieldwork on different topics in four very different societies, widely spaced geographically (Lesu, Mississippi, Hollywood and Zambia). For each case, she discusses how she decided upon her research topic and area for fieldwork, the kinds of problems she faced in getting fieldwork started, of establishing a routine of living and patterns of systematic work, of functioning within the indigenous power structure; the kinds of new problems that arose and how she dealt with them; her successes and mistakes. In each case these problems were unique to the situation. In assessing what she had learned from her different fieldwork experiences she concluded (1966: 287) that the most important conditions for good mutual communication between informants and fieldworker are the physical proximity of the fieldworker to the people studied, a sound knowledge of their language, and a high degree of psychological and emotional involvement. Measured in these terms, she regards her work in Lesu as the most successful and in Hollywood the least successful. Lesu was also her first fieldwork experience: she participated to the full yet it was so divorced from all her previous experience that she was able to observe it from the outside the more clearly. In Hollywood, she discovered personal biases and values which inhibited her total involvement in the culture she was studying while at the same time preventing her from evaluating it with complete detachment. It is this ability to become totally involved and yet to observe with complete detachment, continually to "step in and out" of other cultures, which Powdermaker identifies as the key to successful fieldwork and which she achieved in various degrees in the different societies in which she worked.

It may not be pure coincidence that the greatest sense of involvement she experienced was during her first period of fieldwork, in Lesu. Many other anthropologists (like those contributing to Freilich 1970), who have summarized their experiences working in several different field situations, have also indicated a greater emotional involvement in their first fieldwork experience than in subsequent ones. First fieldwork is always the most emotionally rewarding (Honigmann 1970; Schwab 1970). For Baviskar (1974), a peasant caste ethnographer working in a cooperative in rural Mahrashtra, his first fieldwork was the best period of his life, and so good were relations with his informants that he worried that it was making him complacent because he did not have to face challenging situations and problems. But why do anthropologists apparently get so much more out of their first than out of subsequent fieldwork? It may have something to do with age and one's relative sense of adventure and adaptability. Honigmann (1970) says he did not enjoy fieldwork in Frobisher Bay as much as the first fieldwork he did in Western Canada, where he had identified with the community and had participated intensely "with the libidinous tenor of its behaviour". In Frobisher Bay he felt alienated from the Eskimo youths and treated his informants with a respectful reserve and more sober propriety than he had the Indians. There were other reasons why he liked Frobisher Bay less, like frostbite, but the most important factor seems to have been that greater experience and maturity brought with it a loss of adventure, spontaneity and an ability to enter into the spirit of things. Moreover, when anthropologists first go into the field they are often single, but on subsequent field trips they may be accompanied by their spouse plus children. This has both advantages and disadvantages (§7.11.2). But whatever the gains in terms of access and efficiency, status and morale, the presence of one's partner or family is also likely to remove some of the emotional rewards of fieldwork by lessening emotional dependence on establishing warm reciprocal ties with informants-as-friends, a rapport and involvement which sometimes becomes so strong that the anthropologist regards friendships made in the field as the best friendships ever made. Dentan (1970) and Honigmann (1970) are among the many who have returned from the field with a bittersweet blend of nostalgia and homesickness for the people they studied. The degree of personal commitment of the lone anthropologist is likely to be much greater than that of an anthropologist who divides his love between family and informants, or who is the member of a large fieldwork team.

The first period of fieldwork, usually undertaken as a underfunded graduate student, is also often the longest period of unbroken "immersion" in another culture that the professional anthropologist will ever experience. Once initiated into the profession, monies for subsequent field trips may be more forthcoming, more generous and time at a greater premium. This leads to the employment of field assistants and other labour saving devices. Shorter time in the field and the use of assistants is also likely to lessen the intensity of emotional involvement with informants.

5.2.2 A classified bibliography of writings on fieldwork experience

Among the edited thematic collections on interaction, adjustment and the subjective experience of fieldwork are Béteille and Madan (1975), Casagrande

(1960), Freilich (1970), Kimball and Watson (1972), Spindler (1970) and Srinivas *et al.* (1979). Similar accounts by sociologists include Hammond (1964) and Bell and Newby (1977). Useful bibliographies are to be found in these volumes and also in Pandey (1975) and Wax (1971: 375–388). Personal accounts by the same person of experiences in a variety of different fieldwork situations include Mead (1972, 1977) and Wax (1971). What follows is a selected list of titles arranged geographically using the regional categories employed in the Anthropological Index to Current Periodicals in the Library of the Royal Anthropological Institute. It includes both specific accounts of personal experiences in the conduct of ethnographic research as well as monographs which devote significant attention to such matters. Within each category items are arranged alphabetically.

(A) North and north-east Africa: Cunnison (1960), Rabinow (1977).

(B) West Africa: Bohannan (1954, 1960), Schwab (1965, 1970).

(C) East Africa: Albert (1960), Beattie (1965), Middleton (1970), Saberwal (1969), Turnbull (1972).

(D) Central Africa: Turnbull (1961).

(E) Southern Africa: Holleman (1958), Kuper (1947), Powdermaker (1962), Schwab (1965, 1970), Turner (1960).

(F) North America: Aberle (1966), Adair (1960), Carpenter (1960), Casagrande (1960), Castaneda (1968, 1971, 1973), Dollard (1949), Freilich (1970), Honigmann (1970), Hostetler and Huntington (1970), Keiser (1970), Liebow (1967), Lowie (1960), Marriott (1962), Maxwell (1970), Pandey (1972, 1975), Roberts (1956), Spindler and Spindler (1970), Sturtevant (1960), Whitten (1970), Whyte (1955), Wax (1971).

(G–H) Caribbean and Central America: Gonzalez (1970).

(I) South America: Chagnon (1968, 1974, especially pp. 162–197), Herskovits (1934), Köbben (1967), Lévi-Strauss (1955), Wagley (1960), Whitten (1970).

(J) Middle and Near East: Gullick (1970).

(K) South Asia: Baviskar (1979), Beals (1970), Bellwinkel (1979), Berreman (1962), Chakravarti (1979), Dua (1979), Gupta (1979), Hitchcock (1960, 1970), Joshi (1979), Mandelbaum (1960), Patwandthan (1979), Raman Unni (1979), Ramaswamy (1979).

(L) South-east Asia: Conklin (1960a, b), Dentan (1970), Du Bois (1960), Williams (1967).

(M) Central Asia and Far East: Diamond (1970), Norbeck (1970).

(N) Australia: Hart (1970), Stanner (1960), Yengoyan (1970).

(O) Melanesia: Burridge (1960), Malinowski (1922, 1967), Mead (1960), Read (1965), Watson (1960).

(P) Polynesia: Firth (1960), Maxwell (1970).
(Q) Micronesia: Barnett (1970), Gladwin (1960).
(R) Europe: Bell and Newby (1977), Boissevain (1970), Freidl (1970).

Nicola Goward

5.3 Personal interaction and adjustment[2]

5.3.1 Introduction

Ethnographic fieldwork is subjective both in the sense that the ethnogaphers report selectively what they are predisposed to see, hear and record from the flood of words and events which wash over them every day in the field; and in the sense that the kind and quality of information which comes their way depends to a very large extent (given a certain measure of luck in any field situation) on the kind and quality of relationships between anthropologists and their informants. Some field experiences are truly alarming (Chagnon 1974); all are different. This is not simply because the cultural environment or nature of the research problem may differ, but because each depends upon an interaction between the personality of a particular anthropologist and those of his or her hosts.

If the anthropologist's laboratory is "the field", then the subjects of research are real people who act and interact in relation both to each other and with the ethnographer. Handling them requires such a wide range of social skills (understanding, patience, intuition, subtlety, perceptiveness etc.) that one wonders how academics, of all people, cope (Crick 1982). Scholte (1971: 878–883) has even suggested that our professional training may blunt such critical faculties. Moreover, subjects are usually as interested in finding out about the fieldworker as he or she is about them. What they find out and how they locate him or her in their own social world, as well as the unique personal rapport between the personalities of the fieldworker and host as individuals, is likely to influence profoundly the kind of data produced. It is therefore important to "make the right impression". At the same time, the ethnographer should be aware that informants are also to be concerned about creating the right impression on him or her and on their own people.

---

[2] This section is based on only a fraction of the available literature. The same aspects of personal interaction and adjustment in the course of fieldwork are covered repeatedly in different accounts, and no useful purpose would have been served by citing them all.

Many factors influence the relationship between a fieldworker and object of study. Some of these are less controllable than others. There are the peculiarities of a researcher's own life-history, culture and psychological background in relation to the particular topic or locality of research (Vidich *et al.* 1964: part 2; Agar 1980: 242–262), and the fact that the "self" is bound to alter in relation to the "other" during the course of prolonged fieldwork (Crick 1982). In an attempt to control and develop an awareness of such subtle influences as personality responses and effects, it has long been suggested by some that preparation for fieldwork should include psychoanalysis or psychotherapy (e.g. M. and C. Schwartz 1955: 343–353; Gans 1968; Freilich 1970: 1–38; Gullick 1970: 124; Honigmann 1970), or at least a study of ones own culture (Hsu 1979: 526). In addition, the way in which an ethnographer presents the self and a research project may prove vital, or the fact that the ethnographer (a stranger) has to establish a role and status, and create, maintain and then break-off relationships with uncommon and unnatural rapidity (Nash 1963).

Informants, of course, may have their own ideas about who the fieldworker is and what he or she is doing even before there is a chance for introductions and before work begins. In other words, they have imposed *their* meanings on the situation. The acceptance of the fieldworker will be affected by a host's perception of his or her status among them and the role they believe him or her to be playing. Furthermore, they may have their own reasons for providing the information they do, and may use the anthropologist for their own purposes. None of this should be seen as unusual. In short, the factors must be understood as part of a reciprocal relationship, a complex psychological and social (though unequal) transaction (Wax 1960a: 52; Lundberg 1968: 47; Hatfield 1973; §1.4). It is the character of this transaction which controls both the kind of knowledge which the materials produced will yield "and how we exercise whatever responsibility we may feel to our subjects and to ourselves" (Jay 1974: 372).

Both practically and intellectually, the central problem of most ethnographic research is having to "live as a human being among other human beings yet also having to act as an objective observer" (Middleton 1970: 9; see also Gans 1968) and as a result trying to keep affects and cognition separate (Nash 1963), our knowledge of persons and our "objective" knowledge of social organization and culture (Jay 1974). We may decide in the end that this is impossible, since our data are bound to be the result of a complex interaction beween "self" and "other", where the boundaries between the two are inherently vague and subject to continual renegotiation and reinterpretation (Lewis 1973: 9). Some have sought to turn the autobiographicality and subjectivity of data production to positive advantage. What is important is that we should try to be deliberate and self-conscious in whatever personal approach we adopt (Honigmann 1970, 1976). For most of us it remains a real problem.

Fieldwork inevitably involves a lot more than just sitting around watching things and asking questions. Much of what goes on does so behind closed doors, real or metaphorical, and most of what the anthropologist finds out depends upon being accepted, being invited into people's lives and being volunteered information. This involves winning trust, respect and friendship. Friendship, however, even where it is offered, morally requires reciprocal friendliness. This is where the problem begins, because friendship implies loyalty, and loyalty may have to be demonstrated by taking sides, and taking sides will prejudice relations with other people.

The researcher *tries* to "fade into the background" so that people will behave as if there were no outsider present and confide in the fieldworker as freely as with anyone else in their society (Dentan 1970). In order to achieve this, he or she must try to become as "human" as possible by doing many of the things that insiders do. This often involves learning new skills and overcoming self-consciousness and certain inhibitions which may amount to a second socialization (Yengoyan 1979). Powdermaker (1966) learned to dance in Lesu and to tell the time by the sun. Learning new social, linguistic and technical skills, places the fieldworker in a position of dependence on teachers, and numerous anthropologists have commented on how the pressure of everyday matters and their inability to cope with them led them to be treated like children by their hosts (Köbben 1967; Beals 1970: 49; Dentan 1970). The anthropologist, aware of the need for goodwill and sensitivity, may feel that childlike clumsiness leads to loss of respect and credibility and that innocent mistakes and social blunders may be professionally costly. Veena Dua (1979: 138), an educated middle-class urbanite doing fieldwork in a Punjabi slum, offers us an apt example of how easily misunderstanding can arise. She was given a bicycle, for which she gave the owner cash and proceeded to ride around the village on it. It was only weeks later that she was discretely informed that the owner of the bicycle she thought she had bought it from had only intended to leave it with her as security on a loan.

Despite concern about mistakes made, an anthropologist who is otherwise liked by his informants usually finds that clumsiness and indiscretion are considerably indulged for quite some time. Diamond (1970), for example, explains how, after six months in the field, she was still "inarticulate, uninformed, odd-looking and occasionally quite tactless and mannerless" but had nevertheless been accepted as a "symbol of importance" in the village. Naivety may, moreover, make the ethnographer seem less threatening and dispel suspicions about ulterior motives. It may "break the ice" in establishing relationships with individuals who decide to take the anthropologist under their wing. Futhermore, breaking the rules is often the surest way of learning them (Dentan 1970), and much information about how a community works may be obtained in the form of criticism or advice after the detection of technical, social or moral weakness in the fieldworker (Gupta 1979). Such indirect benefits may be worth risking one's reputation for.

However, what may be accepted as a mistake through understandable ignorance at the beginning of fieldwork may be taken as an insult later on. Only too aware of this, many anthropologists take great pains to adjust their behaviour accordingly. Chakravarti (1979), working in Rajasthan, began fieldwork by ignoring his own caste rules but soon realized that he must live by them in the community he was studying. Hart (1970) remarks that never in his life, before or since, had he been so submissive to the will of others as when living among the Tiwi. The pay-off for conformity to local norms of behaviour is usually acceptance. Even so, some may feel, with Berreman (1962), that while they are "tolerated with considerable indulgence" their presence is never actively desired. Others, like Norbeck (1970), find that – although they make strong dyadic ties with individuals – they are never fully incorporated into the group. Many, however, after being in the field for some time, find themselves flattered by natives saying that they have become "one of us". This can never mean, however, that no difference is perceived between anthropologist and informant. It is simply that his or her presence among them, in a role as fieldworker and friend, has been accepted and is desired. When Powdermaker (1966) was given a clan and kinship position in Lesu, she did not fool herself into believing that she was really part of a Melanesian clan but accepted it as a symbolic gesture on the part of her friends to show that they liked and accepted her.

In many ways, the anthropologist is always an outside observer. He or she is neither undifferentially incorporated into the community, nor usually seeks to be. As Dollard (1949) points out, the fieldworker not only adjusts emotional responses and behaviour to the community being studied but also acts according to his or her own social position and often expects to be treated differently. There always remains a degree of detachment because the role is only temporary and for a specific purpose. At some point, he or she will leave and re-enter the professional world. Moreover, even when participating to the full, there is an ever present possibility that the anthropologist will be taken by surprise by something incomprehensible to remind him that, as a participating member, he still has much to learn (see, e.g. Beals 1970).

So how hard should the anthropologist try to become a native? How far should behaviour be modelled on that of the community? The answer seems to be that every attempt should be made to learn the rules of acceptable behaviour and how far one can choose to ignore them without upsetting informants. In many instances it may well be professionally advantageous to remain "different" in the eyes of informants because this increases opportunities for contributing to his or her own role definition, providing greater flexibility and the ability to do things which ordinary members of the community cannot. The best example of this is crossing boundaries of sexual division (§5.4). However, life may become intolerable if no attempt is made to conform, even if this is ethically repellent.

5.3.2 Problems of adjustment and stress

Most ethnographers entering a new culture find that they experience some degree of "culture shock" (Oberg 1972; see also Junker 1960: ch. 4; Pelto and Pelto 1973: 177–192; Johnson 1975). They must come to terms emotionally with living under quite new conditions, where many of the things they will have previously taken for granted are missing. Many aspects of life may be appealing: Powdermaker (1966: 99) enjoyed living out of doors, Maxwell (1970) enjoyed participating in sports and fishing, attending dances, drinking beer and falling in love with a Samoan girl. But most personal accounts are packed with details about difficulties of adjustment. These may range from problems in dealing with those whose consent is necessary in order to do fieldwork in the first place to poor and unaccustomed sanitary arrangements (e.g. Boissevain 1970; Diamond 1970; Norbeck 1970). There may be a new and extreme climate to cope with. Briggs (1970) in the Canadian arctic, Norbeck (1970) in Japan, and Honigmann (1970) in British Columbia all experienced problems in coping with the cold. There may be new kinds of health problems: Evans-Pritchard had his fieldwork foreshortened in both 1935 and 1936 by contracting malaria; Malinowski (1967) became obsessed with his health and medication; and Dentan (1970) has remarked on how the fieldworker's behaviour may become quite eccentric through attempts to prevent catching things (also cf. Boissevan 1970; Maxwell 1970). The fieldworker may also have to adjust to unfamiliar, unpleasant or boring food (cf. Diamond 1970; Norbeck 1970; Powdermaker 1966: 98; Benedict (Mead 1959: 299)), or even to hunger (Holmberg 1969). There may be problems in finding suitable habitation or clothing (Pelto and Pelto 1973: 180–182, 190–192). Living conditions may be cramped and uncomfortable

(Evans-Pritchard 1940: 14; Norbeck 1970), and bugs and insects may be irritating. The logistics of everyday living may be more complicated than the researcher is used to, and may consume more time than is customary (Honigmann 1970); Gulliver (1966) recounts his problems in simply keeping up with the pastoral Turkana in Kenya. The fieldworker has to adapt to new routines and to use available time in an efficient way (cf. Boissevain 1970; Maxwell 1970). Problems may arise through having servants (e.g. Powdermaker 1966: 95), and lack of independence arising from the non-possession of native skills (cf. Beals 1970), while there may be unexpected technical problems with equipment due to climate (Powdermaker 1966: 99; Beals 1970).

Perhaps the greatest hardships of all, however, are psychological rather than physical and logistical (Wintrob 1969). The anthropologist is usually used to a good deal more privacy than is permitted in the field. Evans-Pritchard (1940: 14) recalls the long time it took him to adjust to "performing the most intimate operations before an audience or in full view of the camp"; similarly, a constant stream of visitors created "a severe strain". At the same time, like Dentan (1970) working among the Semai, Evans-Pritchard was irritated by the way all visitors, particularly children, distracted informants. Diamond (1970) found the constant noise of "Chinese opera, popular songs and emotional radio dramas", broadcast over a public address system or loud private radios, quite unbearable. Benedict (Mead 1959: 301) found the "gregariousness" of fieldwork, of "being in the thick of the herd all the time", a source of tension and Norbeck (1970) reports the extreme measures he took to find privacy and quiet. This is unfortunate since, as Evans-Pritchard has pointed out, his compelled intimacy with the Nuer actually made him feel closer to them than to the Azande with whom his relations were physically less intimate. Whilst the fieldworker may feel the need for privacy, both for work and for his or her own emotional well-being, it is important that this should not disrupt establishing good relationships. The Spindlers (1970) coped by living close to, but outside, the Menomini settlements they were studying and used the privacy of their car for interviewing and writing notes. Others (e.g. Powdermaker (1966) in Mississippi) who have made frequent trips to the lavatory to write notes, have caused concern to their friends over their health.

Life may be public enough in itself in the communities anthropologists study, but the investigator as a rather odd newcomer is likely always to be in the centre of the stage. This generates a constant awareness of behaviour and of conscious acting (Boissevain 1970), of calculations in personal interactions and gift-giving exchanges. It also results in a greater sensitivity to the making of mistakes (Norbeck 1970). Powdermaker (1966: 99) in Lesu and Maxwell (1970) in Samoa, became bored and oppressed by the monotony of life, and found it a strain coping with a dwindling interest in research both in themselves and in their informants.

Most anthropologists have found language learning (§7.5) a source of strain. Evans-Pritchard (1940: 14) mentions the problems of learning a language without an adequate grammar or dictionary. Norbeck (1970) laments how his knowledge of Japanese permitted him to "come through in the gross but never in the fine" and how he craved a steady diet of conversation in his own language. Diamond (1970) tells how her morale suffered as a consequence of not being able to make herself understood. While she was able to exploit the patience of children to practice her linguistic skills, children could provide her with only limited information and she, like many, became frustrated because she could not get on with her fieldwork. In desperation she read English novels and gave English language lessons to show her informants that she was expert in some language. Others have found their linguistic incompetence a source of great embarrassment. Powdermaker's mistakes (1966) in Lesu dishonoured her teacher, and she regrets not having devoted at least the first few months of fieldwork to learning the language.

The lesson seems to be that, where there is time, anthropologists should concentrate their efforts at the beginning of the fieldwork on learning the local languages, not worrying too much about "interviewing", and making detailed notes on what can be observed, including long lists of questions which may be referred to at a later date once the anthropologist is comfortable with the language sufficiently to be able to tackle them. In this way, much misinformation due to incomplete understanding may be avoided. Moreover, an emphasis on desire to learn the language may often provide a useful and acceptable entré into the community (§7.7.3).

A willingness to learn a language may be the most important factor in establishing good relations with informants. Even with the most excellent of interpreters, there is nothing so alienating as to speak another language. Many anthropologists, due to pressures of time and a belief that they can avoid wasting it, in the initial stages of fieldwork at least, employ an interpreter so that they may begin interviewing right away. Hiring an interpreter may also remove some of the strain and frustrations associated with needing to learn a language and at the same time needing to understand and use if efficiently by providing a "check" on the fieldworker's own understanding of information provided. However, having an efficient interpreter can lead to laziness in efforts to learn the language. Beals (1970) has remarked upon how a bad interpreter may not only be a barrier to establishing good relations in the field by repeating unflattering stories about the anthropologist; he may also inhibit the development of linguistic skills by insisting upon translating and explaining everything.

In addition to these problems the anthropologist has to cope with emotional conflicts arising from the contradictory requirements of involvement and detachment (Geertz 1968; Wintrob 1969). Distancing, observes Bourdieu (1977), may have epistemological consequences. He or she may be shocked by local morality or naivety and be unable to take a really detached view. Powdermaker (1966: 29) records how she expected to have to suppress her prejudices in the field, but not those prejudices which were principles as well. Having come from a society which values life and takes care of its disabled and unfortunates, she could not come to terms with native callousness, even when she recognized that it was their way of coping with continually living on the brink of tragedy. Maybury-Lewis (1965: 84) and Chagnon (1974) struggled against similar problems, although the Spindlers (1970) were eventually able to come to terms with it. Even so, problems of adjustment are still likely to create stresses in the fieldworker and may have adverse effects on relationships with informants. Bohannan (1964) vividly describes the stresses she felt in trying to be part of two societies, her own and her adopted one. She felt outside both, as uncomfortable with the Europeans at the Station as with the Tiv, yet at times sharing experiences with and valuing the company of both. Her work made her feel like a trickster, as one who seems to be what she is not and professes faith in what she does not believe.

Since the fieldworker does not usually intend to settle permanently in the community being studied, it is also often important mentally to "escape" from it periodically (§7.8). Most anthropologists, unable to escape physically, take to expressing themselves in their own language by keeping a personal diary separate from their field notes, or writing long letters (Dentan 1970; Mead (1959) on Benedict), though the long time-lapse between writing letters and receiving replies may be another source of discomfort. Benedict (Mead 1959: 299) and Powdermaker (1966: 100) read poetry. Reading novels is also an important means of escape, but the fieldworker rarely carries enough literature. Powdermaker (1966: 100) also remarks how important it was for her to read novels which presented a whole new world rather than short stories with an emphasis on one episode or a quick climax. Anthropological books, though read more critically in the field, she did not find good for escape and relaxation, because she was always comparing her own data with the monograph.

5.3.3 Status

If the fieldworker is studying people in a country other than his or her own, status (see also Junker 1960: 116–120) may primarily be determined by the relationship between their respective countries. In the days when anthropologists often worked in territories which were colonies of their own countries, the political and social system endowed them with high status of a particular kind. The relationship between researcher and informant was often one of patron to client (Colson 1967; Gutkind 1967). Political domination and racial and social distance were reflected in informants' subservience, submission and ingratiation. Moreover, a glance through Malinowski's diaries is sufficient to show that these attitudes were expected of informants (e.g. Malinowski 1967: 272). He afforded himself no moral self-reproach for his deeply-felt aversion to those upon whom he depended for information. Malinowski was himself aware that these feelings often directly interfered with his ability to get information from his informants, as when (1967: 166) one informant was busy supplying him with "valuable information" and he (Malinowski) was overcome with "a violent aversion to listening to him". No doubt it was only by virtue of the same social relations of domination that the Melanesians were prepared to impart any information to him. Where social superiority went hand in hand with direct political domination and hostility, respectful compliance to the fieldworker's demands was likely to be replaced by stolid resistance against having any truck with him. Evans-Pritchard (1940) remarks upon how his own associations with a colonial government towards which the Nuer felt hostile led to his being regarded with "suspicion and obstinate resistance" in the early stages of fieldwork. These attitudes were reinforced by the fact that, because of his difficulties in being accepted by the Nuer, he sought help from the American Mission at Nasser.

Even today, many anthropologists work in countries which have an economically inferior or dependent status in relation to their own, in countries which are the recipients of international aid for example. Others work in sub-cultures of their own or another society whose status may also be racially or socially inferior to the dominant culture with which the anthropologist is identified. Even where this is not the case, the anthropologist is invariably better educated than most informants. Where education is a highly valued resource this may similarly elevate status.

Relative high status *vis-à-vis* informants is often associated with relative affluence and control over all kinds of other resources. Moreover, it is often the case that anthropologists do have certain material resources, access to information and friends in local high places. This not only creates a social distance which may prove a barrier to good communication, but it may also create a moral dilemma about how, and if, to dispense such resources that they can afford. Most ethnographers are involved in gift-giving; to establish rapport, to maintain relationships and to repay moral and material debts. The gifts and services traditionally provided include food, clothing, medical aid, advice, introductions, lifts, photographs and so on. Patwandthan (1979), for example, helped untouchables in a Poona slum to get jobs, offered books and financial assistance to students, made donations towards wedding expenses, hastened an application through the bureaucracy, gave a woman a sari and a *panchayat* a clock. The anthropologist's presence and continual questioning is only tolerated because of compensations he can offer, material or otherwise. Since ethnographers are not usually sufficiently well-equipped to give material gifts to all their informants, they must be careful how they distribute their largess so as not to offend individuals or etiquette. Köbben, (1957), working in Surinam, solved his problem of how to distribute a limited number of gifts among local children by giving them all to the old men to distribute, thus allowing them to increase their own prestige into the bargain. Bellwinkel (1979: 141) avoided the problem entirely by deciding not to give favours to individuals but used his political influence on behalf of the whole community. In many cases, anthropologists have found that their presence alone, perhaps simply as a novelty and welcome diversion from the monotony of village life, perhaps for the prestige their presence has been perceived to bestow on the community, perhaps for their entertainment value (Saberwal 1969), perhaps just because people have liked to talk to someone who was always prepared to listen, has been sufficient reward for their informants' cooperation.

But ethnographers also have to cope with the fact that informants may try to manipulate them to obtain desired resources believed to be in their gift. Diamond (1970) recalls her feelings of guilt at her inability to work the economic miracles expected of her in terms of bringing AID and development agencies to her Taiwanese village; she also felt guilty about her own relative affluence, and felt as if she was boasting when she answered questions about income, television and car-ownership in her native America. Beals (1970), working in an area where missionaries had led local people to expect gifts of food and old clothing, responded to informants' demands for gifts by pretending to be poor. Gift-giving, as a means of creating and cementing relationships with informants, is only of benefit if it is done with discretion. If overindulged in or gone about in the wrong way, its effects may simply serve to emphasize the social and material differences between fieldworker and informant, and therefore represent a barrier to certain kinds of communication (see Firth 1967: 13).

An important part of establishing rapport is to win trust. Simply being white may invoke the mistrust of black informants, as Köbben (1967) found. Where an old colonial government has been replaced by a native civil service which is sensitive to its independent status, as in many parts of Africa, strong suspicion may be directed against foreign scholars (see Gutkind 1967). In political discussions, fieldworkers may be called upon to explain or justify their governments' position and role in native affairs, past or present (Diamond 1970). Even if they are not assumed to be connected with old colonial structures, ethnographers may be identified with particular sponsors associated with outside control or interference (Hart 1970).

The problem of a fieldworker's status being determined for them as, or even before, they arrive in the field is not, of course, peculiar to anthropologists working in old colonial territories. Researchers working in a sub-culture of their own society, like Whyte (1955) in "Cornerville" and Lincoln Keiser (1970) among the Chicago "Vice Lords", often have special problems. As a member of culture *A* – ambassador of the dominant culture among members of the sub-culture, as White among Blacks – the fieldworker cannot hope to be accepted as part of culture *B*, the culture being studied. To be identified as a member of culture *A* involves being caught up in a whole set of expectations about attitudes and live issues in a society in which relations between cultures *A* and *B* are mutually antagonistic – just as they may be in an old colonial setting. Keiser had to cope with being the object of hatred towards Whites and the effect this had on introductions to local leaders, establishing residence, being admitted to fighting clubs and so on.

It would be wrong to assume that inequalities between fieldworker and informant always place the fieldworker in the higher social status. The situation may be reversed when working with local élites. Boissevain (1970) has remarked upon how a focus upon peasants, villages and slums in the literature on southern Europe reflects the researcher's inability to keep up with the local élite as he can in "mud hut" anthropology. The fieldworker may become more deferent and humble in attitude when dealing with élites. This is likely to affect the outcome of enquiries: what is observed, how questions are framed, how utterances and events are interpreted. At the same time, when working with government officials the fieldworker may be questioned and interviewed on his or her own background and present research problems in great detail, and be called upon to defend the policies of governments, show an understanding of current affairs and have views on all manner of issues (Henry 1969).

This raises another problem in relation to the anthropologist's ideological status in the community being studied. In many field situations it is made quite clear that unless the anthropologist's own position is frankly stated in terms agreeable to the informant, the interview will be terminated. Anthropologists usually respond by trying to maintain a neutral position by pointing out the good and bad in both positions or by hiding behind ignorance, both in the interests of "morality" and for fear of losing trust or face with either side. Nevertheless, the fieldworker is often forced to take one position, or at least to sympathize with a particular point of view, simply in order to elicit information from anyone. On issues which factionalize a community, it may be necessary to communicate formally personal and ideological commitment to a particular informant (Henry 1969). Even where this is not the case, the fieldworker almost invariably is, or becomes, identified with a particular social faction, class, caste, political party or whatever, despite any steps taken to "neutralize" status. This is really inevitable, since the fieldworker lives in a community as a person and a social being as well as being a professional researcher, and usually in a community where no social category exists for a "neutral" anthropologist. It is always up to the individual fieldworker to decide how far it is worth upsetting rival factions by associating with both, and how far informants can be expected to accept claims of impartiality when no such impartiality exists in the community. Although the fieldworker may try to minimize identification with any social grouping (e.g. the Spindlers (1970) chose to camp outside the Menomini settlement they were studying largely to avoid residential alignment with any particular faction), it is never possible to remain entirely neutral on every issue if he or she is to be accepted as a normal human being. From the time of arrival hosts will try to locate the ethnographer within their own networks of statuses and social relations.

The problem of being associated with a particular social category or group in the community begins as soon as the anthropologist enters the field. A start must be made somewhere, and as interpreters, assistants and informants are selected or self-selected there is generally little idea of their standing in the local community, except perhaps in the vaguest terms. Yet, since the fieldworker is generally rapidly identified with those with whom he or she associates, this selection of initial collaborators may prove critical in determining the outcome of relations with other potential informants, with the community as a whole, and fieldwork in general. Köbben (1967) found that his early association with one very helpful informant who turned out to be in a power conflict prevented his ever being fully accepted by that informant's opponents. Buechler (1969) is among the many who have remarked upon the way the anthropologist's status may be "assimilated" into that of his interpreter, whose own communication channels also determine those open to the anthropologist. Norbeck (1970) draws our attention to the detrimental effects on fieldwork of being associated with informants who are intensely disliked by others. Other anthropologists (e.g. Freilich 1970: 1–38; Pelto 1970) have pointed out that very often the fieldworker's initial collaborators are almost as socially marginal as the anthropologist. Beals (1970) also notes that those who are able to type, speak the language of the anthropologist and have a social science training, are invariably from outside the community being studied; and the consequences of identification with an outsider with prejudices may be far worse than with an insider with prejudices. For those intending a study of the women of the community, the problem of finding acceptable assistants is intensified. Buechler (1969), working in Bolivia, for example, found that few women were qualified to act as interpreters, except mestizo women who were problematic because of their marginality and their prejudice against local peasants (§5.4).

Informants, then, will always try to place the fieldworker within their own framework of social statuses and values, perhaps even conferring a kinship title. Once labelled and located, the ethnographer is less threatening and more human. But there are often problems. How can an unmarried researcher, working in a society where all adults are expected to be married, be worthy of adult status? (Powdermaker 1966). How can an anthropologist hope to gain respect when working among pastoralists and having to admit to not owning cattle? How can it be that he or she has no clan status among those, in the words of Hart (1970), for whom the saying "white men have no clans" is like saying "white men have no ages"? In many societies the fieldworker working alone is an anomaly. However, although this may be an initial problem, some kind of accommodation is generally reached.

5.3.4 Role

Part of the fieldworker's problem in establishing a satisfactory status among informants arises out of his or her role in the field (see also Paul 1953; Gold 1958; Junker 1960; ch. 3, 120–133; Wax 1960b; Olesen and Whittaker 1967; on role conflicts, see Kloos 1969). Some ethnographers have a pre-defined role, as administrator, missionary, "expert", or even colleague or workmate. However, while ethnographers may assume or be ascribed roles other than that of researcher in times of political or economic crisis (Spillius 1957; Henry 1966), or even periodically in normal times (e.g. as doctor, teacher, advocate), they are principally in the field to record other peoples lives. Roles, of course, emerge from the interplay between stranger and host, but although informants are bound to define an ethnographer in terms which are familiar to them, the ethnographer may also attempt to project a particular image to assist the conduct of research. The various possibilities are well-known (Hatfield 1973): the patron and source of material goods, the culturally incompetent (either as "child" or pawn), and the technical and cultural expert.

One particular problem which may arise is that the fieldworker often appears to have no apparent "job" or economic role. The various forms of "participation" which fieldworkers engage in are to some extent attempts to resolve this problem. In some respects this may be less of a problem for females, where some of the things they do – like going to market – are seen to fit in with the role of females; males living on their own and looking after their own domestic affairs in the same society cannot even begin to conform to their sex role. Other male anthropologists, like Boissevain (1970), who take their families into the field with them, are often a particular enigma since they are seen not to "work" and yet are able to support a family and home. But if informants are puzzled by what the anthropologist does not do, they are even more puzzled by what he or she does, and often leap to their own conclusions regardless of what the anthropologist may claim to be doing. Informants are puzzled by the fieldworker who spends hours at a time chatting to people, wandering about and watching people, asking questions, consulting documents and writing things down – while at the same time doing none of the things they have to do in order to live.

Many anthropologists find it necessary to find a regular task which is comprehensible to their informants, under the auspices of which they are able to conduct their research (Spindler and Spindler 1970). The anthropologist cannot always get away with being "a student" (Diamond 1970). Freilich (1970: 1–38), like Köbben (1967) working in Surinam, was labelled a "collector of evil gossip" in Trinidad, and was accused of lying when he tried to explain what he was doing because they could not believe that anyone would pay him to collect information on such things as kinship, religion and magic. Mohawk steelworkers regarded him as "some kind of scholarly tramp looking for a home". Anthropologists working in areas where there has been missionary influence are themselves often assumed to be missionaries, albeit posing as something else. Köbben was suspected of being a missionary of a rival religion and Diamond (1970) in Taiwan was identified with the missionaries from mainland China. The Berremans (1962), working in the Himalayas, wishing to escape any suspicion that they were missionaries, took care to avoid asking any questions about religion at first, used native Hindu greetings and concentrated their enquiries on the apparently innocuous subject of agriculture. This led to a rumour that they were government agents assessing land for tax purposes (a suspicion similarly levelled at Freilich (1970) in Trinidad) or were investigating the extent of land used in un-authorized areas. Genealogical enquiries were interpreted as a prelude to a military drafting of young men. Other beliefs about their presence included a common suspicion directed against anthropologists, that they were foreign or government spies. Even if the anthropologist succeeds in explaining the purpose of a study, informants may fear an investigation. This is particularly so in politically sensitive places, as in Kenya in 1963, where politics was an area whose "secrets" must be protected from foreigners (Saberwal 1969). Because of the American presence in Taiwan, Diamond (1970) was even visited by a Taiwan internal security agent, who finally decided she was "mad but harmless" in her interest in Chinese culture, and left her alone. Gupta (1979), however, was obliged to stop taking notes at public meetings in Uttar Pradesh after suspicions that she was an intelligence agent. Little wonder, then, that Diamond (1970) felt that she was prying when asking questions, and at her relief when people turned out to be keen to volunteer personal information to a willing listener.

The initial suspicion that the fieldworker has ulterior, hidden motives may persist for a varying amount of time depending upon how quickly the anthropologist is able to establish relations of trust. For Berreman (1962) the turning point came after he introduced his Brahmin interpreter, whose credentials were discretely checked by informants, proved correct and so everything else the anthropologist said could be believed. Sometimes, how-ever, lack of trust and suspicions about real motives have been dealt with more actively by the investigator directly through the prime perpetrators of gossip. Boissevain (1970) coped with his main opponent by drinking him under the table and talking to him. Others have taken less dramatic steps in attempting to convince key individuals that they are doing what they say they are doing. Informants themselves are usually the best people to explain to others what the researcher is there to do since they are able to put it into terms understandable and acceptable to local people, and they are trusted by their own people better than the anthropologist, at least at first.

It usually does not take long for informants to lose some of their initial suspicions. The Spindlers' informants (1970) soon recognized and began to accept them as being different from other Whites simply because they did not begin to teach or preach to them but were interested in finding out about them. Since informants are usually as keen to learn all they can about the anthropologist as the anthropologist is to learn about them, they usually provide ample opportunity for presences to be explained. Time also increases understanding and the anthropologist's capacity to show that his or her actions are harmless and acceptable. Powdermaker (1966) told her Lesu informants that her prestige at home depended upon her obtaining correct and full information about their customs, thus placing an onus on her informants. In a community proud of its culture and history, an anthropologist may satisfactorily explain an interest in the contemporary lives of informants as part of a fascination for their history and the way of life of their ancestors, as Freilich (1970: 185–250) did among the Mohawks. Berreman (1962), in a protracted speech, explained that his interest in Indian culture and livelihood in the Himalayas was to promote better understanding between Americans and Indians, saying he was particularly interested in Paharis because so little was known about them even among Indians. Many anthropologists, including Berreman (1962), find that their informants cannot get over their surprise at the interest taken in them; others find their informants not in the least surprised that the anthropologist should want to study them, like Köbben's informants (1967) who saw their village as the "navel of the world".

5.3.5 Coping with informant presentation

Just as the anthropologist may try to dispel unwanted suspicions and take pains to present an acceptable and professionally useful self-image, so too are informants likely to be concerned about the impressions they wish to create about themselves. Berreman (1962) has remarked upon the way in which people behave differently when they are giving information or know they are being watched to when they are reported by others or do not know they are being watched. Evans-Pritchard (1940: 14) remarks upon the way his enquiries were always carried out in public, either in the open or in his tent with a constant traffic of people going in and out, all interrupting each other and clamouring for attention with little opportunity for confidentiality. Such conditions are bound to affect the quality of information informants present. An informant may not be prepared to divulge personal or community "secrets" in front of others but may be quite open when alone with the anthropologist; alternatively, an informant may by shy when alone with the anthropologist but not when morally supported by friends. This is particularly so if an informant is also trying to impress them with personal knowledge about something. Informants are involved as much in creating impressions for their friends, rivals and relations as for the anthropologist, and an important factor influencing the kind of information an informant is prepared to impart is who else is around at the time.

The fieldworker has to try to discover informants' hidden motivations. Berreman, for example, discovered that some people in the Himalayan community where he was working were more concerned about creating the "right" impression on him than others. Whereas those of upper castes were highly suspicious of being photographed, lower caste individuals were more cooperative. Women, youths and older people were also more cooperative and less concerned about "stage managing" information about themselves. This had interesting consequences for Berreman's fieldwork (1962: 22).

Some anthropologists have become frustrated because they sense that informants are deliberately withholding information from them and assume that it is because, for one reason or another, there is no trust. Köbben (1967) has remarked upon the psychological stress created by having information withheld (even if it proves insignificant in the end), and the problem of coping with informants' lack of a sense of exactness and avoidance in answering specific questions because they do not see the point of them and want to talk about something more interesting. In order to obtain accurate information it is necessary to take into account the motivations and sensibilities of informants as actors. Buechler (1969) has pointed out how embarrassed interpreters were in Bolivia at being asked to translate certain questions (e.g. about conception) directed at other women. Informants may be anxious about having what they say written down or about being photographed or recorded. Diamond (1970) tells how her informants feared that her tape recorder would "steal their souls". Informants who are trying to be sociable may be reluctant to give detailed explanations for fear of boring the fieldworker with things which are, to them, commonplace. Boissevain (1970) has remarked upon the inappropriateness in this respect of prolonged direct questioning and how most information is best gleaned indirectly from many hours of casual conversation. Cooperative informants are generally likely to tell the anthropologist what they think he or she is likely to want to hear (just as the anthropologist does with them) and their assessment of expectations is often as inaccurate as the information they provide. They may simply be unfamiliar with the need for official or academic exactness, or may even find amusement in deliberately misleading the ethnographer, as Köbben (1967) reports his informants misled official census makers. Informants may also deliberately lie or "bend the truth" in order that the listener form a particular impression of them. Powdermaker (1966: 185), working in Mississippi, for example, reports that Whites faked an aristocratic background, and notes that such lies, even when the fieldworker cannot check up on them, often reveal much about the values of a society.

Whatever steps may be taken to present a favourable impression to everyone, the anthropologist usually finds it much better to get on with some individuals than with others. Some are always more helpful, responsive and knowledgeable than others. This is partly because personality types in the field differ as much as they do in the anthropologist's native land – though this only came in a moment of realization to Powdermaker in Lesu (1966: 82). But there may also be other reasons. Diamond (1970) found that, among the women she was studying in Taiwan, she got on far better with old women and young girls, who were flattered to be asked their opinions on things and had the time to answer questions. She became the sounding board for lonely old women to talk about traditional ways and the injustices and difficulties of their present-day existence; for the young women she was a model of freedom and independence. Young married women, on the other hand, were polite but cautious towards her: she was "in their generation but not of it".

Hitchcock (1970), working with his family among the Gurkhas, en-countered particular problems with an uncooperative headman. Why, he wondered, did he have to work so hard at cultivating a friendship with him when it was so unnecessary with others? Hitchcock began by questioning his own attitudes towards the chief and perceived that he might be behaving over-sensitively towards a figure of authority, and suspected he was much less interesting to the headman than the headman was to him. It might also, he supposed, have been due to their religious differences. This leader may have seen the researcher as representative of a new order. Or maybe the researcher conveyed unspoken objections to the caste system through which the head-man conspicuously benefited. The anthropologist is not always aware of why it is possible to develop good relations with one informant and not with another.

Many anthropologists remark upon individuals who have been particularly uncooperative. Saberwal (1969) notes one informant in Kenya who required 15 visits to his homestead and three encounters outside before he would give an interview. Another gave information only when "held captive" on a 30 mile journey in the researcher's car. Margaret Mead, in a 1930 letter to Ruth Benedict (Mead 1959: 313), reports similarly uncooperative informants on an Omaha reservation. At the same time, she also reveals a more specific reason given by informants for their reluctance to give her some of the information she was seeking: a belief that death would follow the divulgence of sacred things.

Clearly, in the early stages of fieldwork at least, Mead had problems in establishing rapport with her informants. Other anthropologists coping with similar problems of uncooperative informants have found a number of remedies. Time, of course, is the greatest ally, since even the most suspicious and uncooperative of informants often mellows when others are seen co-operating with the anthropologist, and perhaps benefiting from the situation; and when the anthropologist's presence has become more familiar and less threatening.

Some fieldworkers have found their field assistants (§7.11.3) – or a particularly helpful early informant – useful in establishing rapport with less cooperative informants. Beals (1970) has pointed out the usefulness of assistants in the first stages of fieldwork in indicating to the anthropologist the proper ways of behaving. Buechler (1969) learned from his interpreter in Bolivia always to greet people, even if one has only known them a short while or knows them only by sight. The fieldworker may also use an interpreter or field assistant to pass on to others autobiographical details which may ease an introduction. If the fieldworker does not know the language of his informants, his interpreter may be a very important agent through which the community assesses the ethnographer. Interpreters and assistants may also be useful in gaining access to settings where invitations would otherwise not be forthcoming. Saberwal (1969) used good informants in Kenya to persuade hostile ones to cooperate, while Norbeck (1970) began work in Japan by asking those who helped him to act as guide and councillors on other suitable informants. Personal recommendations or introductions generally give a far better response than random approaches. Nevertheless, this may lead to its own problems since informants are likely to direct the anthropologists to their own friends and relatives and (as has already been discussed) if the fieldworker becomes involved with a particular network there is some danger of identification with a particular group or faction.

Nicola Goward

5.4 Gender orientations in fieldwork

5.4.1 Asymmetries in the anthropologist's own society

Consideration of gender orientations in social anthropology must begin with social anthropologists themselves, and be sought in their own societies. Selection of anthropologists, which occurs long before fieldwork is entered upon, may itself be gender-biased. As yet little research has been done on who become social anthropologists, but we must assume, until there is other evidence, that their perceptions will at least be influenced by ideas about gender in their own cultures, and that they themselves will not be a random cross-section of their communities.

Probably, for example, in the past the male and female anthropologists who established the discipline stood in different relationships to their respective gender groups, both in their numerical representativeness and in other ways. Anthropologists may still do so today. Okely has contributed a useful discussion on the uses of diaries and autobiographies (1975), where she notes that

> Whereas the female anthropologist, not accompanying a husband, is rejecting her conventional destiny by the act of fieldwork, the white male anthropologist is completing his. (See also Lévi-Strauss (1963: 43) for a discussion of the function of travel among young French men.)

It was suggested recently that early female social anthropologists, unlike the majority of their gender group, were mainly "single and tough", and that most remained unmarried throughout their careers. However, little biographical material is available on social anthropologists at large to substantiate any of these assertions and to enable full discussion of their implications (but see E. W. and S. G. Ardener 1965), and there are notable exceptions to spinsterhood and absence of children – for example, Margaret Mead (1972). In 1964 there were, in fact, only 142 members of the (British) Association of Social Anthropologists. The age and sex structure of the membership then was as shown in Fig. 5.1. The figures for males peak at a slightly higher age than those for women; this possibly reflects the custom in

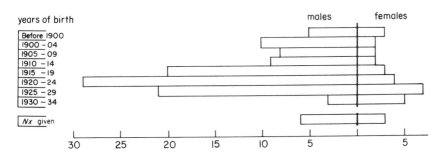

Fig. 5.1 Total ASA membership by sex and dates of birth, 1964. From E. W. and S. G. Ardener (1965: 300).

British society for women to marry at a younger age, which enables some married to established anthropologists to enter the field relatively quickly.

Gender orientations in the home cultures of social anthropologists may have meant that some women anthropologists have had their attention directed to topics thought to be particularly suitable for them, or they may have been discouraged more than men from entering certain fieldwork situations because there were considered more vulnerable to physical harm (although no certain case of attack is known). Thus after Franz Boas gave Margaret Mead the task of studying adolescents, she noted:

> I imagine that my age and physique – at 23 years old I was 5 feet 2½ inches tall and weighed 98 pounds – had something to do with his choice. I had wanted to do my initial field work in some much more remote and 'untouched' place in the South Seas (Mead 1977: 19).

Some women took special precautions. Powdermaker (1966), on the advice of Radcliffe-Brown, carried a gun in Lesu; Ruth Benedict bribed a young girl to sleep alongside her (Mead 1959: 299). Leela Dube (1975: 159) started her fieldwork in 1948 in Gond "under the protective umbrella" of her father-in-law to conform to Indian sensitivities. The presence of a wife and her perceived special needs may occasionally influence the field location of a male ethnographer. Rosemary Firth gives several reasons for the choice of one particular village where her husband worked, including that it was not "too primitive to seem a possible neighbourhood for me as well as for Raymond" (1972: 18). Problems of physical safety worldwide are, in fact, thought to be a more prominent consideration now than formerly, especially in urban locations. The reluctance of women to discuss sexual harrassment makes it difficult to establish whether they actually suffer special problems of this type.

It has been suggested that although there have been fewer professional female social anthropologists than male, they have had greater influence on their discipline relative to their numbers than women in most other disciplines have had on theirs. Nevertheless, whatever the reasons (which probably lie in the cultural priorities of social anthropologists' own societies), female anthropologists have not held as many top departmental academic posts as men and this may have diminished their full impact on students, on research, and on the development of the discipline.

5.4.2 Gender orientation in analysis

The gender orientation of social anthropologists themselves will also have had an influence on the topics selected for study, on the theoretical frameworks which they have employed in the field, and in their later analyses. Susan Rogers (1978: 143) has written:

> If Ardener (E. W. 1972, 1975) is correct in arguing that ideological systems are not necessarily generated by a society as a whole, the question of which members express such beliefs and values becomes a crucial one. That is *men* might regard their activities as predominantly important, and their cultural systems might give superior value to the roles and activities of men. As Ardener points out, most ethnographic data focuses on male cultural perceptions assuming them to be representative of the society as a whole.

That responsibility for such a selective approach might lie with the social anthropologist, rather than with the group studied, is attested to by the early experience of Rattray.

> Expressing his surprise on learning of the role played by women in the formal state hierarchy of precolonial Ashanti, Rattray states: 'I have asked the old men and women why I did not know all this. . . . The answer is always the same: "The white men never asked us this; you have dealings with and recognize only men, we supposed the Europeans considered women of no account."' (Rogers 1978: 145, quoting Rattray 1923)

More recently than Rattray, Iona Mayer, in a 1975 paper on gender relationships based on fieldwork conducted among the Gusii between 1946–1949, noted a bias in the methodology which she and her husband had used:

> Because of our concentration on masculine themes and male informants when in the field, we do not have the material that would enable us to judge either the extent of moral consensus, or how much an appearance of consensus might have owed to a 'false consciousness' in the Marxian sense. Our informants, men, probably thought that women like best to do the service tasks and be 'bought like cattle'. (Mayer 1975: 278)

The great increase in the study of women and in woman-centred studies (which are not quite the same thing), particularly by women, in the past decade or so, is itself a reflection of changes in social and moral consciousness in the researchers' own societies, which have been influenced by, among other things, the resurgence of feminism. Women in particular resisted being deflected from this area of research, as they were sometimes in the past when it was not an important focus of interest in the discipline. (It remains to be seen whether any changes, say in economic support and career prospects, will weaken the present commitment to the study of women.)

Coincidentally, it has been suggested, the interest in symbolic analysis of myth and belief systems played a part in bringing the study of women into more general interest. Symbolism may present perceptions of the universe which do not accord perfectly with predominant explanations of societies, and may give access to certain aspects of thought and organization that are not expressed or are contradicted in natural language. Thus myths and stories may incorporate subordinate, submerged or implicit ideas, including those relating to gender roles and valuations. In any given society we might, in fact, expect a multiplicity of partial models of reality, some of which, depending upon the critical realm of discourse or universe, receive greater prominence. In many (but not all) forums (often indeed in those most easily accessible to most ethnographers), those constructs generated by men seem to become the dominant models.

The possible existence of models which have been overshadowed or repressed, and of "muted groups" which generate them, has been discussed (in S. Ardener 1975, 1978 and 1984, and by others elsewhere). Women, according to this view, have often found themselves muted in contexts where their gender ascriptions are critical. So far there has been little research giving explicit consideration to situations in which men are "muted" by virtue of their gender alone (but work on inter-relationships of this type between two interacting mixed groups – e.g. "gypsies" and "gorgios" – has been undertaken).

The gender orientations of much research are seen then not only as phenomena of the ethnographer's own culture, but as often reinforced by similar cultural emphasis in the societies under study. Sharma (1981: 36–37) confirms that:

> It is not just that anthropologists (male or female) have listened to men and not to women, although this may be true in some cases. In many situations, Indian village women will express ideas and describe events in a manner that does not arise from a specifically female experience. This is partly because of the dominance and public acceptability of the male definitions of the world.

But the problem of eliciting women's alternative views, if any, is not necessarily solved by the presence of female ethnographers. As Olivia Harris has noted (personal communication (1982) confirming E. W. Ardener (1972)) "the gender of the fieldworker does not in and of itself mean that the data collected will reflect gender. It is likely that women professionals trained within the dominant discourse will not produce such a female approach as their gender might imply", in contrast to men who may well produce a male-oriented analysis "since at least in gender terms, they are more fully integrated anyhow in the dominant discourse". Sharma thinks that the village based study, a common basis for anthropological research done by both male and female ethnographers, is itself a model which reflects male interests, being particularly inadequate from the women's point of view when (as is the case of village exogamy and virilocal marriage) they uproot themselves at marriage and migrate over long distances to join their husbands. The difference gender makes to apperceptions of physical and social space has also been considered by Callaway, Wright and Ardener, among others (in S. Ardener 1981). It is also becoming clear to a number of workers (including Sharma and Barnard (1980)) that some female links (between women and their parents, or between women and their sisters and brothers) have in the past tended to be undervalued by social anthropologists; correspondingly, ties to husbands and children have been relatively over-emphasized. That the productivity of women has also often received scant recognition in the past has been well documented since Boserup's pioneering work (1970). Iona Mayer has noted that the Gusii male view (that women who defaulted in their servicing tasks "were naughty, selfish and unreasonable")

> was echoed to some extent by the official view of colonial administrators, and often even [of] colonial anthropologists, that the 'breakdown of marriage' is a symptom of 'social malaise', and that 'stable marriage' is a Good Thing in itself, along with other traditionally sanctified values like respect and discipline.

She adds that in the 1940s "One hardly ever thought of asking, 'good for whom?'" (Mayer 1975: 278–279). If marriage has been studied predominantly from a male point of view, so perhaps have many kinship studies (see for example James' (1978) critique of the "problem" of matriliny, and Ardener (1983)).

Nearly a decade ago it was said that

> we should not be disappointed if, should women's models of the world (or those of any other muted group) be elicited, they were found to resemble in the main those of the dominant structures with which they are associated. It is the small deviations from any norm which may be crucial. Just as the pinch of caraway seed may transform a basic recipe, or a drop of dye may alter a hue, so any small unique differences in world-views may make "all the difference" (S. Ardener 1975: xix).

Sharma (1981: 37) notes that "(unsurprisingly) in many areas male and female experiences do not diverge and there is no specifically male/female model"; to this others would agree. But Sharma continues:

> So it is not just sensitivity to the presence of women which is required of the ethnographer, but sensitivity to the difference between different kinds of situation, and the correspondingly different ideas and experiences which will be expressed within them.

This accords with the stress which has been laid on the significance of identifying the relevant universe or domain of discourse for an understanding of "muting".

5.4.3 Accesibility to domains of discourse

At a recent discussion on the advantages and penalties of an ethnographer's gender in the field, the women present concluded that the lone female fieldworker is usually (but not of course always) better placed than the single male worker. Among the reasons advanced was that, of course, a woman can enter important female spheres where no man may appear. But it was further suggested that a woman is more likely to penetrate a male domain than a man to enter a mixed or female realm. It was generally felt that one woman in all-male company normally has a smaller effect on the group than one man in all-female company. The age of a woman might be more critical here.

Yengoyan (1970: 212), a bachelor, found that Filipino women in Mandaya would not tell him anything about their rituals and activities. Fearing gossip, he also cautiously avoided hiring a female assistant or seeking more intimate female companionship which might have provided some insights into the female world (which, incidentally, led to rumours that he was a homosexual!). John Honigmann (1957) noted that as a male anthropologist he had little opportunity "to directly study women's roles". Hirschon (personal communication, 1982) has written that in Greece gender conformity was demanded and her classification as a single woman student was critical.

> Although the observance of these expectations often meant that I could not be involved in specified 'male' activities or areas (e.g. coffee shops, football match attendance), the limitations were not nearly as severe as they might seem . . . I had the great advantage of free access to homes. Here it was possible to get to know *all* members of the family, male and female. In this respect, a single woman fieldworker has a distinct advantage over her male counterpart . . . (as some (male) colleagues have told me).

Where a woman is seen as less of a threat to the host society, a female ethnographer can move between public and private domains more easily, and she may even gain access to men's sectors which are barred to women in the host society. Barber (personal communication, 1982), for instance, had no difficulty in attending ceremonies at a Nigerian mosque normally only open to men. Powdermaker (1966) in Lesu and Diamond (1970) in Taiwan both found men prepared to admit them to strictly male occasions on the grounds that most of their culture would be overlooked if they neglected the male side of things. Although she decided to decline the invitations, Powdermaker was invited both to enter the seclusion hut of male initiates and to attend their circumcision ceremony a few weeks later. S. Ardener found that in the

Cameroons she could approach secret masked figures to which local women were forbidden access – provided none of the latter were present.

The difficulty experienced by a female anthropologist working in a society which strictly separates the sexes is graphically described by Pettigrew (1981). Starr (personal note 1982) has suggested that both men and women can circumvent gender-restrictions by making use of their opportunities in public settings such as weddings, markets and law courts, to have informal talks with members of the opposite sex. Starr also recommends the use of "*key* informants", and of a local field assistant of the opposite gender to gather material or to accompany the ethnographer. She was able to cross into sex-segregated groups in rural Turkey and enter male political caucuses and into informal discussions of law cases by always being accompanied by "a young poor assistant", native to the town whom she had trained. Tremayne was able to enter restricted circles in Iran under the patronage of a senior male with enough authority to break the rule of exclusion for her. Diamond (1970) used an assistant to interview all males aged between 20 and 40 to avoid unsavoury sexual overtones. Male ethnographers are advised by Starr always to talk to two or more women together. She recommends the use of helpers of both genders where possible, so that everybody's reputation is protected. Sometimes, however, being associated with a local person of either sex may have the effect of inhibiting special treatment for the outsider.

It has been suggested (e.g. in *Notes and Queries on Anthropology* 1951: 30) that among "very unsophisticated natives . . . a woman may find that she is . . . given the status of male". Whether or not a fieldworker can become an "honorary" member of the opposite sex is, however, debatable. Okely (1975) notes that

> To my surprise and perhaps disappointment when I entered the field [among English gypsies] I did not find this so. . . . Increasingly I suspect that women anthropologists are given ambiguous status in the field, not as 'honorary males' but as members of an alien race. So where did this 'honorary male' come from? The idea has roots instead in the anthropologists' own society.

Assimilation in the field meant, for Okely, undertaking tasks (like driving lorries) which in the English milieu from where she came were associated with men. Besides escaping from the strictures of the role "allotted the females of my own kind", she writes of her fieldwork, in addition, "this *rite de passage* made me an honorary male among those back home".

It might also be helpful to consider the special privileges sometimes granted to women fieldworkers in terms of the discussion of "generalized" as opposed to "specified" gender categories. It has been suggested that men tend to "belong" to a general (human) category (which may have representational

functions on behalf of both genders) which can include females lacking
certain criteria which would specify them as of the (social) gender-category
"woman". Such "generalized" females often include virgins and women past
their child-bearing capacity (see, e.g. Hastrup 1978).[3] Outsiders, including
female ethnographers, may appear to lack some of the critical specification-
markers of local women, and hence be "generalized" and classed in the same
category as men. It may be more difficult for a male ethnographer to be
identified by the specified characteristics appropriate to women. The longer
the fieldworker remains in the field and the more assimilation occurs, the
stronger may be the requirement to conform to the host society's gender-
roles. The need to accept gender-based restrictions applies with special force,
of course, to anthropologists doing fieldwork in their own societies, or in
those into which they have (like Pettigrew and Chahidi, for example) entered
by marriage.

The presence in the field of a spouse or children clearly affects gender
relationships. Partners probably influence the choice of field itself to differing
degrees, and there may be a bias towards the interests of the male. Thus
Margaret Mead "left the choice [of Manus] up to Reo Fortune, whom [she]
was planning to marry" (Mead 1977: 61). There is thought to be some loss
for any pair of fieldworkers, because they lack the vulnerability and dependence of the single worker which bind them to the host community. Rosemary
Firth notes that:

> In respect of learning the language, having a spouse present may be at some
> points a hindrance because one is not driven by the desperation of loneliness to
> make outside contact (1972: 24).

Married ethnographers may also forfeit the possibility of a "neuter" or
anomalous role. Olivia Harris argues that a woman fieldworker's classification is "*bound* to be different . . . if she is with a husband" (personal
communication 1982). Buechler (1969) remarks how his wife had great
difficulties in attending meetings in Bolivia because women would usually
only attend if the head of their household was absent or when a complaint
was lodged against a particular woman. Rosemary Firth writes of the effects
of assimilation:

> I remember one evening when I was serving coffee to Raymond and to one of our
> most trusted and respected informants, a very strong urge came over me to retire
> into the back parts of the house after coffee, as a Malay woman would have done,

---

[3] Hastrup sees biology and culture as "mutually affecting spheres of reality" (1978: 54). For
Harris '"Women" as an abstract "biological" category is meaningless: the distinction between
married/unmarried, or childless/with children, or pre-menopause/post-menopause, may be
every bit as significant as the simple male/female distinction' (personal communication 1982).

rather than stay talking to the two men. Then I had consciously to remind myself that I was indeed an anthropologist, not just a Malay woman nor a good Muslim wife (1972: 25).

Gertrude Enders Huntington (Hostetler and Huntington 1970) remarked that she was obliged to maintain a submissive female character in the North American Hutterite commune in which she was working with her husband and family. Pettigrew, who worked in her husband's Punjabi village, gives a vivid account of the problems of being assimilated, as a spouse, into a society which strictly controls the speech and movements of women. She too had to make a deliberate effort to be a social anthropologist. In the village where "it had seemed as if I had lost the facility to express myself through feeling that I had to conform to a system in which shyness in a woman was a value" (1981: 75), Pettigrew states: "when I did go out I soon became terrified of meeting a man's gaze and, worse still, of letting a smile slip onto my face while talking to men" (p. 72); on holiday in Delhi, however, she had "to learn again how to talk to reasonably sophisticated men without being shy" (p. 76). Dube (1975) gives sensitive accounts of her fieldwork relationships in three different Indian communities. On the other hand, a couple may use their gender differences to study different topics, and they may share their findings, while helping each other to keep a sense of balance and of academic purpose. Margaret Mead notes that after she joined Reo Fortune in Melanesia:

> My intellectual life became a cooperative enterprise and the excitement of intellectual discussion was part of my life in the field itself, not something to be shared in letters or much later, after the event (1977: 62).

Rosemary Firth in her perceptive account of her introduction to fieldwork, wrote:

> When two people go into the field together, sensitivity to small signs of strain and the ability quickly to interpret the feelings of the other are obviously important. In the loneliness of the field situation, the presence of another can be a real source of comfort.

She adds that "some possibility of being able to talk over the early feelings of confusion and failure is an important safety valve" (1972: 24). Malinowski's diary (1967) certainly leads one to speculate on how his fieldwork and analyses would have progressed with the presence of a wife! On the other hand Rosemary Firth thinks that

> the obligation to give support to another may also be a serious additional burden on a fieldworker to begin with. I am not surprised then at the anthropologist who wishes to break ground before he brings his wife or family (1972: 15).

Nowadays it is less uncommon for an ethnographer to be accompanied by a husband not trained in the discipline. The effects of his presence have not yet been documented.

The division of labour does not necessarily mean that a female ethnographer only has advantages in her contacts with women; sometimes it is easier for her to discuss sensitive economic matters with men, where the male ethnographer may be seen as a threat, or as, say, ultimately interested in increasing taxation. It is also possible, of course, that women might tell some things to a male ethnographer that they would not mention to his wife. Children may be useful points of contact and their presence implies shared experience between the female ethnographer and mothers in the community (Dube 1975). On the other hand, fieldworkers have reported that "working mothers" may not be approved of in some communities, where it may be felt that they are anomalous or neglectful of their children. Little evidence is available on the advantages and disadvantages for male ethnographers of having children with them in the field.

5.4.4 Conclusion

It is important to remember that gender-based assumptions are embedded in most aspects of social anthropological research, whether the topics under analysis are religious, economic, or other. The study of specifically gender-defined groups and of their world-views and experiences, has been influenced by values and interests derived from the ethnographers' own cultures, and those in the societies they study. Although there is now a greater awareness of the dangers of "generalizing" material gathered only from men or about men, greater sensitivity to the representativeness of such material is often required. Recently there has been a greater concentration on the study of women and some women-centred studies, and many publications have emerged which have gone some way to redress some past imbalances.[4] Indeed there is sufficient output for there to be talk of a "second wave" in recent studies of women (*Choice*, Jan 1982). Already notes of caution are appearing in print.

[4] This is not the place to discuss the findings of all the new work on women. However, among the many publications with implications for methodology, other than those referred to, see Papanek (1964); Bovin (1966); Golde (1970); Chinas (1971); Rohrlich-Leavitt *et al.* (1975); Slocum (1975); Weiner (1976); Harrell-Bond (1976); Lamphere (1977); Quinn (1977); Edholm *et al.* (1977); Tiffany (1978); Caplan (1979); Huizer and Mannheim (1979); Milton (1979); Rapp (1979); Leibowitz (1980); MacCormack and Strathern (1980); Reiter (1980); Rosaldo (1980); Watson-Franke (1980); Bourque and Warren (1981); Leacock (1981); Young *et al.* (1981); Roberts (1981) – to select but a few. Woman-centred studies would ideally include, for instance, a study of kinship where *ego* is female, studies of male activities as seen by women, of cosmologies and so forth; this is still a little explored approach.

Rogers (1978) says that exclusive focus on female forms of power may give the impression that women are relatively more powerful than they are. Strathern (1981) warns against "the manufacture of a sub-discipline". Others might see few signs of that taking place or even of its being an aspiration among social anthropologists, although, as in other fields of enquiry, some specialization is to be expected. Most social anthropologists when they study gender-specific topics might agree with Sharma (1981: 37) when she says that she does so "as part of a necessary stage in the dialectical advance towards a more integrated analysis of society". As is implied here, however, any such advance will depend not merely upon adding more data collected with greater sophistication, but upon using these to find new insights towards a total "re-vision" of anthropology and of society.[5] Callaway (1981a) sees such a re-vision as "correcting or completing the record; ... a deliberate critical act to see through the stereotypes of our society as these are taken for granted in daily life and deeply embedded in academic tradition, and as the imaginative power of sighting possibilities", the effect of which would help "to bring about what is not (or not *yet*) visible, a new ordering of human relations."

Shirley Ardener

5.5 Special problems of fieldwork in familiar settings

The aims and methods of research, especially in the case of intensive field-work, demand, at one and the same time, the qualities of an insider and those of an outsider. Srinivas (1966: 156) has put it this way: the researcher becomes ideally so close to his or her host community that s/he understands their values and assumptions, and must, like a novelist, "get under the skin of the different characters he is writing about"; but equally, the ethnographer must have the detachment of the trained observer, must seek generalization rather than anecdote, must remain faithful to scientific canons which gives validity to his or her presence in these other, often compelling and absorbing, surroundings. The tension between the need for both empathy and detachment is a problem facing *all* anthropologists (§5.3.2) and yet this is often forgotten by people who argue against the "insider" working in his or her own society, as opposed to the "outsider", transplanted to an exotic setting. Aguilar (1981: 22) points out in defence of the "insider", that:

[5] The notion of an "integrated" analysis of society presupposes a coherence that society may not possess, however.

Bias is the human condition, a danger for both insider and outsider researchers. Whereas the insider might labor under a biasing chauvinism, all outsiders, by virtue of their primary socialization in one society, must make efforts to over-come ethnocentric bias. Similarly, the xenophilia of some socially mobile or ethnically passing individuals is also a possibility for the exoticist (outsider) who sees much virtue abroad and little at home.

Anthropologists, of course, have tended to study people whose values and life-styles are different, even in their own society. Far more frequently they have gone to the villages and the slums which have still provided the requisite culture shock of fieldwork for the urban, middle class (or, as in India, high caste) researcher, but without the expense of foreign travel. Even in sociology, Newby (1979: 101, echoing Frankenberg, 1966: 252) has reflected on the fact that British rural community studies have, on the whole, concentrated on family farming in the upland fringes of the British Isles, to the neglect of the capitalist farming areas of lowland England, with their large farms, agricultural labourers and villages divided between commuters and "locals". The very idea of a "Celtic fringe" is, of course, an ethnocentric one, minimizing and distancing the land masses of Ireland, Scotland, Wales and the West Country, from the relatively small area of Central and Southern England, which becomes in metaphor the "carpet" itself. Research, in other words, has tended to focus on those areas least known to the predominantly Southern English, middle class sub-culture of the universities. The rural sociologist, or anthropologist, has sought the exotic at home as well as abroad.

Anthropologists have for a long time studied parts of their own society nearer to home, and made generalizations about their own culture (e.g. Kimball 1955; Mead 1942), but in recent years this area of interest has expanded greatly. In Britain, for example, there have been studies of both suburban areas and villages (e.g. Firth *et al.* 1969; Harris (*pseud.*) 1974; Richards 1980; Strathern 1981a). But these anthropologists had previously done fieldwork abroad (Richards, for example, in Africa and Strathern in Papua New Guinea), regarded by some as a necessary exercise in the acquisition of detachment by the trained observer. Others still (e.g. Lewis 1973: 6) have maintained that it is necessary to work in an alien society in order to produce those sparks of intellectual inspiration which underly the best ethnographic analyses. According to Srinivas, however, the experience of another segment of one's own society can create sociological awareness, or anthropological detachment; it is then that one becomes an "outsider" in one's own place of origin. Might one go further and ask whether, through study, the theories and values of the social sciences *become* one's own society, and the world is experienced through the prism of theory, so that any and every society is viewed with detachment? Is this too much to ask of theory, without the practice of fieldwork? If we take Srinivas's advice and agree that the experience of another segment of one's own society is a valuable part of the process of becoming an anthropologist, we may also reflect that with the spread of higher education, some, especially ethnic and working class students, may experience two sub-cultures by leaving their own backgrounds and becoming, at University, part of the professional classes (see, for example, Jackson and Marsden 1966).

The Essex studies by Robin (1980) and Strathern (1981) exemplify two aspects of research in a familiar setting where, as will often be the case, the researcher comes from a complex literate society. They combine, first, the research methods of the social historian and the anthropologist, and show how important it is for the anthropologist to grasp historical and demographic material where it is available; secondly, the books are the result of cooperation among a number of researchers. The project began as a teaching-aid, and has the benefit of many talents and many combined hours of work. Interdisciplinary work (§7.11.5) is not new, of course: one might compare the striking example of Plozévet (Burguière 1977), a community whose 3800 inhabitants were studied for five years, by not only anthropologists but doctors, psychologists, sociologists, demographers, geographers and historians. In this and the previous examples, the anthropologist is able to take advantage of the fact that he belongs to a society with a well-developed system of higher education, and the availability of on the spot, highly-trained human resources at his or her disposal.

The fact that anthropologists have sought the unfamiliar at home as well as abroad has meant that many of their problems are the same as those of anthropologists doing fieldwork in other societies: Elliot Liebow (1967: 232–256), for example, describes in his own experience of fieldwork, the problems of access, acceptance, language, of the boundaries of the unit to be studied, of the extent of one's immersion in the culture, of one's self-presentation both in clothing and what one said about one's work, and, as in many situations where the anthropologist comes from a richer portion of the world – resentment, and akin to this, the extent to which the anthropologist is to be used as a local resource. Although he belonged to the same nation, he was, "not fluent in their language . . . an outsider not only because of race, but also because of occupation, education, residence, and speech" (Liebow 1967: 252).

A more recent study of Jamaicans in Bristol (England) by Ken Pryce (1979), himself a West Indian sociologist, points to further aspects of the problem. Whereas Liebow, as a Jewish American, could not cross the white–negro divide and be mistaken for one of the blacks he was studying, Pryce took advantage of this ability to merge with other immigrants, and sometimes concealed what he was really doing, in the interests of "total immersion" in the culture. Concealing the purpose of research (e.g. Patrick (*pseud.*) 1973; Humphreys 1975) and taking advantage of informants and well-known institutions and cultural conventions raise special problems when working in familiar settings. Moreover, many people in Britain, as well as other countries, are unacquainted with the idea of intensive fieldwork, although they know about surveys and questionnaires. Coffield *et al.* (1981), studying families in a Midlands town, found that their informants *forgot* that they were researchers, over time, and were treating them as friends; the question arose as to whether they were being told important information as intimates, which would not have been given in other circumstances (§6.3.1).

The fact that anthropology exists as an activity mainly in institutions of higher education and, to a lesser extent, in supra-governmental or government agencies, means that complex societies with more developed education systems train more anthropologists. Consequently, many anthropologists studying their own society experience the problems of an anthropologist in a complex society. More and more, however, anthropologists from the Third World are studying their own societies, and some of their problems are distinct. Aguilar (1981: 21), for instance, points out that the "insider" anthropologist may make informants more relaxed than an "outsider", but if the minority group is in a state of enmity with the larger society, his or her own acculturation may cause resentment. A rather different problem is that of what you are supposed to know already as an insider. The African anthropologist, Nukunya (1969: 19), describes how his informants said they would tell him things which they would not normally tell outsiders, although naive questions might "be considered irrelevant or even impertinent coming from him" (quoted in Aguilar 1981: 21). This leads us back to the debate about bias and objectivity, since it is obvious that even the ethnic minority or Third World anthropologist is an outsider to a degree. The goals of anthropological research distance the researcher from his or her informants – the culture of the university or institute is not the culture of the village. To this extent, unless anthropologists become as introspective as the cinema (with its passion for films about film-makers), there is little likelihood of true insider research ever becoming common: the ethnographer will always be somewhere on the continuum between empathy and repulsion, home and strangeness, and seeing and not seeing.

Jacquie Sarsby

6 Ethics in relation to informants, the profession and governments

6.1 Introduction

Anthropologists, *qua* scientists, citizens and persons, like other social scientists, have to contend with ethical and moral dilemmas arising during and out of their research.[1] In anthropology, as in the other social sciences, there was little concern about ethics among either researchers or their subjects before the mid-1950s and few publications before the revelations in 1965 about Project Camelot, the prime catalyst for ethical debate among social scientists.[2] Today, in the United States in particular, there is an increasing concern with the personal, political and ethical dimensions and implications of fieldwork and a "willingness to address these issues and processes publically, explicitly, and more profoundly" (Emerson 1981: 363). Ethical concerns are now promulgated in codes of conduct,[3] defined by legal

[1] For some anthropological discussions see Barnes (1967a [1963]), R. Beals (1969), Jarvie and Kloos (1969), Jorgensen and Adams (1971), Jansen (1973), Appell (1974, 1976, 1978), Rynkiewich and Spradley (1976), Kimball and Partridge (1979), Social Problems (1980), Adams (1981). For sociological, psychological and general social scientific discussions see Sjoberg (1969a), Social Problems (1973), Nejelski (1976), Barnes (1977, 1979, 1981), American Sociologist (1978), Bower and de Gasparis (1978), Diener and Crandall (1978), Bulmer (1979, 1982b), Klockars and O'Connor (1979), M. Wax and Cassell (1979), Emerson (1981), Wolfgang (1981), Beauchamp *et al.* (1982), Reynolds (1982). For oral history see Henige (1982). Several of these contain lengthy bibliographies.
[2] Project Camelot was a US army sponsored project intended to investigate the preconditions for internal conflict, starting with research in Latin America. See Horowitz (1967), Sjoberg (1969b), Deitchman (1976), M. Wax (1979b).
[3] See Society for Applied Anthropology (1963, 1974), American Anthropological Association (1971, 1973), British Sociological Association (1973), Jansen (1973), Orlans (1973: Chs 2, 3), Reynolds (1975, 1982), Diener and Crandall (1978), Flaherty (1979), Canadian Sociological and Anthropological Association (1979), Long and Dorn (1982).

ETHNOGRAPHIC RESEARCH
ISBN 0 12 237180 1

or administrative fiat,[4] articulated by or on behalf of the people studied,[5] and discussed in fieldwork accounts, textbooks and courses.[6] There is not, nor ever likely to be, however, any definitive agreement about the nature of either the problems or the solutions.

The ethical problems facing social scientists have changed over time and have probably become the more difficult to resolve as the contexts of social research have changed and as the balance of power has altered between social scientist, sponsor, gatekeeper, citizens and government. One cause of these changing power relations is the continually altering social, economic, educational, administrative, legal and political contexts and constraints within which social research occurs; a second is the institutionalization and professionalization of social scientific research. A third contributory factor is the recognition that knowledge is not only a source of enlightenment but also of power and property and, therefore, that it entails the power both to harm and to benefit those studied (cf. Barnes 1979: 22; Cassell 1982: 156–157).[7]

6.2 Ethical concerns and anthropology

The "charter" establishing the interest of anthropologists with ethics begins with a letter by Boas (1919), condemning the use of anthropology as a cover for spying, and stating that persons so using it forfeited their right to be classed as scientists and jeopardized the research activities of honest visiting investigators, for which opinions he was censured by the American Anthropological Association (Stocking 1968: 273–277). Echoes of that controversy and its twin themes of politics and ethics were to recur half a century later in

[4] See, e.g. Nejelski (1976), Solomon Islands (1976), Tebape (1978 – which includes a copy of what must be the only Anthropological Research Act, that passed in Botswana in 1967), Barnes (1979: Ch. 8), Flaherty (1979), Flaherty *et al.* (1979), Klockars and O'Connor (1979), Beauchamp *et al.* (1982: Part 5), Great Britain (1982), Raab (1982), Reynolds (1982).
[5] The literature on this topic is considerable. See, e.g. Deloria (1969), Efrat and Mitchell (1974), Maynard (1974), Gwaltney (1976, 1981), Journal of Social Issues (1977b), Zinn (1979) for North America; Hau'ofa (1975), Mamak and McCall (1978), Keesing (1979), Strathern (1979), Hughes (1980) for Melanesia and the Pacific; Chilungu (1976), Mafeje (1976), Owusu (1978, 1979) for statements by African anthropologists; Grönfors (1982, for Finnish gypsies).
[6] See Appell (1976, 1978), Diener and Crandall (1978), Kimball and Partridge (1979), Spradley (1980), Warwick (1980), Agar (1980), Shaffir *et al.* (1980). A comparison of readers in fieldwork and qualitative methods shows the trend: see Adams and Preiss (1960), Jongmans and Gutkind (1967), McCall and Simmons (1969), Filstead (1970), Shaffir *et al.* (1980), Burgess (1982).
[7] For a detailed discussion of these and other factors and their ethical concomitants see Barnes (1979) who pays more attention to anthropological literature and concerns than is usual in historical accounts of social science. His use of "citizen" rather than "subject" or "informant" is intended to draw attention to "their rights and duties as fellow human beings" (ibid: 14–15), but has some disadvantages in not distinguishing their research participatory role.

the debates about anthropology, Project Camelot and the Vietnam War, but by then the climate of professional opinion was very different and the debate extended far beyond the USA.

In the intervening decades there was little (published) concern with ethical issues. There were discussions among applied anthropologists who formulated the first anthropological code of ethics (see Mead *et al.* 1949; Tugby 1964; Society for Applied Anthropology 1963, 1974; Mead 1978; but cf. Manners 1956); but when ethics were mentioned the main concern was with the effects of interference in the field on humanitarian grounds and, therefore, with the scientific propriety of such actions (see, e.g. Spillius 1957; Gallin 1959; cf. Jarvie 1971; Pelto and Pelto 1978: 186; McCurdy 1976). In the colonial period, too, as Barnes (1967a) pointed out, people, places and groups were not routinely disguised; ethnographer, informants and, to some extent, administrators were neither part of one social system nor shared one moral code; the field of study was clearly defined; governments were legitimate research sponsors and funders; and publications were not read by informants, let alone vetted by them.

In 1965 two events altered the situation: Project Camelot, and the furore which had developed slightly earlier in Mexico after the publication of a Spanish edition of Oscar Lewis' *The People of Sanchez* (see Paddock 1965; R. Beals 1969: 11–15). The American Anthropological Association commissioned an investigation into, *inter alia*, relationships between anthropologists and sponsoring agencies, created a Committee on Research Problems and Ethics, and passed a Statement on Problems of Anthropological Research (R. Beals 1969). In 1971 it passed a formal code of ethics, since periodically amended (American Anthropological Association 1971). Debate within and without the Association was further fuelled by other controversies, notably over sponsorship of research by American defence and intelligence agencies, e.g. in the Himalayas, and anthropological involvement in Vietnam and Thailand.[8]

Radical theoretical and political critiques inside and outside the discipline which directly challenged anthropology and its practitioners also had consequences for anthropological ethics. One set of criticisms arose out of the debate about the relationship between anthropology and colonialism/ imperialism;[9] others appeared in discussions about the legitimacy and uni-

---

[8] See Berreman *et al.* (1968), Saberwal (1968), R. Beals (1969), Wolf and Jorgensen (1970), Jones (1971), Jorgensen (1971), Berreman (1978).
[9] For some of the numerous publications see Manners (1956), Gutkind (1969), Stavenhagen (1971), Lewis (1973), Berreman (1973a, b), Nash (1975), Mafeje (1976), Huizer and Mannheim (1979); also Belshaw (1976: Ch. 16), Anthropological Forum (1977).

versality of western, bourgeois, white, male-dominated social sciences arising out of developments such as feminism and Marxism, or out of discontents articulated by the people studied or on their behalf by spokesmen or by anthropologists who had worked among them.[10] These developments were accentuated by the changing political context as colonial countries attained independence, as minorities and disadvantaged groups became politically more active, and as local scholarship and research developed (see, e.g. T. Weaver 1973; J. Nash 1975; Barnes 1979, 1982; Fahim 1982). Correspondingly, new types of anthropologists and new foci for research have burgeoned. Increasingly, anthropologists are: "studying-up" (Nader 1972), investigating élites, bureaucracies and aspects of industrialized societies and international capitalism;[11] developing an insider/indigenous/native anthropology (§5.5);[12] putting their skills at the service of minority groups and threatened communities through action and advocacy anthropology (§4.4.10);[13] and becoming more deeply involved in client-commissioned research, e.g. applied, development, administrative and evaluation studies on behalf of governmental and other agencies and clients.[14] All these fields present ethnographers with setting-specific problems of method and of ethical and moral dilemmas in relation to informants, clients, sponsors, governments and the profession. Not least is the dilemma that in such settings the assumed normative commitments to the interests of informants and to the public dissemination of findings may no longer be paramount, nor is it easy (even if held desirable) to reconcile "academic", "committed" and "practitioner" concerns.[15]

The themes and studies which have emerged have theoretical, methodological, logistical and ethical implications. Some involve the sociology of knowledge, such as the assertion of the primacy of insider knowledge and experience (see Barnes 1979: Ch. 4, 1982; Hiller 1979; Aguilar 1982; Grönfors 1982); others are concerned with the view of knowledge as power,

[10] See footnote 5, and Fabian (1971), Scholte (1971), Hymes (1972), T. Weaver (1973), Vidich (1974), Mamak and McCall (1978), Barnes (1979: Ch. 4), Huizer and Mannheim (1979).
[11] See, e.g. Harrell-Bond (1976), Britan (1979), Marcus (1979), Galliher (1980), Nash (1981), Messerschmidt (1981).
[12] See, e.g. Kloos (1969), Madan (1969), Shokeid (1971), Brunt (1974), Fahim (1977, 1979, 1982), Posner (1980), Loizos (1981), J. Stephenson and Greer (1981), Aguilar (1981), Wolcott (1981), some of which discuss ethics.
[13] See, e.g Huizer (1973), Jacobs (1974), Schensul and Schensul (1978), Huizer and Mannheim (1979), Jones (1980).
[14] See, e.g. Colfer (1976), Almy (1977), Eddy and Partridge (1978), Britan (1979), Hinshaw (1980), Chambers and Trend (1981), Messerschmidt (1981), Wilkins (1982).
[15] Compare, e.g. Orlans (1973), Rynkiewich and Spradley (1976), Huizer and Mannheim (1979), Cassell (1980), Chambers and Trend (1981), Messerschmidt (1981).

resource and property. The latter include the recognition that much social research may be of no direct use or interest to the people studied (see Huizer 1979; Santos 1981); that studying disprivileged and subordinate groups ignores the real and more pressing problems, such as dominant groups (Saberwal 1968; Nader 1972; Huizer 1973; J. Nash 1975, 1981); that the people studied may be harmed thereby, and that telling the "truth" about minorities or marginal groups may serve further to disadvantage them rather than help to dispel myths about them (Grönfors 1982). Other issues, discussed in Melanesia and elsewhere, require a positive contribution from the fieldworker, as Strathern (1983) demonstrates, including

> 1. Recognition of the rights of people being studied. . . . 2. Contribution by fieldworkers to the interests of the community . . . so as to maximise the return to the community for cooperation in fieldwork; . . . 3. Recognition of continuing obligations to a community after completion of fieldwork. . . . 4. Maximum involvement of indigenous scholars, students, and members of the community in research; . . . 5. Recognition of obligations to make a return to the host community (Keesing 1979: 276–277).

In sum, then, these critiques and developments have questioned the basic tenets and morality of anthropology, its epistemological foundations, its motives, purposes and products, assumed neutrality and objectivity, and its eurocentrism. Even if, as Barnes (1979: 172) suggests, the model of a politically active social science is unlikely totally to replace earlier models, the adoption of a "neutral" or "uncommitted" stance is now seen to be no less political a position than a "radical", "committed" or "antagonistic" stance (cf. T. Weaver 1973; Barnes 1981). The issues cannot be further developed here but they cannot be ignored, not least because while some see a concern with ethics as a way of resolving various dilemmas afflicting anthropology and anthropologists, others see this as an evasion of the basic problem.

6.3 Balancing responsibilities and interests

Ethnographers have to reconcile the rights and interests of the various parties in the research enterprise: informants and other research participants (citizens); gatekeepers; sponsors and funders; themselves; colleagues; their own and host governments; their universities or employers; and the public(s). Moral and ethical decisions occur at all stages of research, from the selection of topic, area or population, sponsor and source of funding, to publication of findings and disposal of data.

6.3.1 Responsibilities to informants and participants

Ethnographers are usually incorporated into the moral community of their hosts, although the depth of their involvement and intensity of interaction vary, as does the mode of researcher–informant relations and hence power relations in the field (see Cassell's schema 1980, 1982; Adams 1981). Long-term researchers (§8.6), in particular, acquire continuing (sometimes life-long) moral obligations and practical responsibilities, which may be personal, professional and civic, in relation to their informants and the host population (Foster *et al.* 1979; Strathern 1979a; Hennigh 1981) as well as to their colleagues and sponsors. It may also be harder for them to avoid active intervention, though the extent to which this poses ethical and methodo-logical problems varies (compare Øyen (1972) and Hennigh (1981)).

Commitment to the people studied is usually assumed; but as A. Beals (1978) points out, identifying, safeguarding and forwarding the interests of informants are not straightforward propositions (cf. Appell 1978) particularly in factionalized, stratified, or otherwise divided or complex settings.[16] Nor is it clear that commitment to the paramountcy of the interests, rights and sensitivities of those studied is a universal norm (*pace* American Anthro-pological Association 1971: paragraph 1), if it ever really was (cf. Brunt and Brunt 1978; Chambers 1980: 337).[17] There is ambivalence about, and some opposition to, granting this professional courtesy to élites, publicly account-able officials and other powerful individuals or groups.[18]

Ethnographers should recognize the rights of citizens to privacy, con-fidentiality and anonymity, and not to be studied; to be informed about the methods and aims of the study, its anticipated consequences and potential benefits, risks and disadvantages, and its sources of sponsorship and funding; to be fairly remunerated for time and assistance; to be given feedback on the results and, where practicable, to be consulted over publications (§6.5.2, §6.5.3); and to have their legal or contractual rights in data respected (§6.4.5, §6.5.2). But such generalized (and idealized) obligations present various problems. Full disclosure has to be balanced against scientific interests

[16] See, e.g. Henry (1966), R. Wax (1971), J. Nash (1974), Mencher (1975), Green (1978), Britan (1979), Srinivas *et al.* (1979), Messerschmidt (1981).

[17] For "traditional" ethnographies apparently lacking this commitment, to which critics have raised ethical objections, see Turnbull (1972) and comments by Barth (1974), Wilson *et al.* (1975) and Reynolds (1982: 84–88); and Hallpike (1977) and his exchange (1978) with Strathern (1978, 1979b). See also Gartrell's (1979) comparison of statements by herself and Slater (1976) about the Nyiha whom they studied within a few months of each other, but note that Slater contends much therein (Slater, personal communication).

[18] See, e.g. Becker (1967), Rainwater and Pittman (1967), Nader (1972), Britan (1979), Galliher (1980). For opposition to these views see Appell (1980), May (1980), and M. Wax (1980).

(Barnes 1979: Ch. 6; Jowell 1982; Beauchamp *et al.* 1982: passim) though these will vary according to the researcher's epistemological (cf. Pinxten 1981) or political stance. Reciprocity, commonly discussed in terms of informal assistance and psychological benefits, also raises questions of what forms of material rewards are appropriate (Foster *et al.* 1979; Gwaltney 1981: 54–55); in the absence of formal rulings (e.g. Solomon Islands 1976) whether, as A. Beals (1978) asks, local standards or potential benefits to the fieldworker should set the rate for monetary payments to assistants; and whether money should be offered at all for information (see Goldstein 1964: 160–173; Walker and Lidz 1977; Weppner 1977: 31–33; Whyte 1979; Henige 1982: 55–57). In some field contexts, too, differences in income and resources between researcher and hosts may be an additional cause for embarrassment, strain and difficulties. Related to such matters are more general financial and moral problems for the anthropologist (see Pandey 1979) and for those studied resulting from the development of dependence upon researchers for economic gains (Foster 1979), prestige (Berreman 1968: 370), or political advantage (Mead 1969: Appendix IV; Goldkind 1970). The issue of financial and moral responsibility for "professional" informants has been raised in relation to long-term fieldwork (Blanchard 1977; Foster *et al.* 1979a: 346); and such "trained" informants present, too, methodological and theoretical problems (Blanchard 1977; Pinxten 1981: 67–68; Santos 1981: 274–275).

Participants' rights, interests and even well-being may be affected by the behaviour of the ethnographer or of his or her assistants in the field. Problematical data-gathering techniques, such as covert research (§6.4.3) or resort to morally dubious practices such as coercion, bribery or promising unfulfillable favours in order to acquire information or to gain access to restricted settings, infringe citizens' autonomy and privacy (cf. Jorgensen 1971; Cassell 1980; Kelman 1982) and may also affect the quality and validity of the data gathered thereby (Cassell 1980: 35–36). Some aspects of personal behaviour, such as sexual behaviour, political or religious activities, the acquisition of art objects, or becoming involved in conflicts, illegal or disapproved activities may create ethical or legal problems for the fieldworker and, particularly in the case of expatriate researchers, affect relations with the host government or authorities and have deleterious consequences for informants and colleagues. Similarly, exiting behaviour (Williams 1967: 59; Maines *et al.* 1980), access to and storage of data, particularly confidential material, and the dissemination and publication of information (§6.5) all need careful consideration if informants are not to be adversely affected.

6.3.2 Responsibilities to and for assistants

Other responsibilities, less adequately discussed in the literature, concern field assistants (§7.11.3). Like key informants (§8.3), aides ranging from school children to fully-fledged professional social scientists tend to be acquired haphazardly and often include local residents (see A. Beals 1970; Walker and Lidz 1977). Yet even detailed accounts, such as those where the ethnographer is effectively a research director (e.g. R. Wax 1971, 1979; Gallin and Gallin 1974), tend to focus on logistic details and interpersonal relations. I know of no detailed discussion of the ethnographer's legal and moral responsibilities for the actions of field aides and other assistants during the research period, or subsequently arising out of knowledge or materials gained through their work. I have no evidence for any abuses (though there are accounts of other ways in which key informants have benefitted, e.g. Goldkind 1970); but discussions on the advisability of malpractice or liability cover for social researchers also suggest this for assistants (Reiss 1979; Trend 1980). Ethnographers may also have to act on behalf of their employees, for instance rescuing them from the authorities should they come under suspicion from some source (du Toit 1980: 277–278).

6.3.3 Responsibilities to sponsors and gatekeepers

Researchers, too, have responsibilities towards interested third parties. Balancing these against rights and interests of informants and other participants is becoming more difficult as ethnographic research is increasingly sponsored and funded by governments, commercial agencies and citizens. Gatekeepers, national and local, who control access to research settings, participants and information, display similarities to sponsors. Both parties may require details of the research topic, aims and methods, and try to influence or exert control over research data and findings and over the activities of the researcher (see, e.g. Efrat and Mitchell 1974; Broadhead and Rist 1975–1976; Barnes 1979). Analogous problems may arise in countries in which government officials, and sometimes academics, argue for or permit only "socially relevant" research whether by national or expatriate researchers (see, e.g. Mamak and McCall 1978; Keesing 1979; Barnes 1982). The fact that field research is seldom as clear-cut as the research proposal formats preferred by sponsors, funders, gatekeepers and review boards is particularly problematical for ethnographers; but changing the research topic because of faulty prior information, changed conditions or serendipitous discoveries ethically may require reapplying for approval from sponsors and renegotiating consent with gatekeepers and citizens (Orlans 1969: 16–17; M. Wax 1977; Diener and Crandall 1978: 170–172; Shaffir *et al.* 1980: 16).

Sources of funding and sponsorship (§7.6, §7.7.1) may be contentious in some political, social or cultural contexts. Frankness may create problems of access or cooperation for the researcher; but concealment may have serious consequences for colleagues, the discipline and participants, as debates over US defence and CIA sponsorship have demonstrated (see Berreman 1973a, b, 1978; Barnes 1979: Chs 5, 6). The researcher may be an innocent pawn and unwittingly mislead informants, but this does not mitigate the consequences for both parties (e.g. Glazer 1972; R. Stephenson 1978; Santos 1981: 281–283). Barnes (1969: 179–181) considers, too, that researchers should be frank about their personal characteristics as these should be important to sponsors but only in so far as they might result in unsuccessful research, render the researcher unacceptable to the host population, or interfere with the fieldwork in some way (§6.4.1, §6.4.2).

The interests of employers, academic and other, have become increasingly overt and pertinent (see Vidich *et al.* 1964; Orlans 1969; Record 1969; Reiss 1979; Warren 1980), the more so as financial pressures increase. Bureaucratic regulations in the United States, furthermore, involve them directly in social research through the requirement that institutional review boards assess federally funded projects and thus in discussions of research ethics and methods (§6.4.4). Particularly unfavourable to the rights and interests of researchers and citizens may be conditions imposed on commissioned research, especially by government agencies (see Orlans 1973; Klockars and O'Connor 1979; Chambers and Trend 1981; Fetterman 1981; and footnote 14).

The recommendations to reject research opportunities "whose results cannot be freely derived and publicly reported" (American Anthropological Association 1971: paragraph 3a) and that "no reports should be provided to sponsors that are not also available to the general public and, where practicable, to the population studied" (ibid: paragraph 1g) begin to seem academic and unrealistic and, it might be argued (in some political and research contexts), themselves unethical.

6.3.4 Responsibilities to colleagues

Fieldworkers also have to remember their responsibilities to, and sometimes for, colleagues, local and non-local, consociates and successors, in the field, the country and the wider discipline. Negative consequences of predecessors' (and contemporaries') fieldwork usually receive most publicity (e.g. Appell 1978: passim; Feinberg 1979; Pandey 1972, 1979) whether or not the allegations are true, or are rationalized justifications (or even inventions) masking a variety of motives. However, for one beneficiary of a predecessor's high reputation see Feinberg (1979). The "offender" need not even have been an anthropologist (ibid; Foster *et al.* 1979a: 347); and in some cases it is publication rather than field behaviour which precipitates problems (§6.5).

Long-term fieldworkers (§8.6) face some specific problems and responsibilities in relation to colleagues: particularly sensitive issues are the division of labour, access to data gathered by others and rights in data, and rights of publication and co-authorship. Careful thought is needed if contention and acrimony, or even abandonment of the project, are to be avoided, though arrangements will vary according to scale and mode of organization (see Foster *et al.* 1979). Similar problems occur in other types of team research (see Whyte 1958; Riesman and Watson 1964; Vidich and Bensman 1964; Bradley 1982; §7.11) and in contract research (Colfer 1976; Trend 1980; Light and Kleiber 1981). Riesman and Watson (1964: 266) proffer the interesting suggestion that in team research individual anxieties may be camouflaged and channelled through disputes over ethical standards. And after a divorce, marital or professional, who gets custody of the data?

Other difficulties and antagonisms between colleagues may arise in relation to the question of proprietary rights, the "my-people" syndrome. "Poaching" and "preclusion" and plagiarism are sensitive personal, professional and ethical matters (see Needham 1974; Appell 1978: passim; Jayaraman 1979; Brown 1981; Strathern 1983); but there are clearly limits to the number of researchers a small community (or even country) can tolerate, simultaneously or over a long period. The high vulnerability of long-term projects makes even more essential the normal professional courtesy of consultation and consideration, as Colson *et al.* (1976) and Foster *et al.* (1979a: 347–348) stress.

Consideration for the interests of local scholars has been a major issue since Project Camelot (see Adams 1971; J. Nash 1975), and that their interests may differ, academically and politically, from those of expatriates should be recognized (see, e.g. Fahim 1977; Mamak and McCall 1978; Barnes 1979: 180–181, 1982; Keesing 1979; Fahim *et al.* 1980; Strathern 1983). Responsibilities to local and other scholars also involve issues relating to academic honesty, accurate reporting, and publication of research (see Diener and Crandall 1978: Ch. 9; Henige 1982; §6.5).

6.4 Ethnographers at risk

No social researcher is spared ethical problems; but it can be argued that some dilemmas are the more acute for ethnographers because of their moral involvement with their informants or are the more likely to arise in anthropological and other participant observation studies. Only a few of the issues can be considered here.

6.4.1 Scientist and citizen

One set of dilemmas, which may create personal, political or legal difficulties, arises out of the differences between a scientist and a citizen. Whether these roles are taken as combined in the person of the ethnographer or as divided between researcher and informants, the dichotomy represents divided and sometimes conflicting interests (Barnes 1979: Ch. 9, 1981; §6.2). That anthropologists are especially likely to conduct research in countries or groups other than their own, among minority peoples and in cross-cultural contexts highlights the potential conflict between their rights, obligations and interests as scientists and as citizens at home and as visitors abroad. This legal dichotomy adds a different dimension to familiar discussions (e.g. Jarvie 1971) of the ethnographer as stranger and as friend or guest. It is crucial in any consideration of the researcher's actions and responses in the field, of relationships with government, with informants and local colleagues (see Fahim 1977; Fahim *et al.* 1980); it renders some researchers more susceptible to exclusion on political grounds (see Jayaraman 1979; Worsley 1982); and it is particularly pertinent for action and advocacy anthropologists working outside their own countries. Insider anthropologists face their own ethical and political quandaries (e.g. Madan 1969); but an outsider or alien status may be crucial not only in influencing the moral and practical responses of the fieldworker to the humanitarian and political aspects of field situations, but also for their actual or potential repercussions (see Jarvie 1971; Rynkiewich and Spradley 1976; A. Beals 1978; Pandey 1979). Furthermore, whether one holds that activities as scientist or citizen and as scientist or person should be balanced or separated, or are inseparable and totally interdependent, personal characteristics of the researcher may be especially salient in such fieldwork contexts. Gender, age, ethnicity, nationality, personality, and political or religious convictions may affect the acceptability of the researcher to the host population, gatekeepers, sponsors or colleagues, constrain behaviour in the field, and may cause moral and ethical (as well as methodological problems (§5.3, §5.4).[19] Since field statuses are interdependent, ethnographers may also be affected by such characteristics in their fellow ethnographers (see Dube 1975; Pastner 1982) or in their assistants (see Berreman 1962, 1968: 360–361).

6.4.2 Stressful, dangerous and illegal situations

Fieldwork is a difficult and taxing activity (§5.2, §5.3), one not for the faint-hearted or self-conscious (Sommer 1971; R. Wax 1971: 370–373; Clarke 1975; Pelto and Pelto 1978: 177–192), as most personal accounts demonstrate. Long-term fieldwork may be particularly stressful (Wacaster and Firestone 1978; Meggitt 1979: 124) and it also becomes more burdensome and time-consuming as the researchers themselves age. Fieldwork may also involve, on the part of researcher or hosts, knowledge or behaviour which threatens the moral code, values or personal integrity of the researcher. Acquiescence, intervention, involvement, and evasion are all problematical in some way, may create further dilemmas and exacerbate field-related stress (see, e.g. Read 1966; Devereux 1967; Bohannan 1972; Appell 1978; Pepinsky 1980).

Some field situations may necessitate (or be thought to require) participation in socially disapproved, dangerous, illicit or criminal activities. Many ethnographers are "accessories" before or after the fact and have to decide how far they will obstruct justice (see Polsky 1967; Sommer 1971; Soloway and Walters 1977), although limits to their participation can be negotiated without necessarily damaging rapport (cf. Polsky 1967; Weppner 1977: 36–41; Agar 1980: 61). Though R. Wax argues that a fieldworker may develop "an overblown sense of his ability to offend or injure his hosts" (1971: 275) in some circumstances participating, recording, publishing or even merely knowing of such activities may place informants and/or fieldworkers in jeopardy of physical, psychological or legal harm: neither social researchers nor their data are protected or privileged under law, nor immune to pressure from powerful agencies such as the police and the courts.[20]

[19] See, e.g. Belshaw (1976: Ch. 17), Gwaltney (1976, 1981), Easterday *et al.* (1977), Journal of Social Issues (1977b), Warren and Rasmussen (1977), Weiss (1977), Jayaraman (1979), R. Wax (1979), Zinn (1979), Santos (1981), Worsley (1982), Strathern (1983).

[20] A political scientist, Samuel Popkin, was imprisoned for protecting his sources, and other Americans have been subpoenaed (Carroll and Knerr 1976); Brass (1982) was imprisoned in Peru; and a South African student was gaoled after police had extracted information about drug-use from a sociological interviewer (J. Barnes, personal communication). For some other discussions and examples see, e.g. Polsky (1967), Keiser (1970), Nejelski (1976), Soloway and Walters (1977), Warren (1977, 1980), Weppner (1977), Patterson (1978), Jenkins (1979), Klockars (1979), Sagarin and Moneymaker (1979), Van Maanen (1981), Wolfgang (1981).

Publicly accountable officials may similarly be at risk from the knowledge and publications of social researchers (cf. Pepinsky 1980; Van Maanen 1981); but some would argue that they should not be protected in their public capacity (e.g. Becker 1967; Rainwater and Pittman 1967; Nader 1972; Galliher 1980). Analogous problems may also arise in relation to activities restricted to one sex or closed group (see Chrisman 1976; Schwartz-Barcott 1981).

6.4.3 Backstage and covert research

Another dilemma for ethnographers is the various ways in which their real interests or research role may be hidden from, or become "invisible" to those being studied. Fieldworkers acquire much valuable information and under-standing in "backstage" (Goffman 1959) settings and relationships about informants and others, on matters private and public, licit and illicit; indeed Berreman (1968: 362) regards entry into back regions not only as an index of rapport but of success as an ethnographer. Participants, however, may be unaware that they are communicating information which may be recorded, have temporarily forgotten the researcher's role or be unaware that an ethnographer is never "off-duty"; and it is impossible in long-term research for participants successfully to maintain a front in order to deceive the fieldworker. These issues are frequently discussed in relation to the field role of "friend", since the instrumental use of an intimate relationship is felt to be immoral or a betrayal by one or both parties (see F. Davis 1961; Riesman and Watson 1964: 258–270; Scheper-Hughes 1982; Grönfors 1982). They are acute for insider researchers (e.g. Chrisman 1976; Posner 1980) and especially for those who are kin, affine or colleague of their informants (e.g. Hansen 1976; Nakhleh 1979; Loizos 1981; J. Stephenson and Greer 1981). The use of data gathered in this way raises, too, issues of consent (§6.4.4), con-fidentiality and privacy[21] as well as of the propriety of publishing such data and, perhaps, even of recording them at all in fieldnotes.

In most field settings anthropologists have clearly been outsiders, and whether or not to adopt a fully covert insider role has not been a major problem (but see Redlich 1973; Murray and Buckingham 1976; Dillman

[21] Privacy and confidentiality present ethnographers with particularly difficult problems, given the legal and cultural variations between societies; and they are much discussed concepts, particularly in relation to the USA. See Lundsgaarde (1971), Journal of Social Issues (1977a), American Sociologist (1978), Appell (1978: 268–270), Bulmer (1979), Flaherty (1979), Trend (1980), Beauchamp et al. (1982: Part 4), Current Sociology (1982), Kelman (1982), Raab (1982).

1977), although all participant observation may contain a covert element, either because the researcher does not fully know and therefore cannot say what he will become interested in (Roth 1962), or cannot explain adequately in a cross-cultural context or is unwilling to reveal on scientistic grounds what his research is about (cf. M. Wax 1979a), or simply forgets to inform newcomers that research is "in progress" (see Davies and Kelly 1976). But as insider research increases so will opportunities (and temptations) for anthropologists to adopt a covert role. Most discussions are by sociologists and psychologists[22] but some anthropologists have commented (e.g. Golde 1973; Barnes 1979: 120–128; Cassell 1980, 1982). Criticisms centre round the impossibility of obtaining informed consent, the ethics of deception and thus of infringing the privacy, rights and autonomy of citizens, the effects on the researcher, the validity of the data thus obtained, and the damage done to the interests of the discipline when the deception is exposed or admitted. Proponents argue that covert research may be the only way to ensure that actors behave naturally, to obtain information about certain kinds of behaviour such as illegal or disapproved activities, or to obtain access to powerful subjects or those in closed settings to which a professed researcher would not be admitted. Some deceptions may involve gatekeepers rather than the host population among whom the research is openly practised. Covert fieldwork is particularly stressful; Murray and Buckingham (1976) argue that it should be the method of last resort, and some proponents have recanted, as Redlich (1973) did for himself and William Caudill.

6.4.4 The regulation of research and informed consent

Among other factors inducing a concern with ethics have been the development of endogenous guidelines and controls through the promulgation of professional codes of conduct by social science associations (§6.6), and of exogenous controls through administrative or legal measures regulating research or affecting it as a consequence of more general social laws (see footnote 4). Most of the discussions and measures refer to the USA, where the situation is now so complex that one field manual suggests including legal consultation at an early stage of project design (Fiedler 1978: 156); but they provide a useful awareness of the problems that might arise elsewhere. US regulations, for example, require federally funded projects to be reviewed by

[22] See, e.g. F. Davis (1961), Roth (1962), Diener and Crandall (1978: Ch. 7), M. Wax (1979a, 1980), Dingwall (1980), Galliher (1980), Beauchamp *et al.* (1982: Part 4), Bulmer (1982, 1982a), Kelman (1982: 85–87).

institutional review boards, and researchers to disclose certain information to participants, to inform them of anticipated risks and benefits, to obtain (except under specified conditions) written consent forms and to protect data and confidentiality. These regulations, based on a biomedical model, have attracted much criticism from social researchers, though some objections were met by amendments in 1981. They are regarded as particularly unsuited to the conditions of ethnographic fieldwork, especially the requirements of risk−benefit analysis and informed consent.[23]

The need for informed consent may present fieldworkers with acute difficulties in relation to cross-cultural contexts, illegal activities or politically sensitive settings; and it may be difficult, impracticable or simply impossible to obtain knowledgeable and voluntary (let alone written) consent from everyone in a field setting. It is also a complex issue in itself. For example, from whom is it to be obtained, in relation to whom, what matters, what events and data? Is obtaining it a single event or, as Warren (1977) and M. Wax (1980) suggest, a negotiated, lengthy and repeated process? If consent is withdrawn, what implications are there for the field data relating to that person, and can withdrawal take retrospective effect (cf. du Toit 1980: 282; Henige 1982: 114−115)? Anthropologists pride themselves on being eclectic in their interests and data collection, and topics for analysis arise long after their fieldwork; but these practices could be construed as unethical (cf. Jorgensen 1971; Trend 1980), not least because obtaining informed consent may not then be possible.

6.4.5 Data protection

Social researchers are also increasingly presented with problems relating to other social laws covering matters such as privacy and confidentiality, freedom of information, data protection and libel, in relation to which neither they nor their data may be protected or privileged.[24] Ethnographers

[23] For discussions of various aspects of federal regulations, see Social Problems (1973, 1980), American Sociologist (1978), Klockars and O'Connor (1979), M. Wax and Cassell (1979, 1981), Jordan (1981), Beauchamp et al. (1982: esp. Part 4), Reynolds (1982). For discussions of informed consent in particular, see M. Wax (1977, 1980), Trend (1978, 1980), Cassell (1978, 1980), Chambers (1980), du Toit (1980), Thorne (1980), Beauchamp et al. (1982: Part 4). See also Cohen (1976), Gottdiener (1979), and, for English views, Davies and Kelly (1976), Dingwall (1980), Jowell (1982).

[24] See discussions and references in Carroll and Knerr (1976), Nejelski (1976), Bulmer (1979), Flaherty (1979), Flaherty et al. (1979), Sagarin and Moneymaker (1979), Trend (1978, 1980), Raab (1982), Reynolds (1982). For other references to privacy and confidentiality see footnote 21.

amass a considerable quantity of personal and/or confidential information, gossip, etc., about individual citizens, groups and organizations in their field notes, diaries, photographs and films, and audio- and video-recordings. That individuals and places should be disguised in publications, and that published data may be subject to misuse and misinterpretation is well-known (§6.5). Less widely appreciated may be the facts that field data, normally assumed to be the private and personal property of the researcher, may as a consequence of legal or administrative measures be widely available, and may therefore pose a potential or actual threat to informants, persons named therein and the researcher. Similar difficulties may be presented by the obligatory, con- tractual or negotiated deposition of data and recordings in data banks, archives or other depositories determined by sponsors, funders, employers, governments or the host community (§7.6.1, §7.7.1), and by developments related to computerization (§9.4). Yet the researcher (and informants) may have little or no say over access to and use of field data under such cir- cumstances;[25] and Foster *et al.* (1979a: 335–336) comment that in certain political contexts long-term data banks may yet prove to be unpredictable "time-bombs".

Anthropologists need to discuss and develop methods and procedural measures in relation to matters such as: styles of recording data and personal identifiers; transcription and processing procedures; lifespan of unprocessed data, or even of all data in very sensitive cases; types and places of storage, and safety of data; rights of access.[26] In addition, more awareness is needed about the extent to which promises of confidentiality, routinely made by fieldworkers, can actually be honoured, a point stressed by Americans (e.g. Carroll and Knerr 1976; Reiss 1979; Trend 1980; Jordan 1981; Beauchamp *et al.* 1982). (Paradoxically, in certain research settings professional concern with confidentiality may itself be the problem (Light and Kleiber 1981).) The issues are complex, vary in different countries, have methodological, legal and political implications, and some are contentious. They have been little discussed by anthropologists, yet their data are less easily "cleaned", dis- guised, separated or aggregated and may be potentially more harmful to citizens and researchers than are, for example, quantitative data gathered by more formal methods for which various methodological safeguards have

[25] See the conditions cited in, e.g. Efrat and Mitchell (1974), Solomon Islands (1976), Mamak and McCall (1978), Social Research Association (1980), Henige (1982: 124–127), Light and Kleiber (1982).

[26] For some precautions, see Warren (1977), Gottdiener (1979), Trend (1980), and compare the approach of Gwaltney (1981) who, although blind, transcribed his own data on the grounds of ethics, trust and confidentiality, with the team-sharing and processing of data in Bohannan (1981).

been devised (see Boruch and Cecil 1979; Bulmer 1979; Boruch 1982; Raab 1982).

6.5 Dilemmas of publication and reporting

Some of the most pressing dilemmas for ethnographers arise during writing-up and publication, which may be an integral part of fieldwork as well as its aftermath (Wolcott 1975, 1981). It has also been argued (e.g. by Cassell 1978, 1982) that if informants are harmed by fieldwork this is more likely to occur after data are published (or subpoenaed) than by anything which happens in the field. The researcher, too, is more at risk as participants exhibit direct interests in publications.

6.5.1 Anticipating harms

That anthropologists should consider the possible repercussions of their reports and publications is now standard practice; but, though they may sometimes be better placed than their informants to appreciate these, not all consequences are easily foreseeable, as Barnes (1979: 148–157) points out (see also Condominas 1973). Many discussions focus on the immediate responses to publications or their effects for subsequent fieldworkers, but the researcher has also to consider other implications and possibilities since, for example, differential interests in the wider society may affect or harm members of sub-groups as a consequence of information being made public or inspiring policy (cf. Barnes 1979; Warwick 1982). In some political contexts minorities may be especially vulnerable and it may be thought necessary to withhold data or even to refrain from studying them at all (see, e.g. Metraux 1969; Mead 1978; Jahoda 1981; Grönfors 1982). The researcher, too, may suffer from becoming publicly identified with certain types of research, e.g. into sexual deviance (see Warren 1980).

6.5.2 Participants' rights

Among issues for consideration are whether informants, and other parties such as gatekeepers, sponsors and clients, should have a right to check the content of publications and, if such rights are contractually or morally recognized, whether participants should be allowed to respond to or alter matters of interpretation and judgement as well as errors of fact. Rights and privileges vary, and practicalities such as distance, language and elapsed time affect the issue; and it is also only too easy to consult only the powerful, the literate and those who can most easily understand the analysis or the language of publication, as Wolcott (1975) belatedly realized.[27] Co-authorship may be a thorny issue in team research (Diener and Crandall 1978: Ch. 9; Bradley 1982); and Whyte (1979) suggests that this should be shared with laymen if the practitioner has provided much case material and ideas and the sociologist had not been actively researching the topic. Copyright and royalties have also become matters for moral or contractual negotiation. Some sponsors retain rights in one or both of these (see Social Research Association 1980); some access agreements require royalties to be paid over (Efrat and Mitchell 1974), or they may be used by the researcher to benefit the host community (e.g. Lee 1979). Henige (1982: 113–114) points out that, although rights vary according to jurisdiction, ownership of interview tapes and transcripts is shared with the interviewee or his/her heirs, and suggests that a "fair use" waiver should be negotiated before leaving the field. Similar moral or legal rights pertain to films, photographs (see Hicks 1977) and video-recordings (Gottdiener 1979). The rights of the researcher in data and publications *vis-à-vis* other parties should be carefully considered; the cautious ethnographer does not concede all rights, as Agar's salutary experience shows (Agar 1980: 186), but this may become increasingly hard to avoid in some contexts (e.g. Efrat and Mitchell 1974), and contract (and other) applied researchers may be especially vulnerable to constraints upon publication.

6.5.3 Feedback and consultation

These are often enjoined upon researchers; but, as Appell (1978: 273–274) notes in his comments on the "social functions of ignorance", they may have negative as well as positive consequences for the researcher (see Grönfors 1982), for participants (Cassell 1982: 157) and for a discipline (see Henige 1982: 81–87). Ethnographers, in the field and through writings, force informants to view their culture and society and the behaviour of individuals through the perspectives and discourse of a critical and explanatory discipline (cf. Hau'ofa 1975; Rabinow 1977: 116–120). This partially explains some

---

[27] Some consultative strategies and responses are described in Whyte (1964), Efrat and Mitchell (1974), Monberg (1975), Wolcott (1975), Cohen (1976), S. Weaver (1978), Van Maanen (1981), Fielding (1982).

hostile or disappointed reactions to publications even if no direct harm is suffered, and even though beneficial changes may have occurred after the first shock has been absorbed, such as those described by Whyte (1964) and Scheper-Hughes (1981).[28] Even if what has been described and analysed is not necessarily secret, confidential, disreputable or normally withheld from outsiders, though it may well be (see Chrisman 1976; Warren 1977; Barnes 1979: 156–157; Schwartz-Barcott 1981; Grönfors 1982), privacy has been destroyed. Beliefs and practices have been exposed to the evaluations not only of the ethnographer but also of other insiders and, perhaps more insidiously or dangerously, of outsiders (cf. Metraux 1969; Morgan 1972; Scheper-Hughes 1981, 1982; Grönfors 1982).

6.5.4 Pseudonyms and disguises (§9.5)

One solution to some of these problems is the now common use of protective devices and omissions, although internal evidence such as geographical clues or vernacular terms may inadvertently negate anonymity (cf. Gibbons 1975; Hicks 1977; Warren 1980: 295). It is impossible to hide all insiders from themselves and their peers and, sometimes, from outsiders, particularly when specific offices or roles are involved (see Vidich and Bensman 1964; Gallaher 1964; Colvard 1969; Harrell-Bond 1976; Wild 1978; Scheper-Hughes 1982), and extremely difficult to protect them from investigative journalism (Morgan 1972; Wild 1978; Scheper-Hughes 1982), or from breaches of confidence by careless or unscrupulous colleagues who may have been given "uncensored" accounts (cf. Gibbons 1975; Hicks 1977). On the other hand, some informants or groups for various reasons want, or would have preferred, some degree of identification (see, e.g. Madan 1969: note 22; Fahim 1977; Foster 1979: 180), and this has been demanded as a control over the veracity of the account (see Fielding 1982). Sometimes disguise seems futile (e.g. Nakhleh 1979) or no longer necessary (Loizos 1981); and disguising individuals who appear in films is particularly difficult.[29] Some data cannot be published at all if promises of confidentiality are to be honoured, if especially pressing moral dilemmas over conflicting interests are to be resolved, or if

---

[28] For some reactions, and effects on later fieldworkers, see Whyte (1958, 1964), Becker (1964), Gallaher (1964), Vidich and Bensman (1964), Morgan (1972), Hau'ofa (1975), Monberg (1975), Finnegan (1978), S. Weaver (1978), Wild (1978), Grönfors (1982), and Scheper-Hughes' exchange (1981, 1982) with Kane (1982a, b).

[29] Films, photography, video- and tape-recordings present some particular ethical problems which cannot be considered here (see Barnes 1979: 144–145; Gottdiener 1979; RAIN 1982; Grimshaw 1982).

researchers are to be protected from libel suits or they and their participants from other repercussions.[30] There may be grounds for not reporting some research data at all (Jahoda 1981); and even those (e.g. Denzin 1978: 326–327) who would not exclude on *a priori* grounds any method or area of observation will protect informants thus. Such measures, however, conflict with standards of truth and other academic canons, and hinder verification and follow-up studies and the communication of results to non-anthropologists.[31]

In disseminating data, as in other stages of ethnographic research, ethical and other considerations require compromises between the interests of research participants and other citizens, gatekeepers, sponsors, employers, colleagues and the discipline, and the researcher's career.

6.6 Professional ethics and personal morality

One response to the increasing recognition of personal and professional dilemmas and to changing external contexts has been the development of codes of ethics (see footnote 3). The aims of such codes vary: some are aspirational, some regulatory, and others educational (Jowell 1982), and their effectiveness as guides to or as controllers of conduct varies accordingly. Furthermore, the adoption of a code engenders yet further problems and dilemmas, for example in relation to the topics for inclusion, procedures and recommendations, and in defining and enforcing standards of "professional" competence and acceptability, policing individual behaviour, effectively sanctioning breaches and redressing grievances. Not all associations have codes of conduct (the Association of Social Anthropologists does not) nor can it necessarily be assumed that all members are fully conversant with or make use of their professional code (cf. Long and Dorn 1982). Anthropological associations do not license their members to practise as do monopolistic professional associations like the Law Society, and not all anthropologists belong to associations; and criteria and standards for membership vary (cf. Mead 1978), as do the interests and concerns of academic, applied and practitioner anthropologists.

Views of the relationship between anthropological praxis, ethics and codes

[30] On libel, see Colvard (1969), Bell and Newby (1977), Wild (1978), Fielding (1982), Henige (1982: 115). For some other risks, see Weppner (1977), Sagarin and Moneymaker (1979), Warren (1980), Wolfgang (1981).

[31] See the exchange between Pitt-Rivers (1978, 1979), J. Davis (1979) and Jenkins (1979); also Gibbons (1975), Hicks (1977), Lynch (1977), Diener and Crandall (1978: Ch. 9), Fahim (1979: 268), Warren (1980), Brown (1981), Jordan (1981), Henige (1982: 113–127).

of conduct vary (Scholte 1971: 787–789; Jansen 1973; Adams 1981); and there is, similarly, disagreement among social researchers about the useful-ness, acceptability of and necessity for such codes. Some writers strongly support them (e.g. Jorgensen 1971); others (e.g. Douglas 1979) argue that the researcher should be responsible only to his own conscience and basic social laws, but this stance ignores possible repercussions for professional col-leagues, argue Dingwall (1980) and Bulmer (1982a: 247). Another view is that attempts to police behaviour will only force "decent fieldworkers" into yet more "morally dubious" behaviour (Klockars 1979: 280); and, indeed, one of the consequences of prior review procedures in the USA has been the development of evasive measures (Reiss 1979). Critics of codes argue that they suggest falsely a normative consensus, that their proponents are trying to impose unacceptable (and probably unattainable) absolutist standards, that they are undesirable and, if linked with punitive sanctions, positively dan-gerous measures.[32] Codes may, however, serve a public relations function (Dingwell 1980) and may also provide some protection against the imposi-tion of external controls; but it is also argued that self-regulation can no longer be regarded as adequate and that external constraints may in some cases protect informants (see Chambers and Trend 1981). Not least of the problems is that, even if periodically amended, such codes fossilize and decontextualize the issues and, it could be argued, enshrine outdated models of the discipline; but, as Barnes (1981) notes, they have served to highlight the issues. Other writers (e.g. A. Beals 1978; Trend 1980), whilst not denying the usefulness of codes, suggest that collections of cases and their outcomes would be the more valuable for discussion, instruction and in assisting the practical resolution of ethnographers' dilemmas, though that proposal, too, poses problems, not least in relation to confidentiality and legal liability.[33]

Opinions also vary about the relationship between personal morality and professional ethics: some commentators hold that the two are inseparable,

[32] See, e.g. Leeds (1969), Jansen (1973), Belshaw (1976: Ch. 17), Denzin (1978: 325–336), Adams (1981); and the regular contributions in the newsletter of the American Anthropological Association.
[33] Appell's (1978) collection of case histories indicates some of the problems that would be entailed, though the difficulties were perhaps exacerbated by publication. He had to use protective devices to protect his sources, some of whom asked not to be named anywhere in the book; and some cases of informants put into political jeopardy by association with anthro-pologists and permanently harmed thereby were too sensitive for publication (pp. 269–270). Only 26 of his 91 cases have attributions; the rest are anonymous and have been thoroughly disguised by Appell, though at least one "new" version closely resembles a non-contributor's experiences (p. 10); and as this is a teaching resource, many of the outcomes have been excluded. Bell and Newby (1977) also had to exclude some contributions, in their case because of the stringent English libel laws.

others that they are separable in the research setting. Appell considers that unethical behaviour in the personal domain must subsequently affect modes of behaviour in the scientific domain and, proferring a view of everyday life very different from that of Berreman (1962), or even of Cassell (1980), argues that

> the anthropologist must refrain from any deceitful, insincere, ungenuine, or false social interactions with *anyone* (Appell 1976: 88; his italics).

He also suggests, though, that the best anthropologists may be those who can tolerate the moral ambiguities characterizing a discipline which involves "cross-cultural inquiry . . . at an interface of ethical systems" (Appell 1978: 3).

In a pluralist world and an increasingly pluralist discipline, consensus about ethical behaviour and research practices is unattainable and compromise seems inevitable. Barnes, indeed, argues that "ethical and intellectual compromise is an *intrinsic characteristic* of social research" (1981: 2; italics added). The social researcher must make compromises in choice of field and topic; between roles as scientist and citizen; between commitment and impartiality; between openness and secrecy, honesty and deception; and between the pursuit of scientific knowledge, the public right to know, and the citizen's right to privacy and protection (ibid: passim). His conclusions reflect the experience of many an ethnographer:

> There is no immaculate praxis of fieldwork. Whatever choice we make we are unlikely to be completely happy with it (p. 21). . . . The competent fieldworker is he or she who learns to live with an uneasy conscience but continues to be worried by it (Barnes 1981: 21–22).

There is no ready panacea for the ethical dilemmas that accompany social research, nor can ethnographers easily abdicate their responsibilities: ethical absolutists and relativists alike agree that the onus for making decisions in practice rests with the individual researcher.

Anne V. Akeroyd

7 *Preparation for fieldwork*

7.1 Fieldwork training and planning[1]

It has long been recognized that "the special requirements of anthropological field research involve the student in a longer period of training than is customary in most other social sciences" (SSRC 1968: 101), and that this training itself generally results in significant contributions to knowledge (ibid: 82–83). Normally, this period is reckoned to be about four years. Despite this, "training" has characteristically involved little formal instruction in methods prior to fieldwork. Often, the preference appears to have been for a "sink or swim" approach, supplemented by anecdotes and advice, sometimes verging on the absurd and eccentric. All this is well known, if scantily documented (but see Beals 1970: 38; Edgerton and Langness 1974: 9). Some of the reasons why this should have been so have already been discussed (§1.1), including the idea that fieldwork should be a tough and mystical initiation, prestige accruing in proportion to toughness and mystery (Nader 1970: 114). However, it is worth providing a few facts.

In British universities teaching social anthropology at a graduate level, there are few formal courses in research methods.[2] A survey conducted in the course of preparing the present volume revealed that at some institutions,

---

[1] Discussions, reports and advice on fieldwork training in social anthropology are to be found in Mandelbaum (1963, part viii). There has been increasing advocacy for a practical introduction to doing anthropology at an undergraduate level (e.g. Myers 1969; Crane and Angrosino 1974; Hunter and Foley 1976), using various kinds of teaching exercises. These are, arguably, more useful in acquiring specific skills (for example, statistical ones), or in the evaluation of source material, than in enabling students to master field methods. Moreover, fieldwork at this level of instruction poses immense logistical problems.

[2] For a defence of the more informal, non-directive, assimilative approaches, combined with theoretical naivety, see Fortes (1963: 406, 429–430). Fortes rejects the apprenticeship model.

ETHNOGRAPHIC RESEARCH
ISBN 0 12 237180 1

graduate tuition in research methods is available as part of joint courses taught with sociologists; at most, tuition involves an informal seminar, or even less. In some institutions it has been thought desirable to precede fieldwork with a library study of original sources for a designated ethnographic area (§7.2) (see Fortes 1963: 430). Elsewhere (e.g. Sussex), special materials have been prepared to accompany a graduate programme in research methods. Undergraduate courses in specifically anthropological research methods remain rare, and many still believe that methods are best taught in the context of courses on theory and substantive issues. It is, however, clear that in recent years increasing attention has been given to research methods. This is partly because of mounting recognition within the discipline that training in basic research skills is important, and partly through external pressure for change exerted through funding bodies, especially the Social Science Research Council. It must be said that the training requirements in social anthropology are not identical to other social sciences, although there is some overlap, and there has been some attempt to resist the imposing of a general social science model for research training.

In sociology in Britain the situation is rather different, though very variable (Wakeford 1979: 1). However, anthropologists might reasonably concur with Wakeford (substituting "social anthropology" for "sociology" where appropriate) when he remarks:

> Particularly when the state of method in the discipline is in such ferment, it is worth asking whether a sociologist with undergraduate and post-graduate degrees should not have a broad knowledge and experience of research processes. It is sometimes a source of some surprise, not only among potential employers, that it is possible to qualify in sociology without such experience.

In the United States the teaching of methods approaches much more closely the pattern of sociology, although in many otherwise highly-respected departments training is still of the informal kind (American Anthropological Association 1981). But as in the United Kingdom, there has been a recent trend towards emphasizing the importance of fieldwork training (Agar 1980: 3). On the position in the United States a decade or so ago, see Eggan (1963: 415).

It is not the purpose of this volume, or the series of which it is part, to provide a training prospectus in anthropological field methods, although the information provided might well be useful in the course of such training. However, training itself must be part of a planned programme of preparation for fieldwork. In this chapter we consider some of the more specific aspects of preparation for fieldwork, other than the research procedures and methods to be used in the field themselves. These include project design and the assembling of background data (§7.2), searching the literature, building up a bibliography and using unpublished ethnographic archives (§7.3, §7.4), learning languages (§7.5), and using audio-visual equipment (§7.9).[3] In addition, we have included sections on practical problems which, while not in themselves methods, are nevertheless part of the crucial logistics of good fieldwork: funding and accounts (§7.6), obtaining research permissions (§7.7), and coping with physical health (§7.10). Other topics appear to represent issues of back-up organization, but by their very character impinge directly, and often fundamentally, on the efficient conduct of enquiries. Among such matters are the choosing of fieldwork locations (§7.7), the duration of study periods (§7.8), time-tabling (§7.7), and working in teams and with assistants (§7.11).[4]

In presenting much of this information, we fully recognize that there is an inherent danger that what we have to say will come over as a set of instructions which give the illusion of being universally applicable, but which inevitably must in practice be modified according to particular fieldwork and project conditions. There is no ideal set of guidelines, no strict body of scientific rules; but what we may be able to provide are some helpful reminders, checklists and hints. With this caveat, and the proviso that speculative ability, freedom of imagination, intuition, adaptability, flexibility, resourcefulness and ingenuity count as much in the field as the most meticulous planning (Honigmann 1976), the following are suggestions for profitable field research.

R. F. Ellen

[3] There are, of course, numerous other skills which the intending fieldworker might wish to acquire, but what these are will depend in part on the character of the research envisaged. One of more general application is computing (§9.4), and courses of varying duration and level are now available at most universities and polytechnics. Among the other more basic skills which are usefully acquired at this stage are shorthand, typing, photography and photographic processing (§7.9). Skills of a more specific utility include the rudiments of statistics, demography and map-making. The various ethnographic applications of quantitative techniques are dealt with in some detail in Volume 4 in this series, but see also §8.8. Demography and map-making will also be the subjects of more detailed treatment in later volumes. For the present, a good general introduction to the practicalities of assembling population data is McArthur (1961); on elementary map-making suitable for anthropologists, see Debenham (1937), Spier (1970) and Monkhouse and Wilkinson (1971).

[4] For a general, non-anthropological, manual on such matters in relation to social research in complex societies, see Fiedler (1978), who covers, in addition, supplies and documentation. Logistically, research in such societies obviously has certain advantages (Gillin 1949); more generally, see Edgerton and Langness (1974: 13–20).

Fortes (1963: 433) was disparaging about the value of many of these kinds of preparation. He regarded them as peripheral, calling them the "public relations" side of fieldwork, and among which he included "establishing a role". Such a position would, indeed, be difficult to sustain nowadays in the light of recent writings on how data are produced and on the wide range of subjective inputs. Crick (1982) provides a useful review (§1.4, Chapter 5).

7.2 Project design and background data[5]

Project designs serve two, sometimes competing, purposes: to promote the effective conduct of research and to meet the demands of funding and administrative agencies. In the real world, it is obvious that the conduct of social research is an organic process which cannot easily conform to simple positivistic models based on the experimental sciences, that is, procedures which follow the sequence: problem definition, theory "construction", operationalization, data collection, analysis and publication. That this is so has sometimes resulted in cynicism concerning the farce of writing-out detailed research proposals and budgets (Wax 1971: 285). Nevertheless, project design is important for reasons other than the satisfaction of bodies which have to examine research proposals. Planning is important, even if research methods are problematical, and subject matter and research procedures have to change for practical reasons (Vidich *et al.* 1964: part 1), and even if we agree with a generalist approach to ethnography which abhors preconceptions and advocates grounding theory in practice (Glaser and Strauss 1967; Cohen 1978).

For others, to understand a study properly, it is necessary to know how it was selected in the first place and why certain approaches were adopted. Anthropologists frequently have been less than open about these matters (Nader 1976). Indeed, it is also a fundamental intellectual necessity for the researcher to be able to identify, define and formulate a "problem" for investigation, to have some idea beforehand of its theoretical and substantive boundaries, to know something of the modes and consequences of limiting fields of study (Gluckman 1964), to elaborate conceptually on the basis of this, and to be able to devise relevant procedures of empirical investigation (c.f. Fortes 1963: 432). Understanding how you get from a puzzle (via a hunch, implicit theory, analytic theory, explanatory theory, hypotheses and research inventories) to research strategies is vital if you wish to evaluate your own research (Ford 1975: 256–257). This is not simply a question of mental hygiene, but rather a means of ensuring that time and resources are used effectively, that the necessary practical preparations are made, that adequate data are generated, and that the results are meaningful in the context of other

[5] On the preparation of research proposals see Krathwohl (1976). British S.S.R.C. procedures for the early 1970s are dealt with by Platt (1976: 208–216). On research design and planning generally see Selltiz *et al.* (1963: 25–144, 502–508), Cicourel (1964), Phillips (1971), Nachmias and Nachmias (1976) and Johnson (1978: Ch. 2); a tough (some would say unworkable) set of proposals are offered by Brim and Spain (1974).

research findings and will stand up to tough critical examination. That research is often undertaken in rapidly changing and unpredictable situations only makes it more necessary for the investigator to be able continually to redesign projects effectively and at short notice. To say that luck and accident play an important role in fieldwork is no excuse for not planning for unforeseen circumstances.

Preparation for ethnographic research requires training in research methods skills, an appropriate intellectual preparation for the problems to be investigated and the acquisition of a data base which will enable new research to be conducted in an informed manner. This, in turn, will enable the better formulation of appropriate theory and hypotheses and prepare the fieldworker empirically for the character of the societies which are to be investigated. "Ideally", says Fortes (1963: 432), the fieldworker "should go out to the field . . . with an apperceptive mass, to borrow an old-fashioned pyschological term, of general regional information". This will help to place the area in its wider ethnographic and temporal context. It takes time: at least six months. It has sometimes been advised that potential fieldworkers should "produce a fairly detailed plan of fieldwork, presenting the significant geographical, historical, and ethnological facts"; at Oxford it has become standard to acquire a familiarity with a body of ethnographic literature by first writing an M. Litt thesis using and appraising it.

The background information generally required by a potential fieldworker consists of relevant existing ethnographic publications and unpublished archive material, various official government documents and statistical compendia (e.g. demographic data), and maps. In some cases it will be necessary also to look more systematically at historical materials of various kinds, to obtain more specialized statistics or cartographic data, or draw on other kinds of documentary sources (e.g. local fiction). In a few cases (e.g. Keesing 1959, for Oceania), there already exist regional field guides which introduce the researcher to the practical problems and facilities for conducting work in particular areas.

R. F. Ellen

7.3 Getting into the literature

The aim of this section is to direct the social anthropologist to those bibliographic tools which will be useful when undertaking a literature search. It concentrates on large, recurrent bibliographic tools in the form of abstracting and indexing journals, but does not attempt to enumerate once-off bibliographies on specific subjects. Some guidelines will be offered on the process of literature searching, as well as some suggestions for locating material for consultation.

7.3.1 Literature searching

The researcher embarking on a study will have already identified and located certain published works of relevance. The bibliographies of these items are often a source of references to earlier relevant literature and further references will no doubt be supplied by colleagues, located from publishers' and library catalogues, etc. It is possible, however, to carry out a systematic search for references to the literature on a particular topic or geographical region by using certain bibliographic tools. Bibliographic tools are defined here as those publications which list, in a systematic way, citations or references to a specific body of literature. Searching through one or more of these works (usually abstracting or indexing journals) is, therefore, often the equivalent of examining, *inter alia*, several hundred journal runs and numerous books, pamphlets and theses.

To ensure that the references identified in a literature search are of maximum value, the researcher should first draw up a list of key subject words which define the area of study as precisely as possible. The list should include any geographical or ethnic names of importance, as well as any synonyms, alternative spellings, and broader, generic subject terms. It may be necessary to amend and enhance this list as the search progresses. It is also important that the researcher decides, before beginning, whether to limit the search by factors other than just subject. For example, it may be decided to record only references to items published since a certain date, or written in certain languages. The place and type of publication may also be a limiting factor since certain publications can be more difficult to obtain than others, e.g. theses written outside the United Kingdom and North America.

It is advisable to commence a literature search using the most comprehensive tool available and to begin with the most recent issue and work backwards. Finer (1982) recommends that the first point of entry should be a cumulation if this is available as, "It is often only in cumulated indexes that full cross-references between subject headings are shown, and this is also a way of gauging the likely quantity of references on the topic in question". Any relevant references found in the indexing or abstracting journals need to be recorded by the researcher, and one tried and tested method is to use 5 × 3 index cards which are both portable and flexible. Alternatively, punched edge cards (§8.12.7) or computer-assisted techniques (§9.4) can be adapted for this purpose. Recording the full bibliographic citation of the item at the search stage will avoid the time-consuming problem of checking references later. Researchers should try, for reasons of accuracy and consistency, to adopt a fixed style of citation, for example that given to authors in one of the major journals, or, in the case of doctoral students, any style recommended by the examining body. It is also useful to record the source in which the reference was found, whether a library location is known for the item and whether it has been seen or not. Remember too that abbreviations may seem appropriate when first used but may be incomprehensible at a later date. Research can extend over a long period of time and the maintenance of full documentation can save duplication of effort. For the same reason, it is important to keep a record of the bibliographic tools used in the literature search including the volumes checked and the search terms employed.

7.3.2 Bibliographic tools

The major, recurrent indexing and abstracting services available to social anthropologists are described below, with the aim of illustrating the usefulness, the coverage and the diversity in arrangement of the different tools. The list of bibliographic tools for use in searches focusing on a geographical area is not exhaustive and students are advised to consult a subject specialist librarian for further advice. The selection of appropriate bibliographic tools for a specific literature search will depend upon the subject and/or region being studied and any limitations which have been imposed by the researcher. Most bibliographic tools provide a description of their scope and organization, and a list of journal titles scanned, both of which can be used to assess a tool's usefulness.

7.3.2.1 *Anthropology*

The *International bibliography of social and cultural anthropology* is the only bibliographic tool devoted exclusively to the field of social anthropology. It covers the years 1953 to date in annual volumes and includes references to books, research reports and articles culled from several hundred journals. The entries in the *International bibliography* are grouped together in classified subject order. In addition, cross-references to related entries are supplied. Each volume has an author index and a detailed subject index which allows easy access to potentially relevant entries. This is an ideal starting point for a literature search covering as it does a time-span of almost 30 years and providing details of thousands of publications in the field. The main disadvantage of the *International bibliography* is the time-lag between the appearance of the indexed items and publication of the annual volume.

One publication which enables the social anthropologist to trace current journal articles in the field is *Anthropological index to periodicals in the Museum of Mankind library (incorporating the former Royal Anthropological Institute library)*. This work, which has appeared quarterly since 1963, indexes the articles from approximately 600 journal runs. These journals are published in many countries and the list includes English and foreign language titles. *Anthropological index* covers the fields of archaeology and physical anthropology as well as social anthropology, and entries are grouped into six major sections. The first section includes articles of general anthropological interest, including theoretical works, while each of the remaining five sections is devoted to a particular geographical region sub-divided by subject, e.g. Australasia – cultural anthropology. Unfortunately, entries on a particular topic (mode of production, for example) can only be located by looking under broad subject headings since no detailed subject index is supplied. The Museum of Mankind library does have one on cards, but only for the period 1963–1967. *Anthropological index* is best used, therefore, when works relating to a specific geographical region are sought. There is no doubt that both the *International bibliography* and *Anthropological index* would be greatly enhanced if cumulative subject indexes were published.

Another long running bibliographic tool which provides access to current social anthropological literature of all types is *Bulletin signalétique 521: sociologie – ethnologie* (formerly *Bulletin analytique*). This quarterly publication has been prepared by the Centre de Documentation du Centre National de la Recherche Scientifique in Paris since 1947. In its present format each issue divides into two sections, one covering sociology, the other ethnology. The ethnology section includes references to books and theses, and to articles from approximately 350 journals. Entries are arranged under broad subject headings, e.g. religion, magic and sorcery, and each issue has two detailed subject indexes (one in English and one in French), a geographical index and one to ethnic groups. All four indexes are cumulated annually. In addition to the bibliographic details many entries also have brief abstracts, in French, outlining the contents of the item. Given the arrangement and indexes of *Bulletin signalétique* this is a more suitable tool than *Anthropological index* when searching for current works of a theoretical nature, although it does not index as many journals.

In the last few years two further secondary services have become available, both of which cover all aspects of anthropology. *Anthropological literature: an index to periodical articles and essays* is a quarterly publication which began in 1979. Based on the acquisitions of the Tozzer library at the Harvard University Peabody Museum of Archaeology and Ethnology, this work also enables the user to search for current items on a particular subject. The first year's issues contained nearly 9200 entries from over 900 journals and 325 books. Each quarterly issue has indexes to authors, ethnic and linguistic groups and geographical areas, but no subject index. *Abstracts in anthropology*, the first issue of which appeared in 1970, indexes less than 100 journals but all the major English language titles are included. Although its coverage is not extensive, the researcher is provided with a brief abstract of the contents of each article. It has a section devoted to social anthropological works which is sub-divided by narrower topics, e.g. symbol systems.

Catalogues of major anthropological collections can also be a rich source of references and some have been published in book form. For example, the 54 volumes of the Peabody Museum of Archaeology and Ethnology *Library catalogue* contains over 80,000 entries for the period 1877 to 1963, and is notable in that it includes periodical articles as well as books and pamphlets. The catalogue is in two sections, author and subject, and a separate volume provides an index to the subject headings. To date four supplements have appeared covering the years 1963 to 1979.

7.3.2.2 *Area studies*

Social anthropologists are often interested in a particular geographical region and will require literature on, for example, its politics, economics and history, as well as anthropological works. While the bibliographic tools already discussed invariably provide for a geographical or ethnic group approach, secondary services exist which are concerned specifically with indexing the literature pertaining to particular regions. Some areas of the world are better catered for bibliographically than others, for example, Africa (mainly due to the work of the International African Institute).

From 1929 until 1970 a bibliography of current publications was published quarterly in the journal *Africa*. In 1971 this became a separate publication, *International African Bibliography*, which is now compiled at the library of the University of London's School of Oriental and African Studies. The aim of this work is "to list all authoritative and serious works within the field of African studies" and it provides references to books, pamphlets and conference proceedings, and to periodical articles from over

1000 journals. The entries are arranged by geographical region and each is assigned subject headings which appear at the end of the entry. The only country which is not covered by this tool is Egypt, but the islands adjacent to Africa are included. Two cumulations are available which greatly facilitate searching. The first is *Cumulative bibliography of African studies* which includes everything in the *Africa* quarterly lists and *International African bibliography* to 1972, as well as entries from other sources such as the *African bibliography series (series A)*. This cumulation is divided into an author and a classified subject catalogue and the latter includes an index to subject headings which provides the searcher with useful synonyms and cross-references. The second cumulation, *International African bibliography, 1973–1978* (edited by J. D. Pearson, 1982), includes nearly 20,000 entries from volumes 3–8 of the quarterly publication plus some 3000 items not previously included.

Publications written in European languages about South, East and South East Asia in the fields of history, the humanities and the social sciences are indexed in the annual *Bibliography of Asian studies*. Published as part of the *Journal of Asian studies* from 1956, this work became a separate publication in 1969. Entries are to relevant monographs, government documents, articles in collected works, and journal articles from some 850 titles. Again there are two cumulations, both entitled *Cumulative bibliography of Asian studies*, one covering the period 1941–1965 and the other 1966–1970. *Bibliography of Asian studies* was preceded by *Far Eastern bibliography* (which appeared in the journal *Far Eastern quarterly* from 1941 to 1955), and before that by *Bulletin of Far Eastern bibliography* (1936–1940).

Of particular interest to social anthropologists working on South Asia is *An anthropological bibliography of South Asia together with a directory of recent anthropological work* (compiled by Elizabeth von Fürer-Haimendorf and later, Helen A. Kanitkar). This work, now in four volumes, covers the period 1940–1969, with a selection of works issued prior to 1940; the volume for 1970–1974 is in progress. *An anthropological bibliography* includes citations to books, theses and journal articles, mostly written in western European languages, arranged by region and sub-divided by subject. Other bibliographic works of use to social anthropologists working on South Asia are described in Kanitkar (1979).

Researchers working on the Islamic world should refer to *Index Islamicus . . . a catalogue of articles on Islamic subjects in periodicals and other collective publications*. *Index Islamicus* covers the period 1906–1975 in five volumes and has been kept up to date since 1977 by *The quarterly index Islamicus*; a supplement covering the years 1976–1980 is planned. Entries are not confined to literature about North Africa and the Near and Middle East, but include items of interest from other countries; they are, nevertheless, restricted to items written in western European languages. The arrangement is by regions and countries, sub-divided by subject, e.g. ethnology. *The quarterly index Islamicus* contains approximately 3000 entries per annum and since January 1983 has provided a subject index for both subjects and places not included among the main headings, or for items which cover more than one area.

The annual *Handbook on Latin American Studies* has been prepared by the Hispanic Division of the Library of Congress since 1935 and is particularly useful to those working on this area. The *Handbook* provides references to journal articles, books and theses and aims to "bring under bibliographic control those works of permanent research value which are published in the social sciences and humanities pertaining to Latin America". A number of factors, including the growth of the literature in the field, led to the decision in 1964 to divide the *Handbook* into two parts, Humanities and Social Sciences, each appearing in alternate years. The social sciences volume has sections devoted to anthropology, economics, education, geography, government and politics, international relations and sociology, each sub-divided by country. Responsibility for each section rests with a contributing editor who also presents a review of the literature. One notable feature of the *Handbook* is that every entry has a brief annotation describing the item's contents. Currency is obviously a problem of the *Handbook* because of the two-year accumulation of literature, but another publication exists which provides the researcher with references to slightly more up to date material, i.e. *HAPI: Hispanic American Periodical Index. HAPI* is an annual publication which began in 1975 and lists, by subject, articles of interest to Latin Americanists from more that 200 major journals in the field.

Some further recurrent bibliographic tools should be mentioned as being of particular relevance to some social anthropologists, including *New Guinea periodical index: guide to current periodical literature about New Guinea*, which began publication in 1968 and is cumulated annually. References to articles about New Guinea are included regardless of where they were published and entries are arranged by subject, sub-divided by place. The Australian Institute of Aboriginal Studies has produced an *Annual bibliography* since 1975, superseding the biannual current bibliographies of earlier years (1963–1975). This bibliography is issued in conjunction with the Institute's *Newsletter* and includes citations to books, pamphlets and periodical articles from some 260 journals, as well as to manuscripts held in the Institute's library. Finally, for those working on the anthropology of the

Pacific area, the annual publication *Pacific history bibliography and commentary* is useful, as is the *Dictionary catalog of the library* of the Bernice P. Bishop Museum in Honolulu.

7.3.2.3 *Human Relations Area Files*

Although not a bibliographic tool in the sense defined above, the Human Relation Area Files (HRAF) should be mentioned here. The files are a data bank made up of a collection of books, articles and some unpublished papers, organized by cultural unit and subject. Duplicate sets of the files are available in institutions throughout the world, although no set is available in the UK. Two publications are used to approach the material, Murdock's *Outline of world cultures* and *Outline of cultural materials*. Part of the work of HRAF has been to compile and produce bibliographies, and the development of HABS (Human Relations Area Files Automated Bibliographic System) has more recently assisted in this area of their activities. Further details about HRAF can be found in their journal *Behavior science research* and in White (1958).

7.3.2.4 *Social sciences citation index*

One bibliographic tool enables the researcher to take a unique approach to literature searching, that is, *Social sciences citation index* (*SSCI*). Researchers have always used the bibliographies in works known to them to identify further works of relevance. *SSCI* works on a similar basis. It is divided into three sections, the citation index, the permuterm subject index, and the source index. The source index lists, in alphabetical order of author the articles which appear in approximately 1400 social science journals each year (including about 70 anthropology and area studies journals). The citation index lists, also in alphabetical order of author, all the works cited in the source articles. Thus, the researcher who has previously identified a publication of key importance to his or her subject can, by looking to see if it is listed in the citation index, discover whether it has been cited recently, and if so, by whom. Ideally the key work, which can be any type of publication, should be an item published some years ago and highly specific to the topic of research; this should ensure greater relevance of the citing articles. The premise behind *SSCI* is that any article which cites the key work is likely to be about the same or a related subject and will also provide a list of further references of interest and thus the researcher will be able to build up a network of citations.

The *SSCI* is a powerful tool once its use has been mastered. The compilation of such a work would not have been feasible without computer technology and users should try to overcome any resistance they might feel towards using it because of its presentation. The permuterm subject index provides the user with a means of accessing the source articles by subject. It should be used with caution since the index terms consist solely of those words which the authors have used in the titles of their articles. Its effectiveness, therefore, is dependent upon the precision of authors when composing their titles.

7.3.2.5 *Theses, conference proceedings and research in progress*

Certain types of publication, such as theses and conference proceedings, are often not included in the bibliographic tools already discussed. Such works, however, can be traced using those secondary services which are devoted exclusively to their indexing. *Index to theses accepted for higher degrees by the universities of Great Britain and Ireland and the Council for National Academic Awards* has, since 1950, provided a list of theses arranged under broad subject headings, e.g. social anthropology and ethnography. Recent issues of *Index to theses* (often referred to as *Aslib index*) have a subject index which extracts specific terms from the titles of theses to assist a search limited to a very narrow topic. North American, and some European, doctoral theses of interest to social anthropologists are covered by the monthly *Dissertation Abstracts International A: the humanities and social sciences* (formerly *Microfilm abstracts*).

Easy searching for theses over a long time-span has been facilitated by the appearance of two retrospective indexes, namely, *Retrospective index to theses of Great Britain and Ireland 1716–1950, vol. 1: social sciences and humanities* and *Comprehensive dissertation index 1861–1972, vol. 17: social sciences*. Because theses are often a rich source of further references these tools are especially useful, as are those once-off publications which direct the user to theses written on a particular subject, e.g. *Theses on the Middle East, North Africa and Islam presented at British and Irish universities, 1894–1978* (compiled by Peter Sluglett, 1983).

Conference proceedings can be identified using such tools as *Index to social sciences and humanities proceedings*. This quarterly index aims to include all relevant published conference proceedings and papers written in both English and foreign languages. The index provides a number of entry points to the listings, including subject category, keywords in titles, author and institution. A recent publication which may set the trend for similar regional or subject indexes to this type of publication is, *Africana conference paper index*. As well as providing information about further literature, such works also help to identify other researchers working on related projects. Some reference tools are designed specifically for this latter purpose, one example being the publication, *Research in British universities, polytechnics and colleges,* prepared annually by the British Library.

7.3.2.6 *Social sciences*

The bibliographic tools named above are those which are of particular interest to social anthropologists, but there are other indexes and abstracts which may be valuable to the researcher working on a specific topic. For example, in the *International bibliography of the social sciences* series there are volumes for sociology, economics and political science, in addition to the volume for social and cultural anthropology. Citations to useful theoretical works may also be found in such publications as *International political science abstracts* and *Sociological abstracts.* Bibliographic tools are produced for most subject areas and specialists in particular aspects of anthropology will need to refer to these. Examples include *Modern Language Association bibliography, Linguistic bibliography for the year . . ., Abstracts of folklore studies* and *Geo Abstracts D: social geography and cartography.*

Because of the breadth of anthropological studies it is impossible to enumerate all potentially pertinent bibliographic tools. It is hoped, however, that readers will appreciate the usefulness of these tools and seek advice from appropriate subject specialist librarians. Information about other bibliographic tools available in the social sciences can be found in Li (1980), Freides (1973) and Walford (1982).

7.3.2.7 *Non-recurrent bibliographies*

This section concentrates on comprehensive, recurrent bibliographic tools since it is impossible to list here all the non-recurrent bibliographies which have been produced on specific subjects, geographical regions or individual authors. Some bibliographies of this type will be discovered in a literature search using the major tools. In addition, over 3200 bibliographies, filmographies and discographies are listed in *Anthropological bibliographies: selected guide* (edited by M. L. Smith and Y. M. Damien, 1981). The annual publication *Bibliographic index: a cumulative bibliography of bibliographies* also lists, by subject, bibliographies published both separately or as parts of books, pamphlets or periodicals. Beckham (1967) and Urry (1977) both give a useful review of some bibliographies relating to the anthropology of selected geographical regions.

7.3.2.8 *Computer searches*

The use of computers in literature searching is now well established and many academic libraries offer facilities for on-line computer searching and usually provide professional assistance in formulating a search strategy and in performing the search itself. If a research topic has been precisely defined this can be an extremely fast and efficient way of searching for references contained in a number of different bibliographic tools. Most libraries will, of necessity, make some charge for this service. One example of a computerized information retrieval service is DIALOG. Among the bibliographic tools which can be searched via DIALOG are Social Scisearch (i.e. *SSCI*), *Sociological abstracts, Comprehensive dissertation index* (1861 to date) and *Conference papers index*. Details of some other useful data bases can be found in Currier (1976).

7.3.3 Locating material

After identifying potentially relevant literature the researcher then has the task of locating the material itself. If the home institution's library catalogue reveals that the item is not in stock there are a number of means available for discovering whether another library stocks it. Some bibliographic tools, such as *A London bibliography of the social sciences*, are based on the acquisitions of a particular library, in this example the British Library of Political and Economic Science, and therefore also provide an immediate location for an item, as do the published catalogues of individual libraries, for example that of the School of Oriental and African Studies, London. Libraries specializing in the literature on specific subjects or geographical regions, or housing special collections, can be identified by using *Aslib directory, Vol. 2: information sources in the social sciences, medicine and the humanities* (4th edn; edited by Ellen M. Codlin, 1980). The *British union catalogue of periodicals* provides a location list to journals held by university and national libraries throughout the country and other union catalogues exist which indicate which libraries hold items. An example is the *British union catalogue of Latin Americana* which is available for consultation at the library of the Institute of Latin American Studies, University of London.

Visitors to libraries other than those of which they are members are advised to write or telephone first to ensure admittance and to check opening hours, etc. It will sometimes be worth spending several days visiting a specialist library known to stock a large proportion of the items identified in a literature search. If no location can be found, or it is impracticable to visit another library, applications can be made via an academic library, to the British Library Lending Division (BLLD) which provides a loan and photocopy service to registered institutions throughout the United Kingdom. In the event that the BLLD cannot provide material from its own substantial stock, it will draw on the resources of major libraries in this country, and collaborating libraries and interlending centres in other countries.

Mary J. Auckland

7.4 A guide to ethnographic archives

Ethnographers can use the records of their predecessors' fieldwork in two main ways. They can incorporate it into their own research, and use it to extend their field of study backwards and to compare past and present. They can also use it critically to re-examine the evidence of that predecessors' interpretations and conclusions. Both approaches depend crucially on the quality of the original fieldwork, the care with which it was recorded and on its survival. Finding and using such material is a difficult process, more like hunting through a large and disorderly market than shopping in a neatly classified supermarket.

The most useful of an ethnographer's unpublished records and the most difficult to use are those made directly in the field. Field notes are the product of observation, often unmediated by systematic interpretation, taken down often in a telegraphic and severely technical style. The user can expect to find a series of small, easily pocketable notebooks, with entries in barely legible handwriting, possibly including shorthand or personal abbreviations, probably including many words in the language of whoever the ethnographer was talking to, and in no kind of subject order; each observation or piece of information was jotted down as it occurred. In parallel the ethnographer probably kept a field diary, more contemplative and subjective in style and written under less pressure. The diaries were used for the recording of more general observations, for the first stages of an interpretation and as a kind of confidant. For the subsequent user, such diaries can be invaluable in interpreting the field notes and should always be consulted if available. They may show the conditions under which the fieldwork was done and establish, for instance, who the ethnographer's informants were, or what languages he spoke. The ethnographer may also have collected texts composed by other people. These can include songs and stories taken down from dictation, autobiographies and local histories and records of court cases or council meetings. Other kinds of material collected or compiled by the ethnographer are lists of words and notes on grammar, drawings, photographs, films and sound recordings. Finally, the letters which he wrote from the field to his tutor, grant-giving body, family or friends may survive and contain usable information even if his own papers do not.

In association with the field records is the material subsequently produced by ethnographers in the process of making their work comprehensible to themselves and to others. It may include typescript copies of field notes. These should, if possible, be compared with the originals to see whether they are complete or contain everything considered relevant in hindsight or just what was needed for a particular book or article; there may also be indexes to the field notes (to which the same cautions as to completeness apply), glossaries and grammars and, finally, material related to the final stage of interpretation, such as research reports, lecture notes, drafts and proofs. These latter groups are of little interest to the ethnographer who wishes to incorporate his predecessor's work in his own but may be of crucial importance in tracing the genesis of a particular ethnographer's analysis.

Apart from understanding them, the major difficulty in using ethnographers' papers is finding them. There is no overall guide to the location of ethnographers' papers as such and no single institution specializing in collecting them. The largest collections at present appear to be held by Cambridge University Library (A. C. Haddon, W. H. R. Rivers), the Royal Anthropological Institute (R. S. Rattray, M. E. Durham) and the British Library of Political and Economic Science at the London School of Economics (C. G. S. and B. Z. Seligman, Bronislaw Malinowski, S. F. Nadel, Margaret Read, Phyllis Kaberry). The papers of any particular ethnographer may, however, end up in the archive of the University where he or she taught, or in the library of the research institute which specialized in the area. The problem for the searcher is that it is possible at present to answer the question, "where can I find X's papers?", with a reasonable degree of accuracy but not, "what ethnographers' papers can I find relating to say, Central Africa?".

The Royal Commission on Historical Manuscripts (§7.4.1, 3–4) set up the National Register of Archives in 1945 to collect information about collections of unpublished manuscripts held in the United Kingdom. Their subject and topographical indices are designed primarily for historians of the British Isles; but their index of persons includes entries for all collections of papers of interesting individuals which they have been able to locate in public institutions and to some in private hands together with important occurrences, for instance a long series of letters, from such individuals in the collections of other individuals or bodies. The index is based on the Commission's own investigations and on information provided by archivists who make annual returns of all new collections they acquire and send the Commission copies of their guides and catalogues. The print-out of the index and the catalogues can be consulted in the Commission's search room every weekday except public holidays. No appointment is needed. The commission's staff will also answer questions about the location of papers by post. The archivists' annual returns of new accessions are published in the autumn of the following year as *Accessions to Repositories and Reports added to the National Registry of Archives*, formerly *List of Accessions to Repositories* (§7.4.1, 3–4). There are likely to be gaps in the Commission's coverage; institutions may not have reported on collections which they received before the National Register was set up or the existence of which they had promised to keep secret for some years after receipt. Bodies which do not have professional archivists, such as departmental libraries and research institutes, may not realize that the Commission exists and is interested in their kind of holdings. Archivists may also have information about collections which have not been formally deposited. For these reasons, if an ethnographer cannot be found in the National Register index it is worth writing to enquire of the archivists and librarians of the institutions with which he or she was connected.

Once the papers have been located, the next hurdle is to get permission to see them. Papers of this kind are very rarely bought outright by the institution; their acquisition is often the result of a delicate process of bargaining and there are a number of possible outcomes. They may be held on a simple loan; this is quite likely if the owner is still alive, does not want to relinquish them entirely, but does not mind other people using them provided that they do not have to do it in his house. In this case, the archivist, as the owner's agent, is responsible for ensuring that the papers are not lost, stolen or damaged and that they are not seen by unauthorized people. The owner, however, will decide who may see them. They may be deposited in the institution as a permanent loan, a gift or a legacy. They are therefore under the control of the institution but the institution may have accepted certain restrictions as a condition of the gift. The restrictions may be on who is allowed to see them, and intending readers may need the permission of the donor or of a committee; they may be attached to the papers themselves and certain sections of the collection may be closed to research because their disclosure would tend to be damaging. Modern field notes in which named or identifiable individuals talk in confidence about sensitive personal or political matters cause particular problems in this respect and institutions and donors may feel that the informant's privacy and safety is the prime consideration and close the papers. Collections may also be temporarily unavailable because they are uncatalogued, or being catalogued, or being repaired or microfilmed. It is therefore wise to enquire in advance in writing, and to make an appointment for the first visit.

Archives usually have, or are preparing, an overall guide to their holdings. This lists each collection by name and reference number if applicable, gives a brief indication of the subjects and dates covered and of the collection's size. This is usually given in boxes, one box holding a stack of paper about four to four and a half inches high. It is always worth looking at guides since they may indicate collections too small and specialized to appear in any published list. Each collection should have its own catalogue; this is a fairly summary list of the contents of the collection usually in book form not on cards, which are awkward to work on for long periods, difficult to copy and easy to disarrange. The catalogue will usually contain a brief introduction and a piece by piece summary of the contents of the collection. A piece will be a field notebook, a volume of a diary or a bundle of loose sheets and will have an individual reference number. The reference numbers are vital, they are used to order documents for inspection and to identify them in footnotes (§7.4.1, 30 for the British Standard for the citation of unpublished documents). The individual entries in catalogues are usually very brief; detailed cataloguing and, in particular, subject indexing are time consuming and therefore expensive occupations, and archivists tend to feel that it is better to list three collections briefly so that they can all be used than to describe one in loving detail and to bury the rest. They do, however, attempt even in a brief list to sort out the material systematically and to arrange it in groups according to origin, subject and date. The value of any piece of information depends largely on its context: who reported it to whom, when and for what reason. It is therefore vital not to remove papers from their context, even if it might seem more convenient for the user, and not to impose on them a classification alien to their originator. It is for the researcher having examined each piece of evidence in its original context to arrange the information according to his or her own needs.

It is wise to work systematically through the catalogues and to note down the references of everything interesting before starting on the documents themselves. Since the documents are, by definition, unique and irreplaceable, you will have to read in a supervised search room, you will not be able to take them away or to xerox them yourself and you will probably be asked to use a pencil to make notes. They may be kept in a strong room at some distance from the room in which you are working, so ask at your first visit what the arrangements for ordering are, how long it takes and whether you can order in advance. You may be able to use a typewriter or tape-recorder if you can do so without disturbing other readers. You may be able to order photocopies to be made for you. Field notebooks which are flimsily put together out of poor quality paper are usually now too fragile to be risked on a xerox machine. Some archives can provide microforms, 36 mm roll microfilm is much less convenient to use than microfiche but much easier to read. Once a negative microfilm has been made of a fragile document, further copies on microfilm or on paper can be made *ad libitum* without touching the original.

A major hazard for users of unpublished papers is the law of copyright as it applies to mechanical copying and to publication. The law of manuscript copyright is extremely obscure. As a rough summary, it is indefinite in duration. An unpublished work, which might be a letter, a note or any other kind of writing, a drawing or a photograph not intended for publication by the original author, remains in copyright until it is published and for the normal period thereafter. The ownership of the copyright does not necessarily go with the ownership of the document. If A writes B a letter, the copyright belongs to A. Although the letter belongs to B, he cannot publish it without A's permission. It is not, therefore, wise to assume that the copyright of a document which you have seen in an institution belongs to that institution. Many archivists hold the view that mechanical copying is tantamount to publication and will not provide copies of anything less than a 100 years old without the permission of the copyright owner. Others take the more robust view that there is no effective difference between copying by hand and other copying and that we should, therefore, provide photocopies if the original will stand it and we have no reason to suppose that the copyright owner objects. However, because it is so easy to make further copies from a microfilm it is usual to sell only a positive copy and to make the purchaser undertake not to publish any part of it or to pass it on to a third party without the permission of the owner of the original.

The major problem of copyright is at the time of publication: you may not quote the actual words of an unpublished text without the permission of the owner of the copyright. You may, however, cite the text – that is, reproduce the information in that text in your own words and tell the reader from where you got the information – provided that you did not, in return for permission to see a particular set of papers, undertake to submit work based on them to the person or body who gave you permission. The archivist should be able to advise you about the copyright problems of a particular collection, though he may not always wish to put his advice in writing.

Although this section has concentrated on the papers of professional ethnographers, other kinds of archival material may contain ethnographic material. For the United Kingdom, the major sources are the records of the British Empire and of the missionary organizations. None of these were primarily interested in ethnographic data and finding such data may require a considerable effort of historical research.

A series of guides to manuscripts relating to particular geographical areas is being published under the aegis of UNESCO by the Conseil Internationale des Archives (§7.4.1, §8–12 and 29). They are primarily lists of raw material, heaped up rather than arranged, and it is wise not to start on them until you have a fairly precise idea of what you are looking for. The principal institutions are the India Office for India, Pakistan, Sri Lanka, Bangladesh, Burma and the British Protectorates in the Middle and Far East; the Public Record Office for the records of other colonial governments; Rhodes House Library, Oxford, and the Centre of South Asian Studies, Cambridge, for private papers of colonial civil servants, etc.; the School of Oriental and African Studies for private papers and missionary archives. Most of these institutions have, or are about to publish, guides, the most recent and important are listed below.

7.4.1 Appendix

I.   Institutions

(1)   *The Royal Commission on Historical Manuscripts*
       Quality House,
       Quality Court,
       Chancery Lane,
       London WC2 1HP
for the National Register of Archives indexes to the location of manuscripts in the British Isles and for unpublished and published archive guides and catalogues of individual collections.
(2)   *The Institute of Historical Research*, and
(3)   *The University of London Library, Palaeography Room*
       Senate House,
       Malet Street,
       London WC1E 7HU,
for printed guides to manuscripts in the British Isles and overseas.

II.   General guides

The Royal Commission on Historical Manuscripts/National Register of Archives.
(4)   *List of Accessions to Repositories* (1957–1971).
(5)   *Accessions to Repositories and Reports added to the National Register of Archives* (since 1972): the annual list of newly acquired manuscripts, published in the autumn of the following year by Her Majesty's Stationery Office, London.

(6)   *Record Repositories in Great Britain, a geographical directory* (7th edn), London: HMSO (1982): a summary list of manuscript holding bodies.
(7)   *British Archives, A Guide to Archive Resources in the United Kingdom*, Janet Foster and Julia Sheppard, London: Macmillan (1982). This is a guide to archive holding institutions, it covers a wider field than number 6 above and gives some information about holdings.

III.   Guides to records relating to geographical areas in British archives

(8)   *A Guide to Western Manuscripts and Documents and the British Isles relating to South and South East Asia* compiled by M. Doreen Wainwright and Noel Matthews under the general supervision of J. D. Pearson, London: Oxford University Press (1965).
(9)   *A Guide to Manuscripts and Documents in the British Isles relating to the Far East*, compiled by Noel Matthews and M. Doreen Wainwright, edited by J. D. Pearson, London: Oxford University Press (1977).
(10)   *A Guide to Manuscripts and Documents in the British Isles relating to the Middle East and North Africa*, compiled by Noel Pearson and M. Doreen Wainwright, London: Oxford University Press (1980).
(11)   *A Guide to Manuscripts and Documents in the British Isles relating to Africa* compiled by Noel Matthews and M. Doreen Wainwright, edited by J. D. Pearson, London: Oxford University Press (1971).
(12)   *A Guide to the Manuscript Sources for the History of Latin America and the Caribbean in the British Isles*, edited by Peter Walne, London: Oxford University Press (1973).

IV.   Guides to institutions

Public Record Office
Public Record Office Handbooks:

(13)   *The Records of the Colonial and Dominions Offices No. 3*, L. B. Pugh, HMSO (1964).
(14)   *List of Colonial Office Print to 1919, No. 8*, HMSO.
(15)   *Classes of Departmental Papers for 1906–1939, No. 10*, HMSO (1966).
(16)   *Records of Interest to Social Scientists 1919–1939, No. 14*, Brenda Swann and Maureen Turnbull, HMSO (1971).

Rhodes House Library, Oxford

(15)   Rhodes House Library, its functions and resources, by L. B. Frewer and Louis Benson, Oxford: Rhodes House (1956).
(16)   *Manuscript collections (excluding Africana) in Rhodes House Library, Oxford*, compiled by L. B. Frewer, Oxford: Rhodes House Library (1970).
(17)   *Manuscript collections (Africana and non-Africana) in Rhodes House Library, Oxford; supplementary accessions to the end of 1977 and cumulative index*, W. S. Byrne, Oxford: Rhodes House (1978).

V.   Guides to archives in other countries

CEYLON, INDIA AND PAKISTAN
(18) *Government Archives in South Asia, a Guide to National and State Archives in Ceylon, India and Pakistan*, edited by D. A. Low, J. C. Iltis and M. Doreen Wainwright, Cambridge: Cambridge University Press (1969).

FRANCE
Archives Nationales

(19) *Guide de lecteur*, Paris: Archives Nationales (1978).
(20) *Inventaire des Archives Coloniales, Lousiane, Martinique, Guyane*, Paris: Archives Nationales (1971–1977).
(21) *La Sèrie d'extrême-orient du fonds des archives coloniales conservée aux archives nationales*, Feireol de Ferry, Paris: Imprimerie Nationale (1958).

GERMANY (FEDERAL REPUBLIC)
Bundesarchiv

(22) *Das Bundesarchiv und seine Bestaende*, Schriften des Bundesarchivs no. 10, Gerhard Granier *et al.*, Boppard am Rhein, Harald Boldt Verlag (1977). (The main collection of German colonial archives is held by the Deutsches Zentralarchiv, of the German Democratic Republic in Potsdam; see Das Bundesarchiv, p. 144).

General Guides

(23) *Die Nachlaesse in den deutschen Archiven*, Schriften des Bundesarchivs 17, Wolfgang A. Mommsen, Boppard am Rhein: Harald Boldt Verlag (1971, 1983).
(24) *Die Nachlaesse in den Bibliotheken der Bundesrepublik Deutschland*, Ludwig Denecke (second ed. revised by Tilo Brandis), Boppard am Rhein: Harald Boldt Verlag (1981).

ITALY
Archivio Centrale dello Stato

(25) *Guida Generale degli Archivi di Stato italiani*, vol. 1, Ministero per i beni culturali e ambientali, Ufficio Centrale per i beni archivistici, Roma (1981).

UNITED STATES OF AMERICA
National Archives

(26) *Guide to the National Archives of the United States*, National Archives and Records Services, General Services Administration, Washington D.C. (1974).

General Guides

(27) *National Union Catalog of Manuscript Collections 1959* – compiled by the Library of Congress, published by Edwards, Ann Arbor, Michigan (1959–1961); Hamden, Connecticut: Shoe String Press (1962); Washington D.C. Library of Congress (from 1963 to date).

VI.   International guides

(28) *Guides to materials for West African History*, no. 1 Belgium and Holland: 1962

no. 2 Portugal: A. F. C. Ryder, 1965
no. 3 Italy: Richard Gray and David Chambers, 1965
no. 4 France: Patricia Carson, 1968
no. 5 the United Kingdom: Noel Matthews, 1973
The Athlone Press, London.

(29) *Conseil Internationale des Archives*
Guides des sources de l'histoire des nations

SERIES A: AMERIQUE LATINE
So far the volumes covering archives in Germany (Federal Republic and Democratic
Republic), Belgium, Netherlands, Spain, Italy, the Holy See, Sweden and the United
States have been published.

SERIES B: AFRICA SOUTH OF THE SAHARA
The volumes for Germany (Federal Republic), Spain, France and Scandinavia have
been published.

SERIES C: NORTH AFRICA, ASIA AND OCEANIA
The volumes for Belgium, France, Denmark, Finland, Norway and Sweden have been
published.

VII. Citation

(30) *British Standard Institution*
British Standard Recommendations for the citation of unpublished documents BS
6371: 1983. London, BSI (1983).

Angela Raspin

7.5 Language learning

7.5.1 Introduction

Anthropologists are normally expected to "learn the language", and while
most try to do so, many of us feel we fail. Since this means failing to measure
up to a publicly required occupational definition, anthropologists have often
taken refuge in silence, instead of thinking critically about how to improve
language learning in the discipline. Increasingly, articulate criticism of the
poor language skills of anthropologists, bias in relation to the use of local
tongues, problems in learning appropriate codes and asking questions (e.g.
Owusu 1978) seem only to have compounded our embarrassment.

   The first step is to break down the delusive simplicity of "learning the
language". This is not a single, albeit complex, activity. No "native speaker",
for instance, commands every style, dialect, technical jargon and so forth
which is included in the English language. And we should not be narrow-

minded about what constitutes a language. Sharing a supposedly common language with informants may sometimes give rise to unwarranted complacency concerning "understanding". It is, however, possible to understand without being able to produce a variety or style. Similarly, everywhere there are specialists who are recognized as particularly skilled in certain kinds of linguistic performance. The outsider might be foolish to attempt work that requires specific, sophisticated or specially rich linguistic knowledge, but quite sensibly aim at achieving certain limited abilities to do certain things, linguistically and socially, and readily seek local specialists to help in anything else. It is as ridiculous to reject interpreters under all circumstances as it is to attempt literary criticism of medieval English literature from scratch, or argue with computer specialists without some preliminary work on their subject.

In learning a language, a child or student abstracts from several kinds of data. It is worth stressing that written and spoken forms of a language are nowadays regarded as constituting interactive processes (Newmark 1981). Speech, however, is obviously tied to specified speakers and interpretable only by listeners within earshot; it can create social context as well as be controlled by it. It is produced as a combination of content, sound, intonation, facial expression, gesture and distance between interlocutors. As well as trying to work out this configuration, which after all partly constitutes meaning and culture (see Parkin 1983), it is necessary to attempt to produce it correctly, because the anthropologist who sends out confused or contradictory signals may be really misunderstood, not just regarded as uncouth.

Conventional linguistics, which developed through detaching the "system" from the speaker, is ill-adapted to understanding how these two are related. Nor has socio-linguistics really tackled the problem, though the developing study of "pragmatics" (e.g. Lakoff 1975) may do so. Besides developing an awareness of cultural variability, as in "proxemics" (Hall 1959), or question and politeness forms (Goody 1978), the intending fieldworker can still benefit from a knowledge of basic descriptive linguistics.[6] This is because it provides models with which to sort out and manage the *patterns* of language flow. One must, however, remember that most writing about language treats it as an autonomous system, so that accounts of language distribution usually sug-

[6] There are many good linguistics textbooks, but also many which would be quite unsuitable. For general introductions to the study of language, and the principles and terminology of linguistics, see Crystal (1971) or Lyons (1981). For aids to description and the handling of data, Allerton (1979), Bollinger (1968), Gleason (1967), Hickerson (1980), Hockett (1958) and Langacker (1972, 1973) are all useful. Some contain practical exercises. As a bibliography, Hymes (1959) is now dated, but it does provide a means of access to some of the older methods literature in this area.

gest they are discrete and mappable onto real space, whereas in terms of *speakers*, they will overlap (Brazil *et al.* 1980).

A language usually has social rather than linguistic boundaries. We talk of different Scandinavian "languages", though they may be more like one another than the "dialects" of Ibo, thereby recognizing political and not linguistic facts. The linguistic status of "a language" is often hard to tell from the literature, partly because the criteria which linguists *tacitly* assume are paradoxically socio-political and not primarily linguistic. Wolff (1959)[7] showed that such criteria can determine inter-intelligibility, which therefore cannot be used as simple evidence of linguistic relationship or difference. The coincidence of "language" and "people" is an ideological fiction[8] if it is not a tautology (because the language is defined as such by reference to its speakers, or vice versa). Any claims to this coincidence should be examined by the anthropologist, and never assumed.

When one thinks about language in terms of use, one realizes that any language is a collection of resources, varying both in number and the means by which they are operated. Individual human beings build repertoires of usages through their lifetimes which frequently cross language barriers. "First World" dominance can be seen in the Third Worlders' need to acquire specific parts of world languages in addition to their own. Put another way, language varieties differ both functionally and in the means by which the functions are achieved. Thus English in England includes closely related varieties which simultaneously distinguish class and region, except for one – "the standard" – which establishes class by the absence of regional characteristics. These distinctions between varieties of English are carried by phonology and, to a lesser degree, syntax and lexis. In addition, there are styles and registers – varieties which similarly distinguish levels of formality and types of occasion.

This pattern is not common.[9] In many parts of the world neighbours can speak quite different languages, and/or the language variety of home, market place and work place can differ from each other and from the official or

[7] Wolff's article is reprinted in Hymes (1964) which is a valuable sourcebook. Even though "language" and "dialect" are not simply linguistic entities, fieldworkers should be (critically) aware of their linguistic aspects: comparative linguistics affords historical evidence if properly used (Ardener 1971a) and information on another dialect can be used for one's "own" understanding, e.g. correspondence rules (so that there may be a correspondence of, say, "k" in a variety A, "ch" in B and a glottal stop in C).

[8] The argument that "wherever a separate language is to be found there is also a separate nation which has the right to manage its affairs and rule itself" (Fichte) shows the ideology clearly but it has deeply infected European thought.

[9] There are many anthologies of sociolinguistic readings, e.g. on multilingualism and social evaluation of styles and varieties, which can introduce these key topics to fieldworkers.

national language, which may also be the only one reduced to writing. A common feature is *diglossia*, which is often the institutionalized, inegalitarian relationship of one or more varieties with another whose ordinary use is limited to the élite and which is the official written mode. For the rest of the populace, the medium of education is not their mother tongue.

The fieldworker has both to work out what the key linguistic variations are in the fieldwork area and to realize that one's own choices will have repercussions. If there are stratified politeness styles, there can be solecisms; if one learns from low class informants, one may present oneself as weak – or insulting – when talking to high class ones. Those who can only communicate through a lingua franca, or the official language, are cut off from many ordinary people, including very often the women. Even the readiness to accept outsiders' attempts at speaking differs, and could be called a social component of language. Where in the past anthropologists learnt well by "total immersion" because there was no alternative, this is paradoxically harder today, when schooling is widespread and the locals are often anxious to learn the standard language from you (see Lakoff 1975).

7.5.2 Achieving basic skills

What preparations can the anthropologist make before fieldwork begins, and what level of achievement can be expected in the field? People vary a great deal in the ease with which they pick up languages and tend to be "soakers" or "analysts" (Healey 1975). Soakers absorb new languages much more easily than analysts, but are less good at analysing them. However, every anthropologist will need to record items of language: this requires some systematic understanding of it and an accurate transcription. In the absence of a local writing system (which in any case would have to be learned) one must make one's own *phonemic* one, using a recognized system like the International Phonetic Alphabet (IPA 1949; see also Pike 1947; Voegelin 1959). All this requires elementary linguistic training, ear training and practice in *phonetic* transcription. Even the "tone-deaf" can acquire them; they are skills to acquire before doing fieldwork, and without their use our ethnography remains embarrassingly amateur.

While many ethnographers work in an unstudied language area, there will typically also be some ability to speak a local language of wider use and/or a lingua franca. This is often a pidgin or creole, perhaps related to the national language. One will need to operate in this, too, in order to cope with officialdom. It can normally be learnt before departure, but beware! Just because the national language is, say, English, do not assume that it will be identical with your own variety. The wider languages travel in the world, the more they develop into local varieties, with distinctive sound systems and with syntax and lexis which change to the point of becoming pidgins or creoles. Often these form the ordinary spoken style, and this has a greater or lesser affinity with the standard, metropolitan or élite variety.

The advantages of knowing the lingua franca include its use as a bridge, both to use with interpreters when trying to learn your "own" variety and as an aid in grasping this when it shares common features (as, e.g. Swahili does to other Bantu languages, or Bahasa Indonesia to Minangkabau). In a city knowledge of a lingua franca is vital, but even here familiarity with specific local tongues may be important. For example, they may serve to maintain ethnic boundaries, promote solidarity, and mark types of occasion.

Given the level of language competence required by a fieldworker it then becomes necessary to consider the resources available to achieve basic skills. The kinds of language help available fall broadly under four headings:

(1) Teaching in the language variety needed by a skilled or native speaker, plus tapes, grammar, dictionary (or any one of these).
(2) Teaching in a related language *or* in the local lingua franca *or* official language as above.
(3) Teaching in basic (especially descriptive) linguistics and phonetics.
(4) Teaching in *how to learn any language* (this normally includes (3), but additional, more academic study of (3) will help here too).

The anthropologist may well find no good resources under (1) and even if a native speaker is offered as a teacher, s/he may not be able to teach or analyse the language in a way which the student finds helpful. Approaches using untrained native teachers are, however, currently very much in vogue (e.g. Krashen 1981). Teaching in basic linguistics and phonetics (3), is of course, always useful. Teachers of (4) train people to "unpick" languages for themselves, working with indigenous helpers – even monolingual ones. The many skills involved are all valuable to anthropologists.

Obviously it's important to start language preparation as soon as possible. Being careful to explain your needs as including (1)–(4), try your own and other local institutions. In Britain, polytechnics and other more technically oriented colleges often have a more pragmatic approach to language learning than conventional university language departments. All language and linguistic departments should be able to offer help in (3), and the Social Science Research Council (SSRC) runs a Pre-fieldwork Language Training Course at Reading University. The Polytechnic of Central London (PCL) claims to offer

(1)-type courses to suit special needs, as do Leeds and Sheffield Polytechnics; there is also the School of Oriental and African Studies (SOAS) and the School of Slavonic and East European Studies in the University of London. The former has a fine library. York, also, is prestigious in this field, and many universities can offer particular lesser-known languages. For example, at Kent, instruction in Swahili, Turkish, and Malay is available.

PCL also runs short courses of type (4), but the leader here is the Summer Institute of Linguistics, with summer schools in Britain, North America[10] and many European, African and Asian countries. They aim to train missionaries: you may also find there are missionary training courses "on site" in your research country. The Peace Corps and VSO increasingly train volunteers in this way. If you can find out about such courses beforehand, you may be able to get permission to attend them, and to budget for time/money to do so in advance, but unfortunately one may only get a response on the spot. The same applies to missionary and volunteer written resources.

7.5.3 Using written and audio resources

The first problem is, often, to find out what the local language variety *is*, especially if you are not able to be sure beforehand precisely in what dialect area you will land up. Besides the bibliographies and standard texts on the language of different areas, there are language maps and atlases, but you must be prepared for these to be wrong (and to try and right them on return). You may find mimeographed resources put out by missionaries (check which societies are working in your area – cf. above) and local language institutions including SIL (or Wycliff Bible Translators) as well as local university language centres. Local churches may have literacy primers which are helpful aids to you. The Peace Corps produces study guides written by professionals, sometimes with cassettes.

Check bibliographies, dissertation titles and specialist booksellers for written materials[11] and consult the linguistic journals.[12] These may also lead

---

[10] The main US address of SIL is 7500 W. Camp Wisdom Road, Dallas, Texas 75236 and their British HQ is at Horsley's Green, High Wycombe HP14 3XL. Theirs is an openly and strongly evangelical organization, which also has impressive linguistic connections and much experience in teaching non-specialists practical learning and analytic techniques.

[11] Catalogue sources include Frederick Ungar of New York (British distributors: Silco Books Ltd) and Bailey Bros. and Swinfen of Folkestone, UK. Johnson *et al.* (1976) provide a good survey of resources for the study of uncommonly taught languages, which comes in eight fascicles.

[12] Useful items occur in theoretically oriented journals as well as journals focused on geographical areas, and on sociolinguistic analysis, and *Anthropological Linguistics*. *The Carrier Pidgin* is a newsletter covering all pidgins and creoles.

you to authors who are worth writing to for help. Many language teaching and linguistic books and general information are available in London at the Centre for Information on Language Teaching and Research (CILT). Its "Culture and Language Guides" so far include some major S. Asian languages, Swahili and Portuguese, as well as many other European ones. An American equivalent is the Center for Applied Linguistics (Washington D.C.). This also publishes a survey of materials for the study of the uncommonly taught languages.

7.5.4 Language learning in the field

There are good guides to learning on your own – study them beforehand as well as taking them with you. Although one can get useful instruction in phonetics from such sources, it's best to try and get practical help and feedback on this before the research starts. A thorough field guide from SIL is Healey (1975).[13] Nida's *Learning a Foreign Language* (1957) remains a thoughtful introduction, since he is a missionary linguist who shows how language learning is socially an end in itself as well as a means of giving (or for anthropologists, gaining) knowledge. It entails humility – willingness to look foolish – and a readiness to try and exist on others' terms, instead of expecting them to conform linguistically to yours.

The normal advice on language learning is: (i) to practise daily and learn daily, in short, structured sessions; (ii) to be open, relaxedly, to as many linguistic expressions as possible, including chat, singing, radio programmes; (iii) to *record* data informally and formally but always systematically (e.g. into a day-book, with date, source, translation if possible);[14] and (iv) to use one's acquisitions, trying out words and phrases. One can also consciously use this regimen for anthropological ends: in particular, *recording* can include systematic enquiry into taxonomies and terminological domains (but look for verbs also, not just the more obvious nominal terms).[15] Discussion of

[13] Most guides have something to offer but Healey (1975) incorporates much of the ground covered by earlier work, e.g. Henry (1940), Gudschinsky (1967), Healey (1964) and Samarin (1967). Books on translation (including literary approaches) are useful pre-fieldwork reading: e.g. Richards (1932), Nida (1947, 1964), Brower (1959) and Newmark (1981). For a very first encounter with the problems of learning exotic languages, Hockett (1957) remains useful.

[14] The day-book is the base record from which one can construct (for instance) the language's phonemes and grammatical features and also set up an ongoing dictionary by classifying data into entries on separate cards (§8.12.7).

[15] One can follow – or learn from – ethno-semantics (e.g. Tyler 1969) and build up culturally revealing accounts through asking informants for calendars, collocations of work actions, etc. This is still a field of research where more of such techniques would be worthwhile reporting. See Howell (1981) as just one approach.

entries, e.g. with a bilingual helper, is also a very valuable ethnographic exercise, albeit in competition for time with straight language work.

Your key need with language, as in all aspects of the research, is patient informants who care about things as they are, not as they think they ought to be. You want teachers who will not fit language into "correct European" grammar, interpreters who will not overbear tellers or interpose their own interpretations. Money to pay assistants is a constraint and so is their availability – suitable aides may be busy people, or in full-time work. You can partly overcome these restrictions – even profit from them, by sharing out different tasks and taking full advantage of skills you find. A research assistant may get interested in working out an orthography; a full-time employee may be willing to transcribe tapes for you in the evenings; a pedantic teacher may prove to be a good cultural commentator. Drills and corrections are skilled but boring tasks: it may be less wearing (but not necessarily time-saving) to devise them yourself, have someone tape them and make your own "language lab". Children can be very good language teachers – straightforward and insistent, more willing to take time than their elders and less embarrassing to the adult learner.

Time is your greatest enemy. Six months is the *least* time it takes normally to get a working ability in a completely new language. Within the last few years, British student research grants have shortened: instead of two years or more, it will get harder to have even a year in the field. It is simply not possible for most people to do linguistically sensitive ethnography or specialized study of language use in this time if they have to start from scratch linguistically too. It would be better to recognize that many kinds of enquiry can be undertaken through a lingua franca or with an interpreter – i.e. to recognize what often actually happens nowadays – while insisting that this work should be accompanied by as much informal language learning as possible and by systematic consistent recording of relevant items (not just translations or glosses of them). These are not ideal aims, but they are realistic ones.[16]

Bilinguals' usage is often a key to first language expression, and can specifically be checked on in order to recover this, besides being important in itself – one is, after all, trying to uncover actuality, not to reconstruct a purist, ideal primitivism uncontaminated by history, state intervention, education or migration.

Much satisfying ethnography exists whose authors' command of their

---

[16] Mead (1939) is often arraigned for arguing that fluency is neither necessary nor desirable (§3.6). As is stressed in this section, the learner's *aims* should be fitted to *need*, and fluency is not simply definable. In fact, Mead seems to have spent much time on language learning and rarely used interpreters (Mead 1972).

subjects' first language is very limited. This guide is addressed to intending fieldworkers who are not linguistic specialists, but who are going to do better work if they have prepared themselves linguistically and achieved a few techniques which even skilled speakers often do not know about (surprisingly few people, for example, understand that it is possible to learn how to analyse and record a language without knowing it beforehand). Of course working through a lingua franca is second best, but it may be the only alternative to understanding five or six other varieties, and it will frequently need considerable learning itself. *Working with an interpreter* can be as difficult and need as much linguistic attention as working monolingually – you have to maintain thoughtful monitoring to make the most of what you hear and not be culturally misled. This is not because the interpreter is trying to mislead you but because s/he is a culture-representative (or unrepresentative) mediator who is trying to carry out new tasks for you, often with limited knowledge of your common language.[17]

7.5.5 Recording data: transcription and translation

Many anthropologists change their interests in the field – or even long after their return – and then wish their data was better recorded. Even when one is not studying the means of communication in themselves, one can at least be steadily conscious of the means as well as the content. It is a tempting mistake, however, to record long stretches of data on tape in the hope they may become useful evidence, without also having the ability to transcribe and translate them. Audio-technology is a very valuable tool, but so long as we *write* anthropology it must be transcribed. Even native speakers will understand a recording better in context and will find transcription and translation extremely slow and skilled work (§7.6.3, footnote 22, §7.9.3).[18]

Properly chosen recording is nevertheless a rich source, often revealing most when studied together with insightful informants. One can sometimes play back material to participants and others for comments. Increasingly, anthropologists are learning to use their technology actively, and not just as a means of examining performance (including political action)[19] but also for

[17] On the use of interpreters in general and in relation to interview schedules, see Phillips (1959–60) and Werner and Campbell (1970).
[18] Casagrande (1954), Phillips (1959–60) and Werner and Campbell (1970) discuss the translation process in ethnographic fieldwork, as well as various techniques (including back-translation) for achieving accuracy. Hymes (1964) provides an extensive bibliography. See also Junker (1960: 27–31).
[19] Stimulating analyses of this kind include Basso (1979), Bloch (1975), Kress *et al.* (1979) and Seitel (1980).

genuinely participating with their subjects. Tapes can, of course, be brought back – and Indiana University will be glad to take copies in their library of sound recordings – for further work either in your home country (where there may also be possible aides) or on a return trip. Unless a fieldworker has special linguistic expertise, local knowledge or access to previous good ethnography beforehand, the most sophisticated understanding of language in anthropology may emerge through such further work.

7.5.6 Conclusion

Three general points may be distilled from what has been said above. (i) Language learning is harder than anthropologists are often led to expect, but it is just as necessary. However, it has many dimensions, it does not solely mean competence as a speaker. (ii) Preparatory training should result in better work generally, if it prepares fieldworkers for all linguistic aspects of their experience (Burling 1970). (iii) Attempts at speaking are socially and humanly rewarding. Trying to talk in others' terms is a surrender of one's own otherness, and generally respected as such. This learning can be properly separated from analysis, although we always have to remember that the greater part of our findings will come through conversation: the more we understand the media of our messages the better.

Elizabeth Tonkin

7.6 Funding and accounts

7.6.1 Funding

It must be said that in the past most ethnographic research has been conducted economically on extremely modest budgets, usually by postgraduate students on subsistence grants. This is due to the self-images of fieldworkers themselves, a tradition of lightweight technology, the low social value accorded to anthropological research by funding agencies, and (at least in Third World countries) the low living costs involved. Although this latter point hardly applies to research in, say, the United Kingdom, even here ethnographic research is cheap compared with large-scale survey research. It is also due to the reflection of this in the total resources available, the small-scale of the projects involved and a professional organization which has not been geared-up to demand more than barely adequate funding. Although increased resources are no guarantee of excellence, there is considerable reason for believing that ethnographic research is grossly under-financed, and that more resources for basic equipment and personnel (including assistants of various kinds and salaries for principal investigators) would improve the quality of research.

In its 1968 report (p. 93), the British Social Science Research Council Social Anthropology Committee estimated that 35% of funds used by fieldworkers originated from foundations, 35% from governments and government agencies, 18% from universities, and 12% from international organizations, industry and personal contributions. There is no reason to think that the position has changed significantly in the past decade, although the proportion of funds obtained from the SSRC has increased to the extent that in 1982 this body provided, in all probability, the bulk of such funds (Annals ASA 1982: 1, 18–19). Nevertheless, social anthropology takes up a very small part of the SSRC budget: around 3% of research grant money. Indeed, it is thought that it will absorb only 0.69% of estimated expenditure by the Council in 1982–1983 (Rothschild 1982: 80).

There now exist a number of guides to the writing and general presentation of research proposals (Agar 1980: 175–178; Krathwohl 1976; Pelto and Pelto 1978: 294–297; Williams 1967: 6). Platt (1976: 14–32) and Field *et al.* (1982: 153–161) are useful for social science applications in the United Kingdom. Much of the advice given in these guides is valuable, but none need be summarized. However, it may be worth repeating here that applications to funding bodies should always be made well in advance of the estimated starting date of a project or fieldwork period, perhaps at least one year. This allows for delays stemming from unanticipated queries, rescheduled plans, the necessity to revise an application, or to submit an application elsewhere. Grant-awarding bodies are usually prepared to discuss projects at varying stages of preparation prior to submission and full advantage should be taken of this facility if it is available.

The next section lists a number of directories and handbooks covering the major British, American and International funding bodies most frequently approached by anthropologists. Some of the sources are restricted to post-graduate students, others to professionally-qualified anthropologists and university teachers, and some to research on particular subjects and in particular regions. The directories do not ordinarily include funding bodies linked to particular universities and colleges, such as the Central Research Fund of the University of London. Few grant-awarding bodies restrict themselves entirely to funding anthropological research.[20, 21] Commercial firms may sometimes be prepared to support fieldwork by supplying their products free (e.g. film and pharmaceuticals), usually in exchange for some publicity. Remember that some sources of sponsorship and funding may be unacceptable in certain countries or settings. Check this out at an early stage in your enquiries. Similarly, funders themselves may impose various constraints on the use of their funds: affecting use of and access to data, royalties, the form in which the research may be reported, whether or not the research can be submitted as a higher degree thesis, copyright, supervision, disposal of equipment, and so on (see Orlans 1973: Social Research Association 1980). You may, for example, be asked to sign an official secrets act. It is important to read carefully any direction which an agency might issue concerning applications for funds.

7.6.2 Bibliography

Annual Register of Grant Support (1980–1). Fourteenth edition. Chicago: Marquis Academic Media.

Awards for Commonwealth University Staff 1981–1983 (1980). Fifth edition (published every two years). London: The Association of Commonwealth Universities.

Charities Digest (1981). Eighty-seventh edition. London: The Family Welfare Association.

Coleman, W. E. (1980). *Grants in the Humanities: a Scholar's Guide to Funding Sources.* New York: Neal-Schuman, London: Mansell.

Crawford, Elisabeth and Norman Perry (1976). *Demands for Social Knowledge: The Role of Research Organisations.* New York: Sage (Britain, France, Denmark: lists some funding sources).

Directory of Funds for Scholarly Communication (forthcoming). Primary Communications Research Centre. Leicester: University of Leicester.

[20] The funding position in the USA, and what is sometimes termed "grantsmanship" is discussed briefly in a number of general works (Agar 1980: 30–39; Freilich 1970: 491–492, 587–589; Williams 1967: 5–6), and at greater length in Hall (1972). Beauchamp *et al.* (1982: part 5) and Reynolds (1982) cover the more recent legal changes in the USA and Institutional Review Board procedures which may also seriously delay starting dates. Some information on funding in the Netherlands can be gleaned from Kloos and Claessen (1981).

[21] Among those that do are the Wenner-Gren Foundation for Anthropological Research (1865 Broadway, New York, New York 10023, USA: Tel. (212) 957-8750), the Emslie Horniman Anthropological Scholarship Fund and the Radcliffe-Brown Memorial Fund for Research in Social Anthropology, both administered by the Royal Anthropological Institute (56 Queen Anne Street, London W1M 9LA). The aims of the Radcliffe-Brown Fund are chiefly to help young scholars, handicapped by lack of finance, to work towards the completion of research already begun; awards are necessarily small, priority being given to those in the final stages of writing-up.

Directory of Grant-making Trusts (1981). Seventh compilation. Tonbridge: CAF Publications.

Hodson, H. V. (ed.) (1979). *The International Foundation Directory*. Second edition. London: Europa Publications.

Lewis, Marianna O. (ed.) (1975). *The Foundation Directory*. Fifth edition. New York: The Foundation Center.

Scholarships Guide for Commonwealth Postgraduate Students 1980–1982 (1979). Fourth edition. London: The Association of Commonwealth Universities.

Scholarships Abroad 1981–82 (1980). September. London: British Council.

Some Awards Open to Graduates of UK Universities and Tenable at Foreign (non-Commonwealth) Universities (1975). London: The Association of Commonwealth Universities.

Some Awards Open to Academic Staff from Foreign (non-Commonwealth) Universities and Tenable at UK Universities (1977). London: The Association of Commonwealth Universities.

Sources of Funding for Scholarly Publishing (1979). Primary Communications Research Centre. Leicester: University of Leicester.

Social Research Association (1980). *Terms and Conditions of Social Research Funding in Britain: Report on the Working Group*. London: Social Research Association.

Studentship Handbook (1981). London: Social Science Research Council (Postgraduate Studentships in the Social Sciences).

7.6.3 Accounting

Adequate funding, successful research applications and forward planning all require sound and realistic budgeting. First, do not ask for absurd amounts: it will be noticed and, even if the project is otherwise excellent, it leaves a bad taste. Secondly, budgeting categories will vary from one project to another, and virement between expenditure heads is usually minimal. Most should include such broad headings as external and internal travel, insurance (for personal effects, health, equipment and travel), the purchase of equipment (professional and domestic), supplies (including medical), accommodation, food, salaries and staffing costs. University (or other institutional) affiliation fees are sometimes necessary and can be quite sizeable. Solomon Islands (1976) mentions a $2000 deposit which would be forfeited under certain conditions. Efrat and Mitchell (1974: 406) quote a $5000 indemnity for research in British Columbia. There are likely to be other kinds of small bureaucratic fees, and allowance should be made for this. If a grant has to be paid through an institutional base, make certain of that institution's requirements. Not only may they have their own rules about what researchers may be paid, what expenses they may claim and such like, but some institutions are now demanding a percentage to cover administrative costs (although British Research Councils will not pay this). In the United States, this may be as high as 100%; for the United Kingdom, 30% is probably common. Perhaps the only safe and near universal recommendation is that prediction of costs is hazardous and that all budgets should allow for unanticipated expenses, including emergencies.[22]

Good budgeting alone is insufficient. The keeping of careful accounts, in the context of overall financial management, is also an important aspect of planning research, of its conduct over time, and its effective completion. Without a clear idea of the pattern of expenses over time, and the relative costs of particular items, vital research time may be cut short and writing periods may have to be curtailed, in some cases threatening completion of a report and its subsequent publication.[23] This, of course, is not to mention the great personal problems and disappointments which arise from monetary imprudence. The larger the project the more complicated and time-consuming budgeting and accounting become, and the greater the necessity for good supervision. In such circumstances, professional advice is always valuable, and often essential. This is usually available from the larger grant awarding bodies themselves, from the financial departments of universities and colleges and from banks.

Some problems may be encountered in the transfer of money to and from fieldwork locations, either internationally or within the countries in which fieldwork is being conducted. Travellers' cheques are not universally acceptable and too much cash in hand is obviously unwise. A local bank account may be useful, but may sometimes pose problems when research is being conducted abroad. International transfers of money sometimes take time and may prove complicated, especially when transfers are being made to remote and small outposts, and when there is a question of negotiating local exchange control regulations. Time should be allowed for such transactions.[24]

[22] The costs of transcribing sound and video-tape recordings may be an overlooked and under-estimated item. Bucher *et al.* (1956: 358–364) suggest that one hour of recorded interviews needs six hours of skilled typing and around three hours of proof reading and checking. Samarin (1967: 105) mentions 11 hours of transcription in longhand by a native speaker for every hour of recording in the field of foreign languages, and 11 hours of checking and identification. Either way, an awful lot of time is involved and at formidable expense.

[23] Certain aspects of budgeting are dealt with at an elementary level by Agar (1980: 178–180), Pelto and Pelto (1978: 178–180) and Williams (1967: 6–7). R. Wax (1971) comments that most people never leave enough writing-up time, and when financial resources are under pressure getting writing-up time can be a problem anyway.

[24] It may also be necessary to check on tax status, not with the revenue services of your own country, but in some cases with the governments of host countries.

Field accounts should be kept simply and quite separately from other notes and financial calculations, and should be kept together. The best way to do this is to use a small hardback account book of the kind available from all commercial stationers. It is not necessary here to discuss accounting techniques in any detail, although there are a number of small points which are worth remembering:

(a) Running credit and debit totals at the bottom of each page assists subsequent checking and calculation.

(b) All major categories of expense should be differentiated, for example air fares, local transport and accommodation. Beyond this, the degree of detail entered into the debit column very much depends upon the effort an individual is prepared to put in. In a sense, the more detail there is the better, but this takes time and a good memory. The advantage of detailed debit accounts is not only that it makes subsequent checking and planning easier, but because it may represent a valuable fieldwork document in its own right, in providing (for example) the prices of goods available in a local market.

(c) On the termination of a fieldwork phase or project, detailed debit items should be collated according to broad categories indicated in (b) and those which appear in the research application.

Some practical guidance in simple accounting methods may be obtained from elementary guides to the subject.

7.7 Getting into the field and establishing routines

7.7.1 Permissions

Once the ethnographer has decided upon a country within which to carry out fieldwork, it is helpful to contact other scholars who have recently done research there. There may be new restrictions, special difficulties, or unusual requirements imposed by the government. This advice may be expected to include the names of local government agents, missionaries, or other useful parties to whom to write for further information. Enquiries must be well in advance of the proposed commencement of fieldwork. In Northern American and West European countries, no prior official permission is generally necessary, although it is usually required that you register as a foreigner. In Third World or Communist countries such permission is required. Some governments require an educational institution, typically a university, to agree to act as official sponsor for the research proposal before they consider granting permission. This institution may demand as a condition of its sponsorship that the anthropologist take one of its own research students into the field to learn the techniques of fieldwork. Given a capable and enthusiastic student, the experience can be profitable for the student, the institution, and for the principal investigator.

Visa-issuing bureaucracies can take up to a year before granting a research permit. Depending upon the country, sometimes one can obtain a type of visa which covers a much shorter period, but is easier and speedier to get. For a pilot study this possibility offers attractions. Relations with officials in the local embassy through which the request is made need to be cultivated. Even though approval will come from the country itself, the embassy staff will initially receive it, and an anthropologist who has been forgotten as an individual may never be contacted and informed that permission has been granted.

Letters of introduction to specific individuals of the "To whom it may concern" type are desirable; they are a politeness as well as being useful. The fieldworker's own university can supply these. Once in the country the fieldworker is usually obliged additionally to register as a foreigner at the local police station or administrative centre. Photographs of passport size may be needed, and a good supply should be taken to the country of research. Copies may be required by other government offices, so 25 is not excessive. Only after registration is the fieldworker technically at liberty to go about selecting a community in which to conduct research.

7.7.2 Choosing fieldwork locations

Among the more usual rationales for field research are: (a) the desire to study certain topics; (b) investigate specific theoretical problems (e.g. the function of exchange, the nature of cooperation in a Greek monastery, the evolution of a particular form of relationship terminology); and (c) suggest solutions to practical problems (e.g. labour disputes on the factory floor, a declining birth rate, a rising divorce rate). These typically overlap. Evans-Pritchard wanted to study magic in central Africa, a region holding a romantic appeal for him. He also wanted to test the validity of Lévy-Bruhl's theories of primitive mentality. He heard, from the Seligmans, of a central African people who practised magic, so he went to the Azande. Years later, Evans-Pritchard's fieldwork among the Nuer was carried out at the request of the colonial government in the Sudan, which was then having trouble with them. The government wanted information.

The choice of a location may be based simply on the fact that its inhabitants have never been ethnographically put on record. Perhaps all that is known about them is their existence. Their prospective ethnographer will get some idea of the social forms which may prevail among them by what is known about their neighbours. This being the case, the wish to record the customs of an unknown people can be matched up with the wish to subject the empirical findings and theoretical conclusions of fieldworkers in the same cultural region to comparative scrutiny. The ethnographer may sometimes want to return to a society previously studied, to see what changes have occurred in the interim (§8.6).

7.7.3 Getting into the field

Once the bureaucratic preliminaries have been completed, and the general location of the research determined, the fieldworker must establish contact with a community (however defined) of a size convenient for the desired research. Specific locations may be chosen for various and often widely contrasting reasons:[25] these may include their representativeness, atypicality, remoteness (for example, where this may be taken to indicate the maintenance of an older, more "traditional" life-style) or accessibility, either in geographical or social terms. Some sites may be selected because they have been studied previously (e.g. Gallaher 1964; Phillips 1966; O. Lewis 1963). The question of size will be determined by the researcher's interests, the nature of the topics selected, the theoretical problems formulated, and the demographic realities of the area. Which reasons become paramount will depend on the subject of research and the practical problems encountered.

Whether the community has fewer than 150 inhabitants, or boasts thousands, or is a factory work group of a dozen members, it must be practicable for research. Transportation may turn out to be a critical consideration. Heavy rains or snows at certain seasons may isolate a community, and for a sick fieldworker be hazardous. Then there is the matter of accommodation. Some communities will have a reasonably priced guest house, or a shop for strangers'to live in, or families willing to take paying guests. A community of this kind may be more suitable than one more exciting for the fieldworker's theories or one more appealing to topical curiosity, but which is

[25] See also Beattie (1965: 13–14), Pelto and Pelto (1978: 179–180), and more generally, Agar (1980: 21–24).

inaccesible at certain seasons and lacks good living-quarters. If a location can be found which is also one of considerable beauty, then this is, of course, an added bonus, but it is hardly a justifiable reason in itself.

Communities of emigrants from the proposed field area living in the anthropologist's own country, in another country close at hand, or in the country of fieldwork can be tapped for language instruction, advice about the best localities for fieldwork, and the names of potential informal sponsors in the fieldwork locality. On these issues advice from the informal sponsor would be welcome. Also, provided the advice is disinterested, a sponsor who knows the area and understands the fieldworker's scholarly requirements can suggest a likely community and even provide introductions to its inhabitants. A sponsor's suggestions should also be invited concerning what rules of behaviour should be observed during the initial phase of research, and what should not be done. If such an adviser is not forthcoming, the fieldworker must rely on his or her own devices, including common sense. It is sensible to stay a week or two in a small town. From this base an exploratory reconnaissance into the area can be made. After this the researcher can make the transition into the community selected with some confidence that the choice is well-informed.

At this stage the anthropologist needs continually to be reminded that he or she is an uninvited stranger. Therefore, the gamut of questions put to informants must be severely limited. Issues which are likely to offend, embarrass, or arouse suspicions, or which may result in the early termination of fieldwork, are obviously unwise. Excessive curiosity about particular individuals, about quarrels, money, tax or wealth are potentially dangerous. Such sensitive areas the fieldworker may learn about from reading, sponsors, first impressions, common sense and helpful informants. In this, the opening stage of research, the fieldworker is generally unlikely, in any case, to know enough of the language to get embroiled in controversial discussions, but will instead rely mainly on visual information. If a language has to be learned then this will obviously be a first priority, but in addition the writing of descriptions of things (objects, landscapes), situations and behaviour, and the drawing of maps, may be a useful first step.

The phase of settling-in may be one of acute attitudinal and personal difficulty, both mentally and practically. Among these is often a sense of inadequacy, uncertainty with the process of selecting a community and an unwillingness to take the plunge. It may be followed at a later stage by an irrational sense of failure. Confidence has to develop and this may take between six weeks and three months.

Most accounts of field methods include a section on arriving, settling-in and on the making of initial contacts. For further advice, and for personal accounts relating to problems of entry and access, deciding on research locations, research settings, beginning fieldwork, getting organized, and negotiating the network in, see Paul (1953), Schwab (1965), Williams (1967: 9–14), Middleton, Uchendu, and VanStone (all in Naroll and Cohen 1970: 220–245), Wax (1971: 15–20), Schatzman and Strauss (1973: 18–33), Edgerton and Langness (1974: 20–26), Freilich (1970c: 492–498), Johnson (1975) and Platt (1976). Berk and Adams (1970) deal with the special problems of establishing rapport with deviant groups. See also §5.3.

7.7.4 Time-tabling

Junker (1960: 12) has estimated that, as a rule, one-sixth of fieldworker's time is spent observing, one-third recording, one-third analysing and one-sixth reporting. However, few plans are more certain to be overturned during fieldwork than those contained in a timetable. Accordingly it must be flexible, to allow for unforeseen contingencies, although it can safely be predicted that several weeks may be "lost" between the time the anthropologist arrives in the country of fieldwork to the time permanent contact with a community is made. With customs clearance and similar bureaucratic formalities, leaving the country may take almost as long. Actual fieldwork may therefore be decidedly shorter than the total period spent in the research zone. This will particularly be the case if one or more breaks are taken from the routine of fieldwork.

And "routine" it should be. The "What-do-I-do-today?" attitude is no substitute for a systematic plan of investigation (§7.2). Enquiries must be circumspect in the early days (§7.7.3), mainly directed towards innocuous matters, and the fieldworker should concentrate on those features of the new environment which can be recorded without much assistance or knowledge of the language. A daily schedule of activities should be set up, commencing shortly after dawn in a tropical climate and allowing for relaxation when the sun is at its peak. The evenings might be spent writing-up the information collected during the day. If evenings are taken up with other activities, then first thing next morning might be the best time to compile notes. At this stage, when contacts are still few, the anthropologist will have more private time

---

[26] Conklin (1960) describes a complete daily routine from his Hanunoo fieldwork. On routines and organizing time, see also Williams (1967: 19–21), Schatzman and Strauss (1973: 39–40) and Platt (1976: 22–35).

than later on, and this is another reason for setting down the less personal and socially intensive elements of the scene.

How long this period of adjustment, contact-making, language learning and preliminary gathering of information lasts will depend upon local circumstances, and on the talents and personality of the fieldworker. After a few months at the latest the fieldwork will have entered its middle stage. Increasingly dependent on the daily round of informants activities, the fieldworker must alter any private daily schedule to fit theirs. By now he or she will know the most favourable times of the day to approach people. At work they will not usually wish to be questioned, so during such times persons unemployed can be consulted: old people who remain at home, and youngsters not at school. After the evening meal might be a good time to engage the workers of the community in conversation, so the lonely evenings spent at the beginning of fieldwork will become fewer.

During this stage questions should focus on matters directly pertaining to research interests. As with timetables, so with questions. Whenever the convenience of informants permits, questions should be organized around specific themes, e.g. death rituals, rules of land ownership, and belief in spiritual beings. In this way time in the field (always too brief) will be spent economically. Once more, common sense is needed. A fieldworker who had decided to ask questions about genealogies in a particular week would be silly to persist with this plan if on one of the evenings his companions chose to talk about matters they had hitherto been reticent about discussing.

After several months in the field is a good time to take a break (§7.8). Fieldwork, after all, is work: arduous, demanding, sometimes boring. Rest is an antidote to the strains imposed by its discipline. The returning fieldworker may find that he or she is no longer regarded as a stranger, but as a former inhabitant of the community. On leaving, people were still probably wary but, by returning to them, friendships are cemented, thereby more obviously integrating the researcher into the lives of the people being studied.

The second stretch of research constitutes the final stage. Questions previously too sensitive can now be broached cautiously. With a more assured command of the language, knowledge of interpersonal relationships enriched by experience, and ever more accepted as a person with roots (albeit shallow) in the community, the fieldworker will discover that the quantity of information acquired now is far more than ever before. Whatever strain this imposes upon personal resources should be made tolerable by the fieldworker knowing that the fieldwork period is drawing to a close.

David Hicks

7.8 Duration of study

David Maybury-Lewis (1967: xix) has remarked that "a good fieldworker may obtain better data in six months than an indifferent one in two years". Evans-Pritchard stayed with the Nuer only nine months; Edmund Leach, only eight months in Pul Eliya; and Maybury-Lewis himself about ten months among the Shavante. Yet the data these three fieldworkers brought back were detailed enough for them to make outstanding contributions to the empirical and analytical capitals of social anthropology. Many fieldwork excursions have lasted longer than a year without adding to the published record. Eggan (1963: 416) has reported that a long period of research was particularly undesirable for doctoral students, suggesting that initial field research might be reduced to six or eight months for thesis-writing purposes. Later field trips might be added post-doctorally.

On the other hand, some have thought it arrogant and deceitful to claim to be able to write about certain difficult topics (such as ritual or secrecy) on the basis of so short a research input, where linguistic (quite apart from other) skills are likely to be limited (den Hollander 1967: 18; Crick 1982: 19). However, the short duration of most fieldwork inputs means that it frequently occurs that many periodic activities and events are not witnessed personally, or only just. Gulliver (1966: 12–15) admits that he was present at only one wedding throughout his entire Jie and Turkana research.

It is clear, however, that the length of time spent living in a community does not, by itself, ensure copious information or contribute to theory. Still, as a rule of thumb, the anthropological profession seems to hold the opinion that unless intended as a pilot study or a follow-up to a previous, more substantial, bout of fieldwork in the community, the period in the field should not be less than one year (e.g. Fortes 1963: 434). The fieldworker should witness his people's annual calendar. Chagnon (1974: 163) has suggested that two years minimum is about right where one has to learn a language, with a break at the end of 12–15 months. Clearly, the length of time to be devoted to initial field research varies according to problem and situation.

A pilot study, if feasible, is always helpful. Contacts on the ground are established, practical details of life in the local field worked out, and desirable locations for subsequent full-scale study pinpointed.

During a full-scale study breaks from fieldwork will be found beneficial for preliminary digestion of material, for rest, and for recreation (Fortes 1963: 436; §7.7.4). In practice, this means the fieldworker actually leaves the community for a more restful place. Though not necessarily restful, a university department not too distant from the field makes an excellent choice. This is especially the case if it harbours other anthropologists; for experiences of the wider region, discoveries and fresh ideas can be exchanged zestfully by individuals sharing common ethnographic interests. Malinowski carried out field research among the Trobriand islanders from 1915 to 1918, taking a break from his labours in Australia after 12 months, and returning in September 1917. Another celebrated fieldworker, the American Frank Cushing, lived on and off among the Zuni for five years (Gronewold 1972). Some ethnographers (e.g. Brokensha 1963) have managed to combine field-work with a teaching appointment; while sometimes frustrating, this does ensure periodic breaks to review and distance yourself from the material.

Once back at home base, the anthropologist still continues the task of interpreting the data produced. What is not quite so self-evident is the desirability that hard-earned contacts with informants, administrators and others be maintained. This is where the training of research assistants will yield dividends, if the fieldworker has managed to inspire these with a lust to document their own society. Information collected by the assistant can be sent by mail and the collaboration can endure. If the fieldworker has done a good job, he or she will often wish to keep up ties of friendship with those who were once friends and neighbours, and with any luck they will wish to as well. No matter that they are non-literate: photographs and other non-verbal tokens of esteem can communicate appreciation and convey good faith as effectively as words on a page. On long-term research see §8.6.

David Hicks

7.9 Audio-visual equipment in general ethnographic studies[27]

7.9.1 Introduction

Pictures of non-verbal communication and of the spatial layout of people can enhance analyses of rites and ceremonies. They can serve a useful purpose by attracting more general attention to the work of social anthropologists and

[27] The use of audio-visual techniques in social anthropology is now well-established, has a long history (Rowe 1953), and a substantial literature. On some general aspects of the use of still photography, cinematography and video-tape, see Collier (1970), Hockings (1975) and Wagner (1980); on the use of aerial photographs and related specialized techniques, see Vogt (1974) and Conklin (1968). On sound-recording, see Polunin (1970) and Samarin (1967).

bring closer to readers the life-styles of different peoples.[28] But the intention to enlighten can also misfire: pictures of an alien people can quickly reinforce stereotypes and invoke prejudiced reactions to an account of, say, religious, political, or economic action that might otherwise be understood as similar to situations in a reader's own society. Social anthropologists have a moral obligation not to stress "the colourful lives of exotic peoples" unless they make it absolutely clear that the customs of their own societies, the Hunt Balls, the Royal Weddings, boxing and football matches, nuclear warfare and religious beliefs, are no less bizarre. They must also point out when the environment of the busy smiling people in their pictures is one of poverty and transnational exploitation.

The need for conversations in social anthropological analysis is, perhaps, less controversial. Interpretations and "thick descriptions" of cultures, analyses of meaning, world views, life-histories, and processes of decision-making in domestic and public situations such as council-meetings, may require verbatim accounts of conversation, in which paralinguistic cues can be as important as the words spoken. Thus, even if a researcher has skills in shorthand, these may not be sufficient for collecting the data that will be necessary for the final analyses.

7.9.2 Visual records

By no means are all social anthropologists agreed about the need to collect conversations, let alone pictures, and the alternative view deserves to be put at some length. Because the case for audio-visual equipment can seem uncontestable, there is a danger of going into the field without the critical appraisal that is essential if it is to be used successfully. For example, a photograph can rarely reveal all the relevant ethnographic detail that an anthropologist had in mind when it was taken; and the time spent on taking photographs can be time not spent on writing the notes necessary to make the photographs useful.

Martin Southwold (1982) has recalled that the occasions on which he was shooting cine-film were those when his notes were grossly inadequate. Even

---

[28] Of course, the pictures you use to remind yourself later of events and people can be of any quality. But the ones you select for publication have to meet technical and aesthetic criteria as "good" photographs: those criteria do not seem to include the representation of stereotypes. Anthropologists might decide to publish only those good snaps which can be tied quite precisely into the main text with a tightly written caption and page references. For some references to literature on the ethics of ethnographic photography and filming, see Barnes (1979: 114–115), Gottdiener (1979), Hanna (1977) and RAIN (1982) (§6.5).

though the quality of the film was good, it was of doubtful value for teaching purposes: "the notes that I failed to take would certainly have been of more value". Moreover, although lectures can be better with some illustrations, for most topics "if one were to show as many as fifty slides one would be in danger of distracting students from the more important analytical issues"; and in any one monograph "we actually use only about a dozen photographs" out of many hundreds taken.

There is the further problem that colour slides are best for teaching but black-and-white prints are necessary for publication. Until recently, one really needed two cameras because satisfactory black-and-white prints could not be obtained from colour slides or negatives. The processing of black-and-white prints from colour slides and negatives has improved greatly, and this problem may already be partly resolved by technological developments.

It is inevitable that an anthropologist with a camera will be asked for portraits by friends and informants in the field, as well as being forbidden on some occasions to take any photographs at all. Martin Southwold felt that on balance the gains of being able to give portraits as gifts to informants were out-weighed by the dissatisfaction caused by delays in printing, by prints of the wrong size, and by his inability to give photographs to everyone. In both Buganda and Sri Lanka, "I felt I would have added more to the sum of human happiness if I had never got into the business of trying to give people portraits". One solution to this problem is to have no camera until the last month or two of fieldwork, or at least to keep one's camera out of the way. During the previous months, one could note what needed to be photographed and then take all the photographs in a matter of days. The only difficulty about this is that one-off seasonal activities and rituals might be missed altogether.

In spite of general objections that can be raised and the fact that photography may be forbidden in some communities, in most cases a camera has social as well as ethnographic uses, and the expense and trouble of giving pictures to informants can be amply repaid by the good relations that it creates and the opportunity that it provides for asking questions about people and events after they have taken place. One should not interfere in a ritual, for instance, but if one has some photographs (taken with permission, of course), they can be a basis for depth interviews and then be given to the participant. A sequence of important events may take place rapidly, or something that involves people and/or a range of material culture may occur only once. Without a camera, such basic information could be lost.

Although professional photographers may keep only one out of every ten or twenty photographs, an anthropologist with an automatic camera ought to be able to use a larger percentage. Carefully shot film of the non-verbal aspects of ritual could turn out to be as revealing and analytically significant as pages of written notes. If film is shown back to informants, they may well draw attention to actions that would have escaped the anthropologist's eye at the time but were none the less important for the people concerned, especially events that they might take for granted and be disinclined to discuss.

Polaroid cameras can be used to resolve some of the difficulties that Southwold encountered. They are relatively cheap, and can be an effective field tool. They remove the problems of distribution, delay, and size of plate, because the results are immediately clear to everyone; but they also raise the problem of duplication if the anthropologist wants a photograph too. One way out of that difficulty is to take polaroid and ordinary shots in quick succession. Several anthropologists have worked with two cameras, especially two simple cameras, without allowing them to intrude on note-taking and observation.[29]

7.9.3 Sound records

A small portable tape-recorder may be essential not only for recording speech at councils, myths and stories, etc., but also for recording ordinary dialogue. The analysis of discourse is so much a part of the interpretation of culture and of people's modes of thought, that the absence of texts, no matter how long transcription may take, would seem to be a major omission in anthropological fieldwork. But if sound-recording is to be used at all seriously as an analytical tool then the importance of immediate documentation and transcription cannot be overstressed. Careful logging of each sound record is essential, and should include the necessary information to identify it once archived: date, time, place and context; the names of performers/speakers; plus full notes and any necessary transcription (§7.5.5, §8.12).

Even amongst those social anthropologists who have reservations about the intrusion of audio-visual equipment on traditional methods of note-taking, there seems to be general agreement that a tape-recorder is at least valuable for four kinds of activity.

(i) For recording meetings, councils, law-cases, rituals, etc., where there

---

[29] Another use for a camera is for copying documents where no photocopying facilities are easily available. Important records are cheaply and quickly copied in this way, and more accurately than in handwriting. The disadvantage is that the copies are not immediately available for consultation, but that may be outweighed by the benefits of having a probably more complete archive when you return home.

are likely to be too many people talking too quickly for it to be easy to take sufficiently full notes.

(ii) For recording important interviews and ordinary conversation where one is hearing more than one could write down at the time, and the speaker's discourse is particularly rich in terminology and ways of expression.

(iii) For learning and improving one's knowledge of the language (§7.5.2, §7.5.4).

(iv) For rhetorical or musical "performances".

(i), (ii) and (iv) not only serve as records in their own right but may provide useful stimulus materials for later conversations with informants. In this sense they serve as a technique to enhance written notes. Nevertheless for (i) and (ii), Southwold comments,

> one must be discriminating. One should take the fullest notes one can at the time, as if the tape recorder was not running, and then after the event ask oneself pretty sceptically whether the time it would take to play the tape over again is likely to repay the bonus one may get from it. Far better erase the tape unheard than waste the time required to listen if one's notes were already adequate.

The risk that one takes, of course, is that after leaving the field new problems may emerge for which some erased tapes might have been very useful! Criteria of adequacy cannot always be settled during fieldwork.

Perhaps the question that we should be asking is not, how can we usefully use audio-visual equipment, but could we manage without it? The value of the critical view that Southwold and some other social anthropologists take is that it urges every fieldworker to ask just how necessary cameras and tape-recorders are for a particular project and what kinds of equipment will be adequate for the task.

7.9.4 Basic equipment

Assuming that photography and recording are acceptable to the informants and communities being studied, and taking into account the reservations made in the previous section, the basic audio-visual requirements for general anthropological fieldwork are an automatic, 35 mm camera and, perhaps, a pocket cassette tape-recorder with built-in microphone.[30]

---

[30] Other equipment which is almost basic are a pair of filters and a light-tight black bag. The filters to use are neutral (they do not absorb light) antihaze: they are cheap, useful for protecting the (expensive) lens of the camera. Keep one permanently screwed onto the lens, never take it off;

Cameras such as the *Olympus XA2*, the *Mamiya U*, and the *Yashica Auto-focus*, have automatic exposure control and a built-in flash system. They can produce good photographs with clear definition for close-up and distant subjects, and they cost between £60 and £70 in 1982. Since light readings are taken from the whole frame, pictures of people in strong light or shade can be unclear and care must be taken.[31]

There are several types of pocket tape-recorder which use the standard size of cassettes. These are generally more suitable than smaller microcassette recorders such as the *Sony M9* (price £40–£45 in 1982) because the standard cassette fits a wider range of playback and transcription machines, and it can also be used to play commercial cassettes for the recreation of the fieldworker and the entertainment of friends. It is essential that a pocket recorder should have cue and review facilities, and preferably a tape counter, but it is not necessary to have stereo or external microphone unless one is recording music or wants especially good quality speech. With an external microphone (a *Sony F-230A* costs £11), one can get very good quality on a pocket *Sony TCM 141* (£55–£60), and it is generally a better buy than the larger, older types of cassette recorder such as the *TCM 757* (£40). The extra £15–£20 required for a pocket tape-recorder is well worth while. Excellent quality can be obtained with even the smaller *Sony WM R2* (£110–£120).

All pocket tape-recorders operate with four penlight batteries, but a rechargeable battery pack (e.g. *Sony BP-23*, £15), and a mains charger (e.g. *Sony AC-15A*, £15) are useful if an electricity supply is easily available.

and keep the other to replace it when it gets scratched. The black bag is for when things go wrong inside the camera: the bag has a light-tight opening for putting the camera in, and light-tight hand-holes. If the film transport jams, for example, the camera can be safely opened inside the bag, any film removed and sealed, the camera extracted and fiddled with, then re-loaded with new film. All other methods short of using a dark-room put the exposed film at risk. Black bags are also cheap.

[31] It is worthwhile to know something about the properties of film. For photographing documents, for example, it is sensible to take a stock of micro-film cut into 36 exposure lengths and loaded into 35 mm cannisters: these can be exposed with flash or other artificial lighting, cemented together again when they have been developed. It is essential to record the speed of the film (the ASA or DIN number). Micro-film is very slow and needs a great deal of light, but is ideal for providing a sharp contrast between what is written and what is written on; which, after all, is what is required. Of course, the aim is to keep these technical matters as simple as possible, but social constraints might make it sensible to carry a stock of very fast film, needing very little light to record an image. Some people consider flash intrusive – acceptable for portraits in a dark room, say, but less so when taking a rapid succession of shots in a public (perhaps solemn) gathering at night. If those circumstances are likely to arise, you should consider taking some Kodak Recording Film (monochrome) or UR1000 (colour), capable of making good images by candlelight. Colour film is slower than monochrome; colour transparency slower than colour negative film.

Finally, a pocket dictaphone, such as *Sony TCM 131* (£30), is useful for recording data when notebooks are not suitable.

7.9.5 Additional equipment for studies of ritual

Anthropological fieldwork that focuses on ritual may require rather more sophisticated equipment, because much of it is non-verbal and careful recording of sequences of action may be necessary without interrupting their flow or disturbing their participants.

Ivan Polunin (1970), has proposed a system for recording events that requires a stereo tape-recorder and two microphones. One microphone is placed close to the theatre of action and the other is held by the fieldworker, who remains out of the way and able to give a running commentary, which can later be transcribed and synchronized with the sequence of rituals. If the quality of the tape-recorder is good, recordings of the music can be extracted without the interference of the anthropologist's commentary by the simple device of playing back only one of the two channels. The *Sony TM 600* could be used for this, together with two microphones, or a more expensive recorder such as *Sony TC-D5M*, the *Uher CR240*, or the *Uher 4200/4400 Report Monitor*.

More expensive cameras, such as the *Asahi Pentax MX* (£130–£140), allows for greater precision in recording ritual, and also for the rapid succession of still photographs that can be made with the addition of an automatic winder (£50–£60). The problem of having the right lens at the right time can be obviated by using a telescopic/wide angle lens, but this costs another £120–£150. Since much ritual takes place at night, a further £50 is required for a good, quick-firing flash gun.

Thus, once the fieldworker abandons the simpler cameras listed in §7.9.4, the cost of equipment increases almost seven-fold, to approximately £350, and a strong case has to be made for the analytical, as well as interest, value of such photography.

The analysis of non-verbal aspects of ritual can be achieved more cheaply, and in some respects more usefully, by the use of a *Super 8 cine camera* such as the *Sankyo* or *Bell and Howell*, which cost approximately £140, and for which an additional good quality, directional microphone can be bought for less than £50.

7.9.6 Maintenance and protection

For recording speech (other than for specialized linguistic work), cheap tape is adequate, but care must be taken not to buy cassettes with poor quality spool mechanism. Moreover, C120 cassettes should never be used on inexpensive or pocket recorders, because the thinner tape requires a more complicated transport mechanism. C90 (i.e. 45 minutes to each side) is the best size, and Fuji, Sony, TDK, Scotch and BASF (especially BASF C90 LH SUPER SM) are all good.

Standard batteries should not be used. The "super" varieties, such as Mallory, Duracell alkaline or at least leakproof "HP" Types, are necessary both for variations in climate and for length of survival after purchase.

All equipment should be insured. Under humid and damp climatic conditions it should at all times be kept dry, preferably using purpose-designed containers. Silica gel is a helpful drying agent. Plastic bags should never be used. If possible, film should at all times be kept cold. Spares (such as a duplicate microphone or flash unit), basic tools and electrical tape may also be useful. Equipment should be thoroughly cleaned and checked by qualified technicians before the researcher leaves for the field.

John Blacking

7.10 Coping with physical health [32]

In the long term it does not make sense to be negligent or take short cuts where personal physical health is concerned. Where appropriate, the same must apply to familial dependents, co-workers, assistants and, up to a point, respondents. It is advisable to bear in mind that all health-care is preferably *preventative* and extends beyond simple clinical measures, to shelter, waste-disposal, nutrition, personal hygiene, clothing and water-use. Nevertheless, any fieldworker should expect to suffer some form of illness or injury in the course of 12 months research.

Particular precautions clearly vary between different research locations, but unless local conditions and facilities are good, familiar and accessible, some will have to be taken prior to beginning prolonged fieldwork. Climatic-

---

[32] You may find the following British addresses useful in connection with matters discussed in this section: The Ross Institute of Tropical Hygiene, London School of Hygiene and Tropical Medicine, Keppel Street (Gower Street), London WC1E 7HT (Tel: 01-636-8636 extn. 213); The Hospital for Tropical Diseases, 4 St Pancras Way, London NW1 (Tel: 01-387-4411); St. John Ambulance, First Aid Training Centre, Headquarters, Greater London, 63 York Street, London W1 (Tel: 01-258-3456); British Red Cross Society, 9 Grosvenor Crescent, London SW1X 7EJ (Tel: 01-235-5454); St. Andrew's Ambulance Association, Milton Street, Glasgow G4 0HR (Tel: 041-332-4031).

ally unaccustomed zones will usually require special forethought in terms of supplies (including suitable clothing) and perhaps a period of acclimatization. Ensure that all vaccinations and prophylactics recommended for the area which you intend to visit have been obtained well before departure dates, and ensure also that you have reliable information about local conditions and diseases which are likely to be encountered. This information should be available from your general practitioner, local health centre or from relevant diplomatic missions; as well as from personal contacts. If in doubt, contact the Ross Institute of Tropical Hygiene. The Royal Army Medical College and Foreign and Commonwealth Office also provide health-care information and briefings for most areas. Freilich (1970: 590–594) provides some useful addresses and information for US citizens. It is advisable to have a medical and dental check-up before you leave for the field. You should also have a medical kit suitable for the conditions under which you will be working and a reliable and practical handbook for situations where self-medication is unavoidable. Adam (1966) is useful in this particular respect, and also provides advice on equipment. Macdonald (1980) is up to date, though shorter and less comprehensive. Two manuals for diagnosis and treatment of common problems used in the training of village health workers, Werner (1980) and King *et al.* (1978), are also useful for the field ethnographer and are strongly recommended. For a broader (but equally comprehensible) coverage of the medical problems of expeditions and of fieldwork under different climatic conditions, see Edholm and Bacharach (1965). This volume also includes advice on dental care; but see also CDF (1965). The *Handbook of Army Health* (Army Code No. 61257) is also useful, as is the *Manual of Army Health*. The Ross Institute publish a series of bulletins on such subjects as anti-malarial drugs, small water supplies and schistosomiasis. These may prove useful in particular cases. On returning from fieldwork carried out under doubtful health conditions a further medical examination is desirable.

It has long been the custom for anthropologists to act as "bare-foot" doctors in the communities in which they are working, at least to some extent. If the fieldworker is a fully-qualified doctor or nurse there is no particular problem arising from questionable medical competence. Non-qualified fieldworkers who envisage, choose or are compelled to adopt this role, might consider obtaining first aid qualifications (such as those recognized by the British Red Cross Society) or attending annual introductory courses, such as those offered by the Ross Institute. The Ross Institute course covers malaria and its control, other tropical diseases, malnutrition and the improvement of nutrition, personal hygiene, sanitation and adjustment to climatic environment. The course is taught by leading experts in each subject, and application for admission should be made to the Director, at the address given in footnote 32. Although such courses can be useful, fieldworkers must additionally be fully aware of the dangers connected with the indiscriminate handing-out of pills, of problems of real and apparent favouritism, and generally of giving advice where competence is limited. On the other hand, it may be that medical assistance from a fieldworker will make the difference between life and death, or between a minor ailment and prolonged suffering. On trans-actional, ethical and other aspects of providing medical facilities, as well as the handling of personal relationships, personality difficulties, stress and depression, see §5.3.2.

7.11 Team research

7.11.1 Introduction

It has long been assumed by some that the intimately personal character of much ethnographic fieldwork does not lend itself easily to team research. The fieldworker who regards the discipline of social anthropology as akin to the art of the novelist is likely to find the presence of other researchers distracting and irrelevant. We have it on the authority of Evans-Pritchard (quoted in Beidelman 1974: 557) that "scholars must wander at their own sweet will".

7.11.2 Couples

Research undertaken by pairs of ethnographers has produced some of the classic fieldwork reports (e.g. Bateson and Mead 1942). Such couples have generally been married or cohabiting, where a decisive advantage lies in a prior knowledge of personal compatibility and of respective skills. This has long been established as a most effective means of obtaining complementary field data, such as on male and female activities and knowledge (e.g. Spindler and Spindler 1970). However, it may also prove a tactical advantage in demonstrating adult status and in providing a further resource in the form of children (e.g. Berreman 1970; Boissevain 1970; Gonzalez 1970: 161; Hostetler and Huntington 1970). Research conducted by pairs other than those cohabiting is less common, although this has now become one of the more successful ways of undertaking interdisciplinary research (§7.11.5).

The main warning to be sounded with respect to research conducted by pairs (and this applies to all teams) is that partners may not react equally to the conditions imposed by fieldwork (climate, cultural differences, health factors). This may give rise to stress and strain which may reflect upon the research itself. The fact that partners share a domestic and emotional relationship only makes it more difficult to resolve the problem by one of the team members backing out of the fieldwork situation.

7.11.3 Field assistants (§8.3)

By these we understand local persons, untrained or semi-trained, who are employed by the ethnographer.[33] Such assistants have in the past been employed most obviously as interpreters, translators, guides and domestic servants; but also (especially where they have attained some degree of literacy) as research assistants for the gathering of basic census, genealogical, textual and other systematic data. In some cases paid assistants have effectively become professional informants.

There is one further benefit in employing a field assistant, though one perhaps not directly part of the anthropologist's fieldwork commission. The field researcher, we must never forget, is rarely an invited guest. Usually, in fact, he or she is something of an intrusive imposition from without. Field-workers are not often requested, and are sometimes unwelcome. The people whose way of life they wish to share possess a humanity identical with that of themselves. In a literal sense, since many such individuals, after all, supply them with information they "are informants", but their moral status is diminished if the fieldworker thinks of them as merely "informants", sources of information for books. It is misleading also, if informants are seen as "actors", "role"-fillers, or "status"-occupiers rather than as flesh-and-blood persons. It is not as modules of a "system", or as niches in a "structure", or as components of a "model" that they involve themselves with the fieldworker. Only rarely does the living reality of individuals come across in modern fieldwork monographs, concerned as most are with these convenient scholarly fictions. Now, the assistant is one such individual, and as the assistant teaches the anthropologist, so should the anthropologist expect to teach the assistant. One thinks in this regard of the fertile and wonderfully productive co-operative labours of the Indian informant, George Hunt, and his teacher and student, Franz Boas.

There are benefits to be gained from research assistants, provided they do

---

[33] Professionally-trained graduate or other research assistants have sometimes been employed. This has generally been in the context of larger teams and projects, with clearly-defined roles (e.g. as census enumerators), or as part of a training exercise.

not detract from the personal commitments of fieldworkers themselves, but provide complementary information to that which they obtain. Naturally, problems may arise. These may be personal. If the research assistant proves unpopular in the community, the disapproval generated will certainly frustrate the anthropologist's inquiries and progress will be hampered. Should the assistant and fieldworker fall out, the latter may suffer local hostilities. Again, it is a question of common sense and maturity on the part of the researcher. There may also be ethical problems linked to payments, contracts and other conditions of employment (§6.3.3). For example, should payments reflect what is locally acceptable or what would be a realistic reward for the same service in the ethnographers own society? Finally, there may be considerable practical problems concerning the competence of individual assistants as researchers. The reliability of data assembled in this way must always be questioned given that assistants must, inevitably, only possess a partial understanding of the ethnographer's requirements and the character of the project. On the other hand, the very absence of complete understanding may serve to instil a sense of reality in data production and interpretation. The amount of attention given to the training of field assistants will affect the kinds of results produced, not simply their quality. See also Beattie (1965: 27), Powdermaker (1966), Platt (1976: 83–97, 142–151), and Pelto and Pelto (1978: 219–221).

7.11.4 Larger teams

The use of field teams any larger than two immediately begin to pose considerable logistical and organizational problems, although these may possess advantages in increased specialization of research effort and the greater amount of work which can be accomplished. There is also an important intellectual problem to be faced: that often no one individual is directly or responsibly concerned with understanding the entire culture or society (Fortes 1963: 435). Research involving larger teams is, therefore, not to be undertaken lightly, and it must always be seriously questioned whether it is the best use of research time.

We must, of course, distinguish between a "project team" and (more specifically) a fieldwork team. A large project may be composed of field staff and home-based staff (secretarial, computing, other research assistants and officers, in addition to the principal researcher). Thus, it is possible to have ethnographers working individually in very different and separate fieldwork locations, but co-ordinated centrally into a project or programme. The intellectual cement for such programmes comes from informal discussions and conferences at home base. The Sussex Highland Europe project of the late 1960s is a good example of this, or the project supervised by Sandra Wallman (1982) on ethnic minorities in United Kingdom inner-city areas. On such "multi-community" projects see also Bennett and Thais (1967) and Pelto and Pelto (1978: 217–219).

The major problems of large-scale team research (Luszki 1957; LeClair 1960) tend to be inter-personal relations within the team: the relative status of collaborators, the issue of effective and agreed leadership, and the delegation of responsibility. A large measure of tolerance is essential. The greater visibility of a team and its impact on the fieldwork locality is likely to put a greater strain on relations between team members and the host community, which may have a very real affect on the kind and quality of data to emerge. Administrative problems may also arise, connected with such practical matters as transport, accommodation, food provision and recreation. On the academic side, cooperation in the formulating of research problems, in the handling and co-ordination of research efforts is necessary, but is seldom achieved perfectly. For example, the necessity to agree on research strategies and interpretations may stifle originality, while the academic content and approach of larger projects is more likely to be subject to interference from funding agencies, especially where these are governmental or quasi-governmental.

For a detailed account of the methods used in one team project, and particularly on the use of periodic team discussion see Hubert *et al.* (1968: 35–37). For further general discussion on the use of larger teams see Douglas (1976), the contributions to Foster (1979), Maxwell (1970), Pelto and Pelto (1973: 276–279), Platt (1976: 66–102, 208–215) and Casley and Lury (1981). Agar (1980: 180–181) deals specifically with the staffing of research projects.

7.11.5 Inter-disciplinary research

Social anthropologists working on specialized topics have increasingly co-operated with researchers from other disciplines or become members of multi-disciplinary teams (Caudill and Roberts 1960).

To date there are examples of cooperation between social anthropologists and botanists (Berlin *et al.* 1974), zoologists (Bulmer and Tyler 1968), psychologists (Lancy and Strathern 1981), biological anthropologists and medical specialists (Chagnon 1974), geographers (Brookfield and Brown

1963), to mention only a few. The reports and other publications of inter-disciplinary teams, as with other teams, can provide unrivalled breadth and depth of authoritative data, but often lack theoretical impulse and analytical focus. From the point of view of practicality and intellectual thrust, fieldwork undertaken in consultation with appropriate specialists is perhaps more satisfactory than full-blown collaboration. It is certainly far easier and perhaps (ultimately) more rewarding.

Inter-disciplinary research requires special and careful planning if it is to be successful, at least six months prior to execution and preferably in or near the actual area of research. It requires "regular support by the holding of inter-disciplinary seminars and conferences, in which the individual research workers report on and discuss their work" (SSRC 1968: 76–77). There must be a sufficiently high level of shared knowledge and an obvious overlap and complementarity in research interests. Moreover, individuals must have some understanding of the other discipline, its problems and research needs. Roles must be clearly-defined, and practical steps taken to minimize differences in professional orientations and the tension which may result from this (see also Pelto and Pelto 1978: 224–227).

R. F. Ellen and David Hicks

8 Producing data

8.1 Introduction

From the language of discussions about methods, it is easy (and perhaps ultimately unavoidable) to treat "data" (or "information") as a material, tangible, resource. We speak of "gathering", "collecting", "sorting", "processing" and "recording" data, as if it were out there waiting for us. Of course it is not: "data" is only a convenient summary term for the documented and memorate results of conducting research, either based on our own first-hand experiences or based on those of others set down in texts. But the situation is more complex than this since we routinely apply the term "data" to representations and constructions at different stages in the complex process of transforming observations of actions and utterances into third- or fourth-hand scripts. This process is indicated rather crudely in Fig. 8.1. As we move through the flow diagram so "data" are by turns simplified, classified and warped in various ways. But the diagram itself is misleading. Particular phases (such as PRIMARY ANALYSIS) themselves involve complex transformations; and DATA may go through a good many more stages than has been indicated here, as it is endlessly interpreted, supplemented and re-interpreted. At the same time, it is no good deceiving ourselves that what we start with are pristine or "raw" mental and physical facts – sense data – which exist in the real world and which are sadly distorted by the inadequate technical means of processing. This is partly true: writing a text is a monstrous simplification of an orally delivered speech. By taping it we may preserve the oral qualities of delivery, but it still remains inadequate when it comes to actions and context. Video-tape improves the situation, but is still highly selective in the information it is able to record about an event. Up to a point we can improve the technical means at our disposal for the assembling of

ETHNOGRAPHIC RESEARCH
ISBN 0 12 237180 1

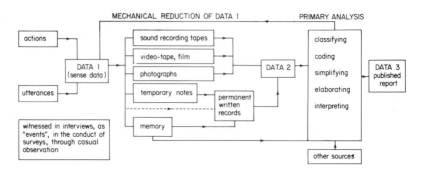

Fig. 8.1 *The transformation of "data".*

data. However, we can never escape from the philosophical dilemma as to what it is we really "observe", how categories affect what we observe and what is "fact" and what "interpretation". Indeed, it is far from clear how we even begin to identify and evaluate the status of an "utterance", an "action" or an "observation" (Chapter 2). In the end, our data are the product of limited and marginal interactions with the people we study and their artifacts. However they are subsequently registered and processed, these data are bound to reflect the uncertainties and ambiguities of the lines of perception and communication through which they were initally generated (Crick 1982). Following Galaty (1981: 90), we might usefully distinguish:

> between data which is anthropologically processed information (notes, census material, photographs etc.), the sources of data which are various modes of observation (visual and auditory channels, recording of discourse, formal and informal interviews, narrative records etc.), and the theoretically constructed objects of analysis and observation (which may be conceptualized as individual events, human motivations, cultural models, folk models, biological types etc.).

In the end, however, we are bound to adopt a woolly epistemological compromise which rests on the convenient fiction that while our perceptions and representations are biased, selective and socially-constructed, the bias is sufficiently similar among professional social anthropologists such that this can be treated as a serviceable basis for analysis and comparison (Ellen 1982), even though the degree to which this furnishes such a serviceable basis depends upon the kinds of phenomena we are dealing with.

The present chapter is concerned with means of producing data in the course of fieldwork in the most efficient, useful and accurate way possible. During the conduct of fieldwork, apart from material varying according to substantive fields, it also varies – as we have just noted - according to its epistemological status. The greater part of ethnographic fieldwork is usually reckoned to comprise of observations of actual actions and utterances which accompany them. The prominence of this approach often amounts to a professional declaration of faith (SSRC 1968: 84–85). Much of this may occur simply through watching and listening to (sometimes little more than eavesdropping) people interacting with each other. The fieldworker, therefore, is just hanging around, although this is not to say that all of this does not require skills which have to be cultivated (see especially Phillips 1971; Schatzman and Strauss 1973: 52–93; also Hilger 1954), or that it has no influence on the situation being described. Some have suggested that, at this level, something can be learned from the systematic and detached character of ethological approaches to observation (Pelto and Pelto 1978: 103), or from the use of other aids to rigorous and detailed description, such as activity charts (Whiting 1970).

Most observation, however, goes hand-in-hand with elicitation. Observations seldom stand independently and are often both the precursors to questions and the means to follow them up. You can prepare yourself for an interview by careful pre-observation. Also, compare informants' statements in an interview with actual practice after the event. So, in addition to simply digesting information which we have played little or no part in stimulating physically, we actually solicit statements about what people generally do (second-hand generalizations), what the respondents themselves may do, what other named individuals do (indirect access and extrapolation of current practice), what people think they should do (contemporary ideals), what people have done in the past (life-histories, oral histories), and the meanings which people employ to order, interpret and negotiate their way through life. Each of these kinds of data are subject to distinctive and often very considerable difficulties of investigation. As has been noted, even watching and listening to unknowing actors is fraught with problems; as, indeed, is the inference of behaviour through unobtrusive measures.[1] And these techniques are among the most problem free. In all of the others we are usually obtaining information in the highly artificial context of the interview, whether formal or informal, and often asking questions which make little cultural sense, are rare, or in other ways inappropriate. On occasions, we may even purposively (and quite legitimately) go one step further and simulate events which might not otherwise be witnessed, or conduct "experiments" (§8.10).

[1] While anthropologists have used unobtrusive measures from time to time (as in the evaluation of physical traces), and while archaeologists use them all of the time, there is little discussion of them as a distinct set of methods. More generally, see Webb et al. (1966).

Because of the general character of this volume, there is very little in this chapter on the use of photography, sound recording equipment, taped transcripts, video-tape and film (§7.9). We provide only an outline and a guide to further literature. It is hoped that some of the future volumes in the series will be devoted to particular technical aids, or to subjects (such as ethnomusicology and oral literature) which make specialist use of them. Additionally, there are further techniques not covered in this volume at all for much the same reason. Thus, we say nothing of the special methods necessary to analyse verbal materials, or for eliciting certain kinds of semantic and cognitive data (such as card-sorting tests and attitude measurement). It is planned that these should go into a separate volume. Similarly, details on techniques for mapping will be covered in various ways in projected volumes on land tenure, economics and subsistence.

It is perhaps worth remembering that in the end the sources of our data are seamless and fragile webs of information. In this fabric, fine distinctions between different substantive fields become somewhat irrelevant. Labels such as "kinship behaviour" are invented to give meaning to what we observe and produce, and to simplify it for the purposes of handling. Moreover, it is now common knowledge, through a variety of carefully-controlled experiments, that the human mind and its faculties provides only the most imperfect of instruments for observation and recording. In discussing the technical minutiae of research practices, it is a sobering experience to reflect upon this.

R. F. Ellen

8.2 Participant observation

8.2.1 On the meaning of the term

Participant Observation is an oxymoron, a form of paradox which generates meanings as well as permitting different – indeed contradictory – interpretations. In sociology, it first proclaimed the possibility of a species member observing its own species, so that people-watching could be as detached, unobtrusive and scientific as bird-watching. In this positivist extension of the naturalist's paradigm, the aim is to interfere as little as possible, to be a "fly on the wall".[2] Interesting variants of this approach were attempted by Mass

[2] Halfpenny (1979) sets out the combinations of practice and presupposition possible in "the handling of qualitative data" in a special issue of *Sociological Review* on this topic. Collins (1979) has attacked the theoretical justification for "flies on the wall".

Observation in the 1930s and 1940s.[3] Other sociologists have used the term to argue the theoretically opposite view that observers have to be total participators in order to gain shared meanings (see Schutz 1972). Both perspectives can posit an observer who is indistinguishable from the observed, while the interpretivist tradition of sociology also encompasses the neo-naturalism of ethnomethodologists.

Social anthropologists are often surprised to hear of these meanings since for them "participant observation" has become almost a definition of their work, a label which somehow validates anthropological practice. In fact there is no substantive theoretical treatment of the term in British social anthropology[4] and it has become generally used only in the last twenty years.[5] In default of a literature to review, this section explores some of the theoretically significant aspects of anthropological practice and focuses first on the term "participant observation" because it must have been adopted to satisfy felt needs. The change from the sociological meanings may have occurred because anthropologists are frequently unable to blend into the wallpaper. The phrase also evokes some of the contradictions of fieldwork practices and, by linking them together, affirms there is an anthropological mystery. The conjunction of terms does not solve this mystery, and to talk of "the standard methods of participant observation" is even more contradictory and innacurate (§1.1).

[3] T. Harrisson's simplistic views (he was himself a birdwatcher) are not representative of the Mass Observation project, which used many techniques; Stanley's interesting account (1981) is also methodologically stimulating.

[4] Perhaps the first published use of "participant observation" to describe anthropological practice is by Frankenberg (1963, reprinted in Burgess 1982), specifically reflecting on the greater "group consciousness" of anthropologists compared with sociologists. The phrase is used almost as a synonym for the experiential, and this in turn is said to be the experience which transforms the social outlook of the fieldworker.

[5] Comment rather than theory has emerged so far in the large, scattered literature for which Burgess (1982) is a useful bibliographical guide, besides offering an excellent anthology. Through giving some emphasis to sociologists' and urban centred accounts, Burgess offers examples of the wide range of fieldwork practices. Here too, as in anthropologists' accounts, theory is implicit in the (very often valuable) commentary on findings. Apparently theoretical discussions of participant observation are usually on methods, methodology and/or morality; if by sociologists they normally presuppose a positivist approach or try to justify a "humanistic" mode within it. This is generally true of textbooks, including the anthropological ones (e.g. Pelto and Pelto 1978). Apart from other citations in this section, see also Devons and Gluckman (Gluckman 1964) and Jongmans and Gutkind (1967). Koepping (e.g. 1976) has written from a phenomenological approach, while Lowry (1981) valuably inaugurated the theorizing of "observation".

Many of the points raised at the 1979 Social Science Research Council Workshop on Participant Observation are incorporated into this section, which also draws directly on the contribution by Tonkin, as well as those cited. Firth (1972) is an illuminating and insightful personal account which corroborates many of the arguments voiced at the workshop.

If using the label "participant observation" permits the holding of apparently incompatible theoretical positions within one discipline, it also expresses for anthropologists generally a reliance on *experience* and a need to affirm that this is a valid source of scholarly understanding. But so far there has not been much effort to separate the proper study of empirical understanding from empiricist presuppositions and to advance from the innocent pragmatism, which Leach discerned in Malinowski, that "culture consists in what the fieldworker himself observes" (and makes intelligible) (Leach 1957: 120).

8.2.2 Observation through participation

Ordinary life with its ongoing, changing social relations and practical demands is varyingly "observed" and "participated" from moment to moment. Any sustained fieldwork, however conducted, will be the same, and much of it will consist simply of watching and listening. These activities should never be regarded as residual or unproblematic. Different kinds of research will obviously demand different levels and kinds of involvement with human beings,[6] but even the archival researcher has to gain access to documents and may well trade on shared understandings to do so. What the anthropologist writes is an outcome, successively worked over, recombined and distanced from disparate sources. Experiences of different kind and quality are cumulatively synthesized.

Distinctions between account and action often dissolve imperceptibly as the anthropologist, like any other human, digests, infers, abstracts and has in turn to act or react, even while trying to record. Like one's "subjects", too, one is simultaneously trying to make sense and to make an impression. Fieldwork in any unfamiliar circumstances – they need not be geographically remote – challenges traumatically one's confidence in these abilities. The

---

[6] It has become fairly standard in the sociological literature (although in some respects this is profoundly misleading) to classify kinds of role involvement according to their approximation to three foci on a continuum: "complete" participation, participation-as-observer and "complete" observer (Gold 1959; Junker 1960: 35–38; Spradley 1980: 58–62). The practical problems of participation, especially where investigators attempt to pose as ordinary members of a community, have been extensively discussed (e.g. Festinger *et al.* 1956: 237–252). On occasions, attempts to "participate" have gone to the absurd lengths of involving minor surgery and other modifications of physical appearance (e.g. Griffin 1960). Deception of this kind raises profound ethical issues (§6.4.4). In situations where it is not possible (even if it were desirable) the *practical* problems of finding an acceptable participatory role may be considerable (§5.3.4). Others have argued that participation of any kind is best minimized, for moral, practical and intellectual reasons.

additional effort continually to gloss one's experience and turn it into notes,[7] the time, effort and self-conscious unself-consciousness necessary, all increase the stress, while participant observation often becomes a paradox in the sense that the more you observe, the further you are from participation; while the time you must take to keep up your notes threatens to preclude either activity.

8.2.3 Acceptability (§5.3, §5.4, §5.5)

This is a personal as well as a professional need, for without a measure of support from others, as Evans-Pritchard pointed out, "disorientated craziness' results (1973: 4). The investigator is not just a social being but one bereft of most familiar props. People impose participation on you in their terms, you cannot help but try to achieve it in yours. Again, while short-term researchers can, effectively, "hit and run", especially if they get assistants to carry out their work for them, the anthropologist has to keep on maintaining good relations, to be tactful, socially sensitive – and one result is to feel never off-duty, overstretched and over-suppressed. It is then difficult not to believe that one's first-hand knowledge, so hardly gained, is important. The quality of one's knowledge is also an aspect of one's personality, and not only in academic circles. (He is stupid, she is old-fashioned, they are open to new ideas . . .)

Fieldwork is the experiential teaching of what "social" means. Because it is exceptional to undergo such intensely effortful, raised-to-consciousness experiences, it has a professional mystique and effects the solidarity attributed to a rite of passage – which it is often said to be (§1.1). It is also sometimes compared to psycho-analysis, which has been recommended by some as a pre-condition for doing it (§5.3.1). Rosemary Firth has argued that such experiential understanding of otherness is not only a precondition of grasping much social analysis but should be a part of formal education (1971).

---

[7] Just as participant observation is wrongly used as a unitary label, so are "notes" (§8.12). Critics of anthropological reluctance to allow access to notes are treating these as a uniform "data base", whereas they may vary from hurried details of observation to preliminary general conjectures. Confidentiality is also a major problem, especially for those who may have to deposit notes as a condition of their research project (§6.3.4). Since the activity of informed insight is a selecting, sorting and generalizing one, the notes will not all be "raw data", but we must avoid the sort of professional territoriality which leads to suspicions that the results are cooked.

8.2.4 Insight and interpretation

In trying to define participation, A. Strathern has argued that "it is only by entering a realm of meaning that we can make it properly meaningful for ourselves" (1979b; cf. also Willis 1976). Such receptive recognition develops unevenly over time, and it cannot be acquired through standardized enquiries alone because there are always things that people do not say publicly, or do not even know how to say. They live them as their common experience. "One's way into this experience", says a commentator who can speak for many,

> is through noticing the villagers not only as they speak but as they behave at
> different times . . . these details may be slight but they add up slowly over time,
> and point the way to other elements . . . which otherwise may go unobserved (du
> Boulay 1979: 25).

Observation, to be of value, must lead to insight, the noticing of apparently insignificant points, the making of connections, the discovery of what Henry James (1962) called "the figure in the carpet". The knowledges involved in this activity may vary from everyday to esoteric, but either way they come from individual encounters and include a mass of informal background which cannot be acquired without personal involvement, chance, and all the characteristics which are rejected within the positivist tradition.

Insight, like other "creative jumps", cannot be comfortably accounted for in that tradition even though it is central to scientific discovery (cf. Koestler 1971; Schon 1969). It is not fieldwork-specific, indeed there seems to be no intrinsic difference between inferences linking one observed activity with another or with a reference in a primary document or with the recollection of a point of theory. Moreover it can occur at any time in a process stretching through fieldwork and for years after it: "the decisive battle is not fought in the field but in the study afterwards" (Evans-Pritchard 1973: 3). And once there, one moves just as when doing the "practical" research, through realization interrupted by error, with retrospective illumination and the sense that there was surely something else very important if only you could remember what it was or where you wrote it down!

Anthropologists have rightly been criticized – and have criticized one another – for allowing time to turn their once vivid understandings into untrustworthy memories. But in so far as human understanding is constituted through experience, glossing is recurrent. Redefinition is also part of research (repeated *search*), a process not finite in any discipline but contingent upon social constraints, like requirements to publish. Lack of completeness is likewise characteristic, so there must always be missed cues, and key points not understood. Whether back from the field or from the library, it is only when one tries to synthesize one's material and create an argument (i.e. do anthropology) that one finds the gaps in the enquiry as it is coming to be constituted, much less the gaps that would exist in another sort of enquiry, with answers sought to the questions one has not yet thought to ask.

8.2.5 Experience as method

Participant observation is not, and cannot be, a method. Rather, we should look more closely at the characteristics that may exist behind the label.

This becomes even clearer if one considers the numerous research procedures involved but the label may evoke for its users a sense that anthropology is characterized methodologically by the research worker's being the *medium* as well as the recorder and interpreter of his/her research. This means – for instance – that *access* becomes an end in itself – which has to be interpreted – as well as a means which enables further interpretation. Since trying to become a member is everywhere an activity as soon as one moves from an ascribed group, much fieldwork activity is directed to it, so as to find out what the rules of access are, as well as to gain access to information. Acceptability (§8.2.3) is not the same as identity, and the participant fieldworker is not primarily aiming for empathy but to use himself or herself as the medium of research. One uses experience to further observation, by experimenting on oneself. Bereft of a tap, the fieldworker grasps the time and labour needed to fetch water, or – more significantly – learns the number of skills needed by gypsy women by going Calling with them (Okely 1983).

Learning by making mistakes is as valuable as it is painful – and we should not be trapped by scientism into concealing our methods (cf. Okely 1975). But the level of one's achievement is not germane to the investigation. The competence of introspection, the use made of achievement is what matters. Our inability can also be disarming (others are happy to have a chance to give expertise and to be superior) while, for practical people, respect is *enacted* when one shows by poor imitation that one realizes what skills their lives demand. A less comfortable advantage of imperfect identification is that it can make you feel safer to your subjects, who are also very capable of using you, while you are pretending to yourself that you are not really using them.

If the anthropologist is in one sense the method, it follows that the age, sex and indeed the personality of the investigator will in some degree direct the findings (§5.3, §5.4). "Long-term" fieldworkers are especially aware of how growing into a locally understood role both limits and enlarges one's opportunities (cf. Foster *et al.* (eds) 1979) (§8.6). But there are ways of countering the ill effects, e.g. by quickly establishing one's right to go elsewhere and to do different kinds of research, as on short visits for surveys and specific enquiries, and by expanding one's personal network. Since fieldworkers tend to rely on a small bunch of key informants and one or two research assistants, one must escape from their prejudices by deliberately cultivating others'.

8.2.6 Many methods, many roles

Anthropologists get accused of limiting their understanding methodologically. It is interesting to consider the limitations of other methods. Anthropologists usually employ several methods and techniques, not just one. Maps, diagrams, photographs, surveys, reportage, all form first-hand documents which will have a historical importance besides affording objective evidence or systematic analysis. Language work, like "acceptability", can be an end as well as a means (§7.4). Yet again, one can have a diversity of social relationships ranging from "total participant" to "schedule filler" by way of "just hanging round", "buying food", "going to football matches", "initiating group discussion" or "conducting formal in-depth interviews" and so on. Findings about social organization and action, symbolism and praxis, can be inferred from elements as diverse as choice of clothes, order of seating, formal genealogies, market prices, recipes, law cases, others' surprise at one's remarks or comments on one's figure – the list is endless.

The anthropologist attempts to acquire a diversity of knowledge of different kinds, in order to deduce and extrapolate explanations of a reality which has the same multifarious character. Other kinds of research tend to be more limited in the kinds of reality they investigate. By using a diversity of research procedures one can increase one's knowledge, but the risks are equally greater. It is easy to be amateurish, or to proffer as evidence what is really an inappropriate or ill-founded generalization.

The redundancy of social life is our salvation, because with even patchy understanding of one domain it is often possible to add corroborative evidence from another, or use it as a clue to the meaning of a third. This is anthropology's strength, and it enables, as it demands, serious and scrupulous cross-checking. But such validations are not always easy to explain, because they can be experienced as part of fieldwork – that is, as part of a social activity in real time. The western academic tradition with its reverence for literacy, scientific experiment and intellectual property (citations) has literally little space for the evidence on which our accounts, so often not "true or false" but "better or worse", must rest.

Elizabeth Tonkin

8.3 Informants

8.3.1 Introduction

Social anthropology is a discipline uniquely suited by its methodology to the discovery and description of culture. That, indeed, is its proper task. Its many styles and theoretical postures vary according to the ways in which "culture" is defined and associated with all of the other elements and processes which are regarded as constituting society. Throughout much of the first half of the twentieth century, most anthropologists who addressed the question of the meaning of culture tended to treat it as a keystone, directly influencing the configuration of all social institutions, through which it was then reproduced. Culture was thus regarded as the mechanism by which members of society were conditioned to play their roles properly: hence the widespread use of the term "actor" to refer to individuals. Following the assumptions of old-style anthropology, the ethnographer could retrieve the monolithic culture of the society studied through the form of its major institutions and from the interpretations of its authoritative spokesmen. The chief could tell us about politics and war and hunting; the priest or shaman or witchdoctor about religious, magical and mystical affairs. Therefore, the ethnographer's principal tasks were (a) to discover the authoritative voices, and (b) to record cases to validate, or invalidate, the accounts offered by these "official interpreters". Modern social anthropology cannot proceed in this way, for our revised and more sophisticated conceptualization of culture also leads us to the view that there can be no definitive rendering of culture: there are versions of it, and these versions will vary according to who offers them. Moreover, they cannot be regarded as having a truth value, as being "correct" or "false". Rather, they are "indexical": they say something about the relationship of the speaker to his society and its members. The interview, then, is not a simple recording of objective fact: it is an exploration of meaning (§8.4).

"Exploration" may be a fruitful metaphor for, just as the pioneer traveller follows the course of the river in order to map it, so too the anthropologist is led along by the nature of his social field. A map would be worse than useless if, instead of representing accurately the topography of a landscape, it recorded arbitrary figments of the cartographer's imagination. Similarly, the anthropologist seeks to draw the contours of the culture studied without imposing fictional features upon them. Interviews and observations are part of a trek along the river. But, just as the requirements of cartography and the principles of geographical science discipline the explorer's record, so the theory and methodology of social anthropology must place certain requirements on our ethnography. The "raw" data of our investigations – our informants' words and behaviour – have to be mentally digested, re-assembled, scrutinized and *interpreted* – so that they make sense (anthropological, as well as "common") to us.

8.3.2 Types of informant

Fieldwork may, to a large degree, consist of watching, hanging-around and listening, but its essence lies in *interaction*. We solicit information directly from particular individuals, prompt them to speak generally about special subjects of interest, pose specific questions, and benefit from them through more diffuse and subtle exchanges. In all cases, these are our *informants*, respondents and subjects of our research.

While almost every person in a population being studied represents a potential informant, the information each has to offer will differ considerably. Thus, we might identify "key" informants on the one hand, and casual informants on the other. Using a different axis we might also distinguish informants in terms of those who provide us with specialist information on particular subjects, as against those who have a more general role. The categories, of course, are not mutually exclusive, and may conceal important differences. Thus, among key-informants, we should separate those who become so because of their formal statuses from those who are significant by virtue of their informal positions. These latter may arise from their being foci in networks of gossip or from the social relations and relations of friendship we share with them. These do not always coincide.

Most ethnographic research characteristically relies heavily on key-informants, and at one extreme this may verge on biography and auto-biography (e.g. Radin 1926, 1927a; Wadel 1973). Most of us rely on a large number of casual informants, but focus on a relatively small number of key-informants (Edgerton and Langness 1974: 33–37; Tremblay 1957), upon whom we often become very dependent. Moreover, the relationships we enter into may be very close (Casagrande 1960; Lewis 1973: 14; §8.3.4).

Ethnographers may acquire key-informants in a fairly haphazard way, through a combination of their structural significance, knowledge, social visibility and the ease of making their acquaintance. There are, however, good grounds for taking some care in selecting informants, systematically seeking out key individuals in relation to envisaged research, while reflecting all the time on the implications of relying on particular persons. On selecting and using informants, see Hilger (1954); on informants in relation to linguistic and quantitative fieldwork see, respectively, Hale (1965) and Campbell (1955); and, on the special problems of handling the "well-informed" informant, see Back (1956). The use of interpreters and translators is discussed in §7.5.

8.3.3 Talking to informants

The subject of interviewing is treated at some length in §8.4. However, the place of informants in ethnographic research cannot be separated from a discussion of the character of the conversations we have with them. To begin with we must dispel the suggestion of any similarity between the ethnographic interview and the interview conducted by the journalist and the opinion sampler. The latter two are imposing arbitrary features on the respondent's mind: the respondent is required to provide an answer to a question that is none of his making. The anthropologist, by contrast, is using conversations with informants largely to discover the appropriate questions to ask. We have to navigate the river in order to discover its interesting features. Were we simply to pursue a schedule of our own devising we should then merely be displaying the contrivances of our *own* minds, rather than discovering the minds of those we want to study (see Glaser and Strauss 1967). Our problem is not only to search for the correct idiom in which to conduct the study, but to search also for the appropriate *subject* for study. Frequently, this means jettisoning the projects we had thought prior to fieldwork we would undertake. We try to "tune in" to local discourse in order to discover its germane issues and, thus, to render ourselves competent to ask questions which will be meaningful in our informants' terms. This process is a haphazard affair and is usually accomplished by making and learning from mistakes – saying the "wrong" things; being unintelligible; realizing that one has unwittingly given offence. There are many instructive accounts of this painful learning experience: guaranteed to make the most unsqueamish wince are Beals (1970), Whyte (1955) and Briggs (1970). These mistakes later become occasions for hilarity when the anthropologist regales students with an early naivite. But they are much more important than that: they are indications of the way in which ethnographers use their own consciousness to refract the character of the culture studied. It is, essentially, a subjective and introspective learning. This being so, interviews are only partly with informants; for we are, simultaneously, talking silently to ourselves. A fine and explicit case of such soliloquy is Gearing (1970).

The proper ethnographic interview is a conversation in which ethnographers risk the appearance of naivite and ignorance in order continually to satisfy themselves that they have understood what is being said; and risk wandering up blind alleys in order to confirm the validity of the ways in which they are beginning to make sense of their data. This painstaking business is sometimes glossed in methodology textbooks as "the open-ended interview", and is contrasted with the more or less structured questionnaires used in sample surveys. There are, of course, occasions on which the ethnographer has to put "matters of fact" to informants: to check a geneaology or some other item of detail, to record a myth or the interpretation of a ritual, and so forth. But, if the ethnographer is to exploit the opportunity for close social relationships with informants which is afforded by a long and intimate immersion in the field, these exchanges should not have the formality of the "interview": they should, rather, be conversations between friends, or didactic transactions between cultural teacher and ignorant but eager pupil (Rabinow 1977). Frequently, the imperative requirement to test one's success at making sense requires that one plays up the naive role by putting questions to which you think you can already anticipate the likely response. A classic illustration of such role play (although one informed also by a genuine scepticism) is to be found in Evans-Pritchard's account of the confrontation between Azande mysticism and European "rationality" (Evans-Pritchard 1937: esp. pp. 314–315).

The conversations we have with our informants are instruments, first for stripping away the ballast of expectation and assumption which we take with us from our own cultures into our fieldwork; and secondly, for consolidating the understanding which we progressively acquire through greater acquaintance with the field. Ironically, we often seem to use the stated principle as expressed in interview to confirm our impression that there is a significant gap or contradiction between principle and practice (see, e.g. Pitt-Rivers 1971: xviii; Fox 1982: *passim*).

8.3.4 Involvement and detachment

Culture, in our present understanding, is a field of meaning; in Geertz's (1975) phrase, "a web of significance". The discovery of meaning lies, as Wittgenstein so persuasively taught, not in the lexicon but in *use*. In our search for meaning, then, we are not so much concerned with matters of fact or with some objective representation of reality, but with the more elusive topics of the perception, cognition and expression of reality. The interview is a mechanism through which we learn to talk, and thus to think and understand, within the idioms of the people we study.

It is partly for this reason that anthropologists attempt to subject themselves to the indigenous regime of conventions and norms of behaviour. One of the most persistent problems we confront is how to so subject ourselves and yet maintain the degree of "detachment" necessary for us to analyse our observations: in other words, to be anthropologists as well as participants.

By "detachment" I do not mean that we have to distance ourselves emotionally; that, I think, would be perverse. But we do have to attempt to maintain some *intellectual* detachment. In extreme cases, some anthropologists have found it temporarily impossible to do so, for the circumstances of their studies were, by their very nature, outside their control. The works of Castaneda (1970) and of Favret-Saada (1980) provide particularly dramatic examples. It has to be recognized that ethnographers, if doing their work properly, are never merely observers: they are also an integral part of the field of study (cf. Bruyn 1970). They are allocated identities, have a social character imputed to them and, often, become important figures in the locality (Cohen 1978: 55ff; Pelto 1970). This aspect of research is celebrated by Hiatt (1981), bemoaned by Malinowski (1976) and revealed in Frankenberg (1957) and innumerable other cases in modern anthropology.

Fieldworkers in the colonial era often had quasi-governmental or official status. Even those who were not thus distanced from their informants would have distanced themselves, with their household retinues of interpreter, cook and so forth. However sympathetic, however skilful they may have been, the resulting distance must have been extraordinarily difficult to bridge, both for themselves and for their native informants, especially since they were associated by colour, education and known contact with the colonial power (see Asad 1973; Goddard 1972). Although they may now lack the accoutrements of superior status, the contemporary fieldworker is still inevitably distanced from those studied – by virture of colour or class or education or merely by the fact that they are doing the study and their informants are being studied.

But now they recognize their informants' deference to them as an obstacle they must surmount: they must sell themselves to their informants as individuals, as friends, rather than as the "Gentlemen Professors from England". This struggle for intimacy is not just a matter of rapport (Wax 1971): it is also a struggle for a different kind of information. Anthropology's task is not just to map the structure and process of social organization; to collect myths; to discover the morphology of religious systems. It is to achieve, what the phenomenologists call "intersubjectivity": to be able to think, emote and cognize with one's informants and, thereby, to come close to their perceptions and understandings of their social realities.

In this respect, anthropologists may differ from other investigative scientists in that they have only a limited interest in managing sources of information: much of their knowledge will come from being managed by their informants – or, at least, appearing to them as being managed by them (see Buechler 1969; Sansom 1980; Henriksen 1973). The contemporary ethnographer now increasingly experiences the requirement to reveal competence as a member of the society studied, or to suffer the social consequences. This is particularly true of anthropologists working within western societies. Where once anthropologists "studied down", they have now learned also to "study up" (Nader 1972).

8.3.5 Conclusion

The conduct of ethnography, of anthropological fieldwork, is an intensely personal exercise in which the ethnographer pits personality against the odds of incomprehension, rejection, sometimes extreme physical conditions, sometimes outright hostility. If successful, winnings are not to be thought of merely in the clinical terms of "informants" and "data": they are friends, often the closest and most loyal friends we may ever have. In the course of fieldwork, anthropologists learn a great deal about themselves, however much they may manage to learn about other people. Ultimately, the ethnographer's success does not depend upon intellectual mastery, but upon the competence with which s/he can interact socially with the members of the field studied, and on the help provided by informants. The latter is of crucial importance, for the anthropologist *is* a nuisance. We need help, not only to negotiate the tortuous social paths of the field, but often also for our very physical well-being. We intrude upon people and require them to bear the burdens of our presence. Our intellectual task is to represent them fairly. Our moral responsibility is to approach them with humility and integrity. Our use of interviews and informants is not, therefore, merely a matter of procedural and methodological principles. It should also be informed by these essentially humane values, for it is in their proper application that there lies the special competence of anthropology to discover and describe other cultures.

A. P. Cohen

8.4 Informal interviewing

8.4.1 Introduction

The basic tool any fieldworker uses to discover what is going on and the meaning it has for those studied, is speech. And like any conversation back home with the family or in the local "pub", little may be revealed and much concealed. Conversation is by no means a neutral activity; as we all know from our own cultures, "It all depends on who you talk to", "We don't talk to *them*", "One doesn't speak of such things", "Well, between these four walls . . ." and so on and so forth. It must also be noted that an ability to talk and listen to others in a variety of circumstances does not normally play any part in that complex selection process by which the would-be anthropologist becomes trained, obtains funds, and finally ends up doing fieldwork. An important key to good research, then, is development not just of conversational skills but in the ability to judge and manipulate circumstances to maximize both the amount and, above all, the quality of information so yielded.[8]

8.4.2 Learning the rules

The newly-arrived fieldworker, labouring hesitantly and painfully in a very alien and difficult environment, is not likely to be over-concerned with such thoughts, and this is understandable. Yet some sensitivity must be retained right from the start lest a whole series of rebuffs and blocks on avenues of

---

[8] On interviewing in general, see Kahn and Cannell (1967); on informal interviewing in particular, see Nadel (1939b), Paul (1953), Hilger (1954), Merton (1956), Madge (1957), Whyte (1960), Denzin (1970), Agar (1980) and Burgess (1982). On structured interviews (but with good advice for ethnographic conversations), see Gordon (1969) and §8.8.5.2 and §8.8.5.3. On some of the special problems of interviewing children, see Rich (1968).

investigation come into play. These may well be exercised by informants in response to premature and tactless inquiries but the researcher is also easily trapped by convention and convenience. Habits of conversation are rapidly learned and a growing awareness of the fieldwork situation may make the flouting of social conventions a psychological barrier for the investigator long before the informant is faced with the impropriety or discomfort of such queries. Similarly, there is always a temptation to go back to those with whom discussions can be relied upon to be rewarding, even fun. Such pressures are considerable given the limited time usually available for completion of field investigations.

Above all, then, remember that the interview is a social process, and that conversations are governed by a variety of cultural conventions and expectations which have to be learned. The language, codes and registers in which the interview is to be conducted have to be carefully decided upon (Fabian 1979); rules of conversation have to be discovered and adopted, including the use of cues (Junker 1960: 106–116). All this requires the development of general insights and social perceptiveness. But in addition to learning the appropriate strategies to adopt in the interview it is important to adopt the same care when it comes to details of context, arranging times and generally setting-up interviews, even the most casual. You do not, as it were, interview the mayor as he stumbles out of a brothel.

8.4.3 Types of interview

At one level it is perfectly feasible to construct a technical guide to informal methods of interviewing,[9] but to do so is in many ways so restrictive as to miss the point of much of this kind of investigation and disguise the underlying issues in its employment and relative merits of the data yielded. The whole point of not fixing an interview structure with pre-determined questions is that it permits freedom to introduce materials and questions previously unanticipated (Whyte 1960: 352). Moreover, on any reflection, it is apparent that people mean so many things by the "informal interview" that in fact we are dealing with a wide range of techniques, with a residual category of those

[9] For example, see Spradley's *A taxonomy of ethnographic questions*, which distinguishes "Descriptive", "Structural" and "Contrast" questions, specifying some 17 major types, some of which are, in turn, differentiated into up to four subvariants (1979: 223). Such a formal approach with its emphasis on "frame elicitation" may indeed be useful in certain kinds of linguistic fieldwork (e.g. Harris and Voegelin 1953), and as part of the corpus of techniques to elicit cognitive and semantic data. It is, therefore, perhaps not unexpected that these approaches are associated with those who propose that ethnographic description should resemble the description of language (see, e.g. Frake 1964; Burling 1969).

things left over from the "formally structured questionnaire" with its precise delineation of questions and procedures by which replies are coded (§8.8.5.2).

Types of interview vary along continua according to a number of criteria. One is the degree of pre-determination in the questions asked: thus we move from formal questionnaires through standardized agendas and checklists to questions arising on the spur of the moment. A second criterion is the degree of directiveness (Whyte 1960: 354–355): from neutral, vaguely encouraging prompts to the most specific of questions on particular subjects. A third criterion (obviously related to the second) is the degree of openness or closedness of questions asked; that is, for example "How are you?" versus "Are you suffering from malaria?" A fourth criterion is the length of the interview (brief encounter versus in-depth probe); a fifth the degree of prior arrangement (from a pre-set appointment to discuss a special topic to a totally unexpected meeting and subject of conversation); and a sixth the interview setting (group versus dyad; subject's house, ethnographer's house, neutral territory, and so on).

Given such a range, it is important to remember that just as claims for the objectivity of formally structured interviews can be false, in that this kind of discussion is alien to all (except perhaps the professional survey worker!), so, too, can less structured approaches themselves be unfamiliar to informants and subject to repudiation and more subtle forms of non-response and distortion. Morris Freilich's experience among Mohawk steelworkers is an extreme but none the less revealing example. Even once accepted as an habituee of the Wigwam Bar and as "Joe's friend", any attempt to overtly direct conversation as an anthropologist towards a specific topic, let alone note down any particular points, immediately met with a hostile response (Freilich 1970: 185–206). In these circumstances, the only way open was to identify as completely as possible with the Mohawks and their culture and then periodically slip away to make brief notes which served as an aide-mémoire for the main ethnographic record constructed later.

Freilich's experience of fieldwork is clearly at the extreme end of the formal–informal continuum. More generally Michael Agar observes that, with the informal interview "everything is negotiable. The informants can criticize a question, correct it, point out that it is sensitive, or answer in any way they want to" (1980: 90). Such a description clearly reveals some of the usages involved in this approach. It is also relevant to note that if everything is negotiable the ethnographer can expect to be questioned himself, and that when this occurs it is important for rapport[10] or even for the very continua-

[10] As Freilich points out, "rapport" is not necessarily synonymous with *good* relations, but is rather "a conditional agreement to communicate" (1970c: 540).

tion of the interview[11] that he be prepared to *exchange* information. It can also be a salutory warning to the anthropologist to then consider the honesty, degree of generalization and simplification of his own replies. In a rural Thai village, to translate British research grants and salaries into Thai Baht gives a grossly distorted picture of relative wealth, and explaining a "riceless" culture to those of one where rice is in so many respects central is well nigh impossible. Of course, one of the problems is that Thai informants lack a background knowledge of British society and this serves to emphasize two important points. The first is that informal interviewing must in no way be equated with lack of preparation and, secondly, that an initial period of rather general and unfocused inquiry can, if used sensitively, serve to train both observer and participants and provide the former with a grounding in the techniques of inquiry appropriate to that culture.

Even where one is going to be using essentially the same pool of informants throughout fieldwork, resort to a less focused or "passive" approach is advantageous to start with. The knowledge sought by anthropologists is frequently a resource not to be dissipated indiscriminately or to individuals who are strangers and clearly cannot understand its value and appreciate its significance. Prematurely detailed, but poorly formulated probings of a particular area are thus likely to generate poor responses. Once some familiarity has been achieved, at least one set of barriers tends to come down. There then exists the possibility of more detailed probing and, here again, by leaving things for a while the ethnographer will have gained the opportunity to learn the appropriate techniques for cajoling or provoking answers and even unsolicited information. The humble seeker after knowledge might well "have his head kicked in" by the members of a street gang for his appearance of weakness or ignorance. In contrast, the hard, challenging responses to statements which worked so well for Agar in his work on the streets might prompt an elderly ritual specialist in some rural milieu to turn away in disgust rather than provoke further elaboration and validation.

If there is some observable degree of progression in the methods appropriate in most fieldwork situations it is in the move from a "passive" to more "active" investigation, as research becomes more directed and focused. The latter can certainly include the use of formally structured interview schedules, but more often consists of non-scheduled standardized interviews where the same kinds of question are asked, or the non-standardized interview where an issue is pursued in some depth. Where the emphasis is on qualitative research, problems of sampling are not prominent yet it remains important to contact

[11] See Henry (1966, 1969) for a discussion of situations in which an explicit expression of personal and ideological commitment is required.

informants who are in some way representative of their society and the cleavages within it. Those most open and helpful to the newly arrived ethnographer have often been described as deviants or somehow marginal. Whether or not this is the case in any particular situation, it remains necessary to persevere with those who are not so readily available and it is in these circumstances especially that preparation, relevant background knowledge and possibly an introduction through intermediaries can be crucial (§8.3).

One often thinks of the interview as essentially being between two persons but this is by no means necessarily so. At the more informal end of the continuum especially, the ethnographer might question groups of people, or others will cluster round and start to chip in. Just what one's response should be depends on the situation. Group interviews can develop their own dynamic and the discussion can open up new fields, though it might also be impossible to control the direction of conversation. Clearly, though, such investigations can directly reveal important cleavages of opinion, the way in which consensus is achieved, and other valuable things. What is often more problematic is what is intended by the ethnographer to be a one-to-one inquiry but which results in questions being shouted to neighbours, or other people wandering in and giving the answers. The replies themselves might be extremely informative, but where one is looking for consistency, for the views and extent of the knowledge of one person, then the situation becomes difficult. It is not always diplomatic to tell a husband to "shut up", and even the mere presence of others, deemed by the informant to be of senior status or more knowledgeable, can affect the quality and form of information. For an excellent ethnographic account of the dynamics of the unstructured interview using verbatim illustrations from taped exchanges, see Chagnon (1974: e.g. 62–63).

8.4.4 Asking questions and interpreting answers

Prepared lists of questions (§8.11) may help the interview to flow, for it is frequently the case that questions you wanted to ask are forgotten; they are also useful when conversation grinds to a halt. Phrasing questions, even if they are not pre-determined in questionnaires, schedules and checklists, is something which must be undertaken with some care. This is particularly so for the beginning fieldworker, where linguistic and cultural competence is generally very limited. It is important to ensure that questions are not too complicated, but then not too short to invite unsolicited ambiguity and hedging. Try to ensure that questions mean the same to the recipient as they do to the sender; that they are culturally legitimate and answerable. Some questions, of the "When did you stop beating your wife" variety, may contain answers to questions which you have yet to ask: avoid them, and unless employed strategically, all leading questions as well. As a rule, big questions – those providing an infinite number of answers (e.g. "How are things today?") – are safest. Show you understand what is being said, and do not agree or disagree (even by using apparently neutral interjections) unless you are certain of the significance of so doing. Cultivate appropriate conversation styles, sitting positions and eye-movements.

If conversation disappears it can sometimes be rekindled by indicating that you know something already, by proffering a different opinion or by providing a calculatedly wrong assertion. Such tactics, however, require care and experience. If a conversation is wandering or stagnating it can be pushed in the right direction with an appropriate probe. Prompting is an art which has to be cultivated, and a certain amount of effort must initally be put into pump-priming (that is, encouraging informants to speak freely and informatively on subjects which interest you). Never interrupt accidentally, but rather with due consideration for the consequences, and do it gracefully. In some situations it may be appropriate to play dumb, that is, to make tactical use of statements of disbelief; elsewhere, however, this may be counter-productive. Indeed, it may make sense to say you have understood, even if you have not. Repetition of a question, or what an informant has just told you, may serve to concentrate the mind and perhaps provide a means of confirming understandings. Certain kinds of questions may be clarified by the use of non-verbal stimuli, such as artifacts, show cards, pictures or even photographs (Whyte 1960: 368ff). Projective aids of this kind may take on a special significance in the context of systematic cognitive and psychological tests (§8.10.5). On the art of asking questions, see, in particular, Payne (1965); Bulmer (1977) is also useful.

Interpreting answers requires, if anything, even more care than asking questions. Even unsolicited statements require interpretation. To begin with, an ethnographer must always bear in mind the limitations of linguistic competence and, no less important, cultural competence. Meaning, after all, resides ultimately in cultural context, and is only expressed transiently and situationally in language. Certain answers may owe more to courtesy than to a desire for accuracy (e.g. Jones 1964), others to impatience, others still to political, moral and other social constraints. At all times you must be on your guard for figurative, rather than literal expression. One way of approaching the problems of interpretation is systematically to envisage all the possible questions, or interpretations of a question, which might have elicited the actual answers which you get. When employing professional translators or interpreters (Phillips 1960: 297–301), such problems are obviously compounded: such third parties are not translating machines, they often have a limited vocabulary in any lingua franca, and may be culturally marginal. In the final instance, the meaning which we attach to statements of informants arises from, and is constrained by, the wholly complex and somewhat artificial exchanges between ethnographer and subject (Fabian 1979; Clifford 1980). On translation in general, see Malinowski (1923), Kluckhohn (1945), Casagrande (1954) and Phillips (1960). Bohannan (1967) is perceptive on cultural barriers to translation.

Numerous factors may influence the "validity" or "reliability" of informants' statements; or, in terms of a positivist paradigm, represent "sources of error". These include involuntary error, intentional error ("lying"), the personality of the subject, the dynamics of the interview situation, and the various cultural conventions and social constraints mentioned already. The psychology of the interview situation, interviewer effect and the control of error have been much written about (e.g. Hyman *et al.* 1954).

Careful formulation of questions and perceptive interpretation of answers is the sure way to ensure that data are "valid". However, there exists a considerable sociological literature on formal means for establishing the accuracy of statements and their reliability; on verification, validation and triangulation; and on proof and inference (e.g. Vidich and Bensman 1954; Becker 1958; Dean and Whyte 1958; Whyte 1960: 358–365; Denzin 1970). Implausible statements in the area of descriptive data (e.g. answers to the question "How many cows does Tom own?") can be checked through internal triangulation (eliciting the same data in a different way from the same informant), by comparing accounts given by different informants, or by including your own first-hand observation. Checking implausibility in evaluative data (e.g. "How do you feel about Tom?") is altogether more difficult, and must for the most part be interpreted in the light of the consistency or otherwise of other remarks or actions of an informant. One of the advantages of informal rather than highly structured interviewing is that further sensitive probing can often produce internal evidence for validation. Chagnon (1974: 67–83, 89) writes perceptively and engagingly about unreliable informants, and in particular on the problem of lying (1974: 92–93, 101–113); on which subject see also Salamone (1972). Manning (1967) deals with some general problems of interpretation.

8.4.5 Recording (see also Whyte 1960: 365–368; §8.12)

Assuming, then, that some of the techniques work some of the time, within a short period a lot of information is going to be floating around and one must decide how best to record it and what to record. There are three obvious possibilities: (1) taking notes during the interview (which can be disturbing), (2) taking notes after the interview (which lessens accuracy and fullness), and (3) tape-recording. The flow of a conversation can be destroyed by the appearance of a large notebook or by the time taken to write down entries. On the other hand, while a small notebook into which key words or brief statements are immediately entered is often adequate, occasions do arise when fuller recording is necessary. Indeed, such efforts on the part of the ethnographer can be a useful demonstration of his seriousness to the informant. Furthermore, the presence of maps and genealogies can themselves be of interest to informants and stimulate exchanges. However, the tape-recorder (§7.9.3), given its technical capabilities and reliability, now has an immediate appeal but is still by no means the ideal technique. Whatever the response of informants to its appearance, or the ethics of concealment, the fact remains that turning tape into a readily accessible record for further inquiry is time consuming. Full transcription by a good typist, working with a clear tape at the rate of six hours for every one of recording, is clearly a problem, especially when one has to make the transcription oneself in difficult physical and social conditions (Bucher *et al.* 1956; §7.6.3, n. 22). Nevertheless, where note-taking is impossible, or where the interview is relatively structured and includes stories or descriptions of rituals the tape can be used to great effect, in that details can be checked and questions asked at the time of playback, without the original flow of information being disrupted.

Even so, in many cases it does seem better to let some of the nuances of conversation go, to make shorter more limited written records, and so be able to spend more time actually talking to people. This way the relevant data will eventually be discovered and retained, together with a better appreciation of its generality and extent of variation. It is also pertinent to note that, in returning to an informant for further clarification and development, one reveals one's interest in the subject and informant, and might thereby gain access to the informant's own doubts and uncertainties which would never be revealed in a single "hit and run" interview. In this sense, even if fieldwork is not always the "long conversation" it has sometimes been held to be, it is wholly reasonable that we should try to make it so.

Jeremy H. Kemp and R. F. Ellen

8.5 Case studies

8.5.1 Introduction

Case studies are the detailed presentation of ethnographic data relating to some sequence of events from which the analyst seeks to make some theoretical inference. The events themselves may relate to any level of social organization: a whole society, some section of a community, a family or an individual. What distinguishes case studies from more general ethnographic reportage is the detail and particularity of the account. Each case study is a description of a specific configuration of events in which some distinctive set of actors have been involved in some defined situation at some particular point of time. Ethnographic reportage tends to be general in form: the analyst makes statements about the overall pattern of behaviour or belief derived from extensive observation. These statements are in effect summary statements reflecting modal or molar forms of behaviour or belief. Case studies, on the other hand, as van Velsen (1967) describes them, are concerned with the imponderabilia of everyday behaviour. Because of the special circumstances surrounding the events as they occur, the principles underlying behaviour or belief will almost certainly never be as clear cut as in more general ethnographic statements. This indeterminacy is crucial in using case material effectively and we will need to return to this again.

In setting out a case study, the analyst must decide in advance at what point to enter the ongoing flow of events and at what point to withdraw from it. For the purposes of exposition, a set of events must be lifted from the ongoing stream and presented, as it were, isolated from antecedent and subsequent events.

The ways in which case material may be used were distinguished by Gluckman (1961). The simplest, he argued, was as "apt illustration". The anthropologist, having produced a general account of the culture or the social system of a people, proceeds, as Gluckman puts it, to use "apt and appropriate cases" to illustrate specific customs, principles of organization, social relationships, etc. Each case is selected for its appropriateness at a particular point in the argument; and cases coming close together in the argument might be derived from the actions or words of quite different groups or individuals (Gluckman 1961: 7–8). The aim, presumably, is to impart a sense of concreteness to an otherwise overwhelmingly abstract account. Obviously, when using case material in this way, the anthropologist so selects the instance from field notes as to present the most "typical" case possible.

The analysis of a connected set of events or a "social situation" as a "case" constitutes a different use of case material (see Gluckman 1958; Mitchell 1957; Garbett 1970). Here the emphasis is on the theoretical connection *between* the events rather than in the attributes of the events themselves. The aim of the analysis, in Gluckman's words, is to "exhibit the morphology of the social structure". But the morphology of the social structure in social situations is reflected in the behaviour of the protagonists and their inter-pretation of the behaviour *as conditioned by that situation.* The selection of the particular situation for analysis is therefore a crucial tactical considera-tion for analytical purposes. Clearly the analyst chooses a situation precisely because it exhibits the "morphology of the social structure". There is thus a duality between the analysis on the one hand and the situation selected to support the analysis on the other.

A third way in which case material may be used in anthropological writing is in extended case analysis (see Mitchell 1956; Gulliver 1966; Turner 1957; van Velsen 1964). Whereas the analysis of a social situation is limited to a single situation or at most to a restricted set of events located in the same situation, an extended case analysis, by contrast, typically covers the same actors over a series of different situations. This implies that an extended case analysis recounts events over a relatively long time period, typically over several years. This allows the analyst to present material which contributes an historical or dynamic dimension to the account. As Gluckman phrases it: "[extended case analysis] treats each case as a stage in an ongoing process of social relations between specific persons and groups in a social system or culture" (Gluckman 1961: 9). This is the most sophisticated use of case material.

8.5.2 Typicality and the case study

Misgivings are sometimes expressed about the use of case material for analytical as against illustrative purposes. Eggan (1961: 22), in response to Gluckman's early article, for example, writes:

> If we are to concentrate our attention on fewer people over a longer period of time, the sampling problem becomes important, since the question of how typical the group is of the society becomes crucial. Gluckman recognizes this problem and thinks it may be partly solved by the use of statistics: but the essence of the method is in the cooperation required for intensive study, and this can seldom be achieved with cases selected at random.

This criticism, however, is founded on a misunderstanding of the basis upon which analytical inferences are made from case studies. The confusion seems to have arisen from the inappropriate extension to case studies of the procedure of making a particular kind of inference from quantitative data. When quantitative data are used to support theoretical interpretations, two different and unconnected types of inference are involved. The first is that of inferring that relationships observed in a sample of instances available to the analyst exist in the wider population from which the sample has been drawn. Sampling theory in statistics devotes itself to providing numerical estimates of the likelihood that the population values be within some defined range of that established from the sample – provided that the sample has been chosen in such a way as to meet the mathematical conditions to justify the computation of the probabilities concerned. The sophistication and elaboration for choosing a "representative" sample in this restricted sense has overshadowed the other kind of inference involved when analytical statements are made from associations uncovered in a statistical sample. This is the inference that the *theoretical* relationship among conceptually defined elements in the sample will also apply in the parent population. The basis of an inference of this sort is the cogency of the theoretical argument linking the elements in an intelligible way rather than the statistical representativeness of the sample. The inference is based on *analytical* induction rather than *enumerative* induction (see Znaniecki 1934: 221 ff). These arguments are more fully set out in Mitchell (1983).

It should be obvious that the inference from case studies is based on analytical induction. What the anthropologist using a case study to support an argument does is to show how general principles deriving from some theoretical orientation manifest themselves in some given set of particular circumstances. A good case study, therefore, enables the analyst to establish theoretically valid connections between events and phenomena which previously were ineluctable. From this point of view, the search for a "typical" case for analytical exposition is likely to be less fruitful than the search for a "telling" case in which the particular circumstances surrounding a case, serve to make previously obscure theoretical relationships suddenly apparent.

It follows from this that the particularity of the circumstances surrounding any case or situation (or set of situations) must always be located within some wider setting or context. Any general statement which links theoretically relevant events or phenomena must always assume that "other things are equal". Case studies allow analysts to show how general regularities exist precisely when specific contextual circumstances are taken account of. When it is difficult to do this, then it is likely that the theoretical formulation of *the regularities* underlying the regularities needs some revision.

Case studies used in this way are clearly more than "apt illustrations".

Instead they are the means whereby general theory may be developed, since it is through the fieldworker's intimate knowledge of the interconnections among the actors and events constituting the case study or social situation, that the fieldworker is strategically placed to appreciate the theoretical significance of these interconnections.

8.5.3 Data collection for case studies

Case studies are, as previously stated, detailed accounts of a particular series of events or actions of actors. As such, the case study is essentially a *configuration* rather than a summary or sort of "average account" of events. The analyst tries to present as complete a set of information bearing on the set of events or action as can be mustered. The "case" may be, as earlier stated, a nation, a people, a district, a village, a family or some particular set of actors involved in a series of social actions. Obviously, the procedures for assembling the configuration of data necessary to describe, say, a nation going through some crisis, are quite different from those concerning, say, a family. To provide the necessary background to the events befalling a nation will require recourse to national accounts, statements by politicians, accounts of debates in parliament. Some of these data may be statistical, such as economic trends for example, some may be formal statements by politicians in structured situations, others may be personal accounts by personalities themselves involved in the events. The data needed to document the passage of a family through a crisis will naturally be very different. While the accounts of social workers, health visitors and other professionals may be an important part of the data, it seems likely that the statements of the members of the family will constitute a major part of the evidence, as will the personal observations of the fieldworker.

Any technique of data collection may be used to assemble the information to provide as complete an account of the course of events as possible: no one technique takes precedence over any other. Just what techniques are used to assemble the data depends partly on the scale of the set of events being described and partly on the theoretical purposes of the fieldworker.

Usually the fieldworker has much more information than needs to be included in a case account. Since the purpose of the case study is to present sufficient evidence to demonstrate how events and actions are linked to one another in theoretically significant ways, data beyond what is strictly germane to this purpose are redundant. This is why some anthropologists (Gluckman 1961: 14; Garbett 1970: 217) prefer to operate with "social fields" rather than notions such as "culture" or "society". Social fields are events and relationships so chosen because they may be linked to one another in some theoretically cogent way. But at the same time, one of the striking advantages of well presented case studies is the possibility that the information recorded in the account may be reanalysed by others either to deepen the analysis or to present an alternative interpretation.

J. Clyde Mitchell

8.6 Long-term research

Long-term research is at the heart of the anthropological method. Malinowski is credited with originating the close, intensive study of a small community which became the foundation of the ethnographic method (Kaberry 1957): "To grasp the native's point of view, his relation to life, to realize *his* vision of *his* world" (1922: 25). The value of Malinowski's Trobriands fieldwork and many-faceted analysis of Kiriwina society and culture as he found it is demonstrated in the many studies which have examined or re-examined aspects of Trobriands society and its changes (Powell 1969; Weiner 1979; Damon 1980, are but a few).

The long-term studies of Firth in Tikopia and Mead in Manus which, including return visits, extended throughout their professional careers, have shown how these communities have experienced some 50 years of internal developments and external influences. The involvement of other anthropologists in these field sites has resulted in studies of particular questions and issues (Firth 1936, 1959, 1971; Mead 1956; Schwartz 1962; Tuzin and Schwartz 1980).

A fieldworker who plans to conduct a study in a community usually prepares for a sojourn of a year or longer, and should consider the possibility and feasibility of an extended stay, return visits and the involvement of other fieldworkers to carry on or develop new lines of interest. Every new community is a challenge to the anthropologist's training, adaptability, language and cultural facility, patience and commitment. The investment made by the anthropologist in establishing good working relations, in being accepted by the community and privy to personal and community affairs, gaining general knowledge of people, place, language, institutions and values, is worth following-up with further or time-depth study of a great range of problems and subjects. The accessibility of community informants to an accepted fieldworker can be a great time-saver and advantage in pursuing other or later information. Most fieldworkers develop close relations with many members of the research community and region, which may be continued over many years in correspondence and visits. The involvement of the anthropologist in local affairs varies with personality and interests; its implications for future research should be evaluated carefully.

On a first field visit, the anthropologist gathers much general information on the community. When this has been collected and recorded in a systematic form that can be corrected, expanded and updated, it becomes a valuable resource for later study of many special subjects. Certainly, it is to be expected that new fields of specialization will develop, and no initial research can lay the groundwork for all future interests. However, many particulars about the community and its residents recorded in a census, such as that prepared by the Rhodes-Livingstone Institute and later developed by Colson (1954), and discussed by Scudder and Colson (1979), is invaluable as a database. The census should at the least include personal information, information on kinsmen, residence, land, personal history. Genealogical information is a database for the study of kin groups, family formation, marriage patterns and residential history. The *Outline of Cultural Materials* has been a useful catalogue for collecting and filing base data. A standard minimum core database is summarized by Foster *et al.* (1979: 333) and reproduced here as Table 8.1.

Table 8.1 *Standardization of minimum core data (from Foster* et al. *1979: 333)*

Accelerating rates of change and increases in complexity require reliable reference points for successive and cross-societal comparison.

From the start, each investigator should define his sample(s) and parent populations in space and time.

Defintion of terms (e.g. household) must be consistent and unambiguous.

Minimum core data should be recorded for each individual in the sample according to the following categories:

  Name
  Sex
  Date and place of birth and of death
  Marital and Parental status (including number of all children with their dates of birth and death)
  Unit(s) of affiliation, past and present
    (a) Social
    (b) Residential
  Occupation(s) and their locations, past and present
  Education
  Religion
  Resource base (minimum descriptive data)
    (a) Ecological categories
    (b) Economic categories
  Sociopolitical differentiation (e.g. rank, title, caste, office, political roles, etc.)

Practical research programmes may try to assess attitudes and behavioural changes and the effects of introduced political, economic or other programmes. The census and database can provide accurate, dated information for studies of such matters as population change, income, employment, economic change, education, political participation, changes in kinship, law and religion. For these studies, historical information is essential.

The new fieldworker can better provide for his or another's future work by careful recording and orientation at the beginning. Often some written reports, by naturalists, administrators, visitors, residents and observers can be located to provide a background for the study. It is often possible to select a community of which something is known through these early reports and, if there is no compelling reason to settle elsewhere, some historical depth is already gained by this. If long-time residents can be interviewed in the field or in retirement, photographs examined etc., these may be helpful with cyclical or changing phenomena. Nowadays, a fieldworker may have so much information about the past that some aspects of his work may be said to constitute a restudy, allowing him to concentrate upon a particular problem rather than a general ethnography.

As anthropologists have developed interests in social and cultural processes, they have increasingly realized the importance and value, indeed the necessity, of a long, continuous or intermittent period of observation. For example, we now recognize many phenomena to be cyclical: life-cycles of people, livestock, plants, trees, aspects of land use (garden rotation and fallow), ritual cycles which depend on seasonal or calendrical events, astronomical sequences, and age grades, human growth, marriage, childbirth, old age. These all may be on a different time-scale, but cannot be fully understood until at least one cycle has been observed and recorded.

Ritual cycles involving growth or maturation, which may be defined by initiation, and long-cycle growth of livestock or certain slow-growing plants, etc., should be studied over the whole life-span. Some information can of course be gained by observing people or communities at different points in a cycle, but it is always difficult to account for incidental events. While sampling of such cycles is possible, the combining of data from communities at different points in a cycle requires assumption of regularities that may be inaccurate. Seasonal ritual cycles may vary from year to year: a single series may not be reliable as a basis of generalization.

For the study of agricultural cycles, a regularly recurring series of visits, every 2, 3, 4, or at most 5 years, is most desirable. Also, a full sequence of seasons is essential for any phenomena linked to land, crops, growth, sunlight, rainfall, etc. When land and crops are rotated, fallowed, rested or replanted on a variable cycle, only repeated thorough surveys of a defined area will reveal the patterns. Soil variations, seasonal and year-to-year differences in rainfall and temperature may affect plant growth. For example, the growth and maturation of trees may be considered an indicator that fallow has been long enough and new cultivation may proceed. The actual observation of soil, rainfall, temperature, crop rotation, fallow planting, the use of these indicators, with time depth information on the land use of the area, is essential to really understand the land use pattern. Where local people do not record or document their affairs, their time estimates are often unreliable.

Understanding of trends, fluctuations and reverses, external influences, and other historical processes are the other main field for long-term study. These studies are less easy to plan, for the fieldworker's opportunity to return may not be predicted, another person may be able to conduct a study, and the community may be inaccessible to return for continued study at the optimal time. Yet our best information on trends and processes of change has resulted from such studies by one or several anthropologists. Restudies of communities are valuable not only to understand change processes, but, as the interests of anthropologists expand to new research problems, these can be examined in communities for which basic information is already available (Kemper 1979).

Foster *et al.* (1979) reported upon long-term field research comparing a variety of practices and experiences. The volume cannot be summarized easily, but suggests certain distinctions, such as that between a long-term study directed at examining processes of change, cycles or trends by continuous or intermittent fieldwork, and the restudy, planned or fortuitous, which is usually directed at measuring change between the first and second research periods (e.g. Epstein, in Foster 1979). Despite careful documentation of behaviour and events during the periods of study, one must make assumptions about the intervening times. A restudy, even if planned at the time of the first fieldwork, is more likely to search for one-directional trends rather than the flux which some continuous studies have observed. Scudder and Colson observe that their long-term study is a "stream of events which occur in response to and in a feedback relationship with a variety of stimuli, rather than . . . trends which can be extrapolated from one time to another" (in Foster 1979: 241). Similarly, long-term "local study can produce a multitude of trends, reversals, continuities, discontinuities, cycles, upheavals and changes, stimulated by internal and external events" (Brown 1974).

Several long-term projects have had teams (§7.11) of fieldworkers and students over many years (Foster, Helm, Lamphere, Lee, Mangin, and Vogt, in Foster 1979). These certainly can both deal with many subjects and serve a valuable training function, but are likely to become artificial environments for research. Fieldworkers who join teams or follow another anthropologist into the field may find that some members of the community who have worked closely with preceding research workers have learned expectations and response patterns which are utilized in relations with other fieldworkers, regardless of their field of interest, personality, or expectation of relationship with informants. This can seriously restrict the work of a newcomer.[12] It also becomes difficult for familiar fieldworkers to direct their inquiries into new areas or escape from the categories in which they have been placed. The difficulties associated with age and sex identification are obvious, of course, but anthropologists are often identified with local groups, factions, and subjects of interest as well. Long-term fieldwork brings opportunities to see old acquaintances in new contexts, to delve more deeply into subjects concealed from superficial visitors, such as sorcery or infanticide, and to trace continuities and trends. But it may also frustrate the exploration of new subjects and relationships with persons and groups outside the familiar range.

Many long-term and team projects evolved without prior or early-stage planning as opportunities and expanded interests led to further research (Lee 1979). Once the initial basic data and fieldworker relationships are set up, new workers and subjects are sometimes readily incorporated. The advantages are obvious, but perhaps some caution should be advised. The diverse personalities and interests may clash in the field setting: the community may be confused or divided, there may be competition for resources. Certainly a small group can be strained by the presence of several visitors, and members of the visitor group may find the fieldwork situation trying.

Fieldwork is often a solitary, individual undertaking, written up by the only anthropologist who has knowledge of the community at the time concerned. Such ethnographic works are at least in part unverifiable. When another ethnographer, at another time and with a different orientation describes the community differently, the validity of ethnographic research is subject to question (Lewis 1951; Redfield 1960). The same problems arise whenever conflicting reports are presented by two or more witnesses. Independent studies are to be welcomed as a way of evaluating research tech-

[12] Ben Finney related, years later, the difficulties that he and his wife Ruth encountered in the Chimbu (Bamugl and Mintima) field site where Harold Brookfield and the author had worked. Ben was addressed as "Ari". He was expected to examine landforms and conduct land-use surveys, while Ruth was told about all recent marriages, children born, new houses, and residential changes. They found this so frustrating that they moved to another area to pursue their studies of commercial coffee growing and education.

niques and ways of testing conclusions. Thus team research and field teaching forms can use data recording forms (Vogt 1979: 209; Foster 1979: 169). However, each anthropologist works in a somewhat distinct theoretical context, local and temporal circumstances, which make exact duplication impossible.

A long-term commitment to a community may have a goal of studying the continuities (Vogt 1979), long-term processes and changes (Brookfield 1973), as well as a change in some interests, when the investigator feels some questions have been answered, some have lost interest, and some new ones developed (Meggitt 1979; Lee 1979). There may well be some subjects whose study is best pursued early in a field involvement, others which can only be broached after a long period in which confidence has been built up and others whose study is fortuitous. For example, in Chimbu spatial distribution and marriage data were readily obtainable, while witchcraft and some other beliefs and practices that were disapproved of by officials were revealed only after several seasons. The occurrence of an unusual event, be it an epidemic, a rare ceremony or a dispute between clan leaders, may not be amenable to any planning.

Certain events and processes are to be understood through historical observation; inter-group relations and interpersonal relations always have a history of incidents, quarrels, alliances, divisions, grievances, unresolved differences and uncelebrated agreements. End-point observation and inter-viewing reveal only some of the conscious factors. But experienced field-workers frequently recognize the re-emergence of an old dispute in a new event. Community realignment (Turner 1957) may only be explicable through an historical explanation based upon long-term observation.

A basis for future work is a careful defining and recording of the particulars of a community: census, mapping, territories, land use, housing and local population distributions, cultural inventories, genealogies. Perhaps an ex-ample of some work, which continued for some years but had not been planned as well as I would now advise, will help to demonstrate this. The 1958 season in Chimbu (Brown and Brookfield 1959) involved a new collabora-tion between geographer and anthropologist. While population, kinship and land use were among our interests, we did not have a satisfactory survey map to begin with; our knowledge and definition of the area went hand-in-hand as we explored and surveyed. Brookfield mapped with field observation, com-pass, altimeter and pacing. Brown gathered land ownership, land use, residence and genealogical information. The end-product of that first year was an oddly-shaped swatch of territory, the land and houses of parts of two clans, nearly complete for three subclans but incomplete – we did not know how much – for several others. Genealogies and population data covered a larger group, some of whose land lay outside the area surveyed. Only on the next field trip (1959–1960) did we see how this might have been improved and co-ordinated. The 1958 data could never be complete. By April 1960 we had nearly a year-round of field observation, including changes in land use and residence over a two-year period. Subsequent fieldwork was mainly in periods of 4–10 weeks. Many subjects were closely followed in subsequent visits. Our understanding of the cyclical and change processes expanded greatly in the next years (Brookfield and Brown 1963; Brookfield 1973; Brown 1974, 1979). On inter-disciplinary research generally, see §7.11.5.

The actual experience of long-term fieldwork demonstrated that the distinction of cycle and change trend can only be theoretical. In our long-term study we found that cash cropping and other change processes had a great effect on the subsistence and livestock land use and food production cycles. Social and ritual cycles were also affected, and some special short-lived problems arose to affect both. Thus in 20 years, the agricultural cycle was so disturbed by social and cultural changes that some conclusions must remain more speculative than demonstrated. However, it is only through such prolonged study that the relationship between cycle, trend and flux can be examined at all. A repeated static study, or a series of such, would probably force one to some incorrect conclusion.

Paula Brown Glick

8.7 Collecting life histories

8.7.1 Introduction

The life history method has been claimed by many of those who use it to be "autobiographical"; thus it is worth quoting a perceptive comment from the poet Kathleen Raine on the subject of her autobiography: "What at some other time I might have recalled, and called my story", she writes, "I cannot say; for we select in retrospect, in the light of whatever present self we have become, and that self changes continually. Some new experience may awaken a whole sequence of past memories, or some mood shed on the past its own colour". And she adds, in a comment which for the historian or the anthropologist is both provocative and disarming, "There is also a mechanical memory, but who cares for its records?" (Raine 1973: 5). Autobiography is thus not a simple matter, for these comments make clear at once that there is not one possible story, but several, and that of these several stories some are more revealing than others.

In this lie both the attraction and the snare of the life history: for the self portrayed in the story lies less in the "mechanical memory" than in the individual's capacity to select events, reflect upon them, and assign them values; and the pattern of these reflections may, and often does, change over time. Correspondingly, therefore, the selection and evaluation of these events by others, too, when they seek to describe some fragment of the protagonist's story, is bound to vary. For anthropologists and their readers, finally, there is a further source of variability in their own understanding: the story only reads truthfully when what is taken for granted in it by the teller has become well understood by the reader or hearer.

There are therefore many versions of a life history, and examples which deal adequately with the questions which arise thereby are rare. Nevertheless, for all these difficulties, biographical material retains a peculiar fascination for the anthropologist. Thomas and Znaniecki (1920: 77) hold the extreme view that the biographical method is "the only one that gives us a full and systematic acquaintance with all the complexity of social life". Langness (1965: 4), more cautious but still enthusiastic, says, "It is not completely far-fetched to assert that all anthropology is biography"; and while this comment must be understood in a very broad sense, it testifies to the fact that, in one vitally important respect, the reflective sensibility of individuals about their past and their present actually comprises their culture, since it is the individuals in a society which receive, recreate and transmit that culture over time, and it is their evidence on which the anthropologist, to a greater or lesser extent, relies. This native experience can, however, be elicited in various ways: by extended observation of conversation and behaviour, by the use of historical documents, or by interviewing which is structured more by topic than by chronology. It is relevant to ask, then, what differentiates the life history? And how far is it dependent on the employment of these other methods to fill in the elements taken for granted by the protagonist?

The nature of this relationship of life histories to other methods thus defines one crucial set of questions about data collection; the other set of questions, already apparent from what was said earlier, turns on the variations between different versions of the same life history – variations both between tellers and over time – and how the anthropologist should control them when gathering material or, more to the point, make use of them. These two sets of questions are considered shortly; but first a brief account must be given of what sorts of biographical material are commonly collected, and how they are used in analysis.

8.7.2 Types of biographical material

The "life history" has been the subject of recent discussions (Burgess 1982; Shaw 1980; Faraday and Plummer 1979), and these may serve as a guide to the literature which is formally so described. The main characteristics of this "formal" life history are (1) the sheer quantity of raw data presented in the informant's own words, and (2) the emphasis on long-term continuities in individual experience, rather than on broadly based thematic and comparative topics. However, Thompson (1978: 203 ff) correctly notes that *all* biographical material provokes a tension in the analyst between fidelity to individual continuity on the one hand, and the logic of thematic argument on the other, and that all biographical data, however apparently "raw", are acquired and used selectively with interpretive purposes in mind. Formal "life histories" thus, in varying degrees, simply maximize the individual aspect and minimize the thematic.

The "formal" life history, therefore, needs to be seen as a particular polarity in the use of biographical material. Amongst other uses, for example, which emphasize the thematic aspect, are the analysis of migration histories (Philpott 1973), health histories (Blaxter and Paterson 1982), occupational and religious histories (Moore 1974) and so on. Then finally, at a level where the individual biography becomes almost completely absorbed into the theme, come those uses of informants' recollections, amidst other material, to construct the contents of a normal ethnography: the life-cycle, for example, or the processes of social change.

These differing presentations of biographical material correspond, in part, to the analytical purpose for which the material is collected; and thus we next consider these analytical purposes and their implications for data collection.

8.7.3 Analysis of biographical material

The type of life history which emphasizes continuities in individual experience is sometimes, we have seen, deliberately contrasted with an emphasis on thematic organization; but despite this view of pure biography as being at the opposite pole from thematic analysis, the belief also persists among anthropologists and others that a biography should be susceptible to special forms of analysis which can be case-specific. There are thus three typical options in relation to analysis: (1) comparative analysis of biographical material by theme, (2) biographical presentation as contrasted with analysis, or with thematic analysis only as background, and (3) suitably biographical forms of analysis. These options each pose different requirements for data collection.

Biographical material which is relatively closely governed by theme (occupational histories, migration histories, etc.) can be analysed by comparative methods which do not differ essentially from those used in reaching other ethnographic generalizations. Similarities and differences are discerned between groups of informants, or between the fieldwork population and other populations, and features associated with the differences are analysed. Where the necessary depth of information can only be obtained from a proportion of the fieldwork population, some form of sampling becomes appropriate; and in all cases studied the researcher needs to be sure that comparable information has been obtained.

Where biography comes to be contrasted with this type of analysis, on the other hand, the contrast is often "the result of a characteristic phase of the anthropologist's own life experience". As he writes his account he finds the people he knows "dissolved into faceless norms" (Mandelbaum 1973: 178). To rectify the balance the author then turns to biographical presentation, a "means of presenting almost 'raw' fieldwork data" (Shaw 1980: 231), as a complement to analysis. The main problems posed in collecting data then concern the selection of subjects, and the nature of the additional materials required to provide a true and balanced picture.

Finally, there remain various claims about forms of analysis which are capable of illuminating the individual biography or case history even in the absence of case comparison, although it must be said that these analyses are much more powerful in a comparative setting.

One such approach – which has a long history – is that of analytic induction. This tradition, which goes back to Znaniecki (1934), is usefully summarized by Mitchell (forthcoming) and by Denzin (1970: 194ff, 238ff). Essentially, it consists in a serial testing and reformulation of a proposed explanation, by moving sequentially from relevant case to relevant case in search of negative instances; but it is implicit in this procedure that each case on its own provides the possibility of a case-specific causal analysis. An example might be Mandelbaum's (1973) study of Ghandi, where he shows that the emergence of Ghandi's policy of active non-violence was made possible by a wide range of cumulative experiences in which some elements of Indian, British and South African culture were rejected and others retained. The tendency of such analyses thus to identify the necessary conditions which make an event possible, rather than the sufficient conditions which would precipitate it, has been noted by Robinson (1951) and Turner (1953). It is probably this aspect of the method more than any other which has occasioned critiques over the years from the standpoint of statistical inference – a recent example being Brown (1973).

A second approach with, similarly, long theoretical antecedents, is probably capable of a much more extensive application to biographical material than has so far occurred. This approach is that which seeks to analyse the logic of cognitive and moral rules (Winch 1958; Lukes 1967), and to draw from them practical inferences about behaviour (von Wright 1971; Hollis 1977). While this tradition has hitherto drawn mainly on history and general ethnography, one biographical application has been attempted in relation to the logic underlying people's accounts of their health histories (Williams 1981).

From the point of view of data collection, these two methods present somewhat different requirements. While logical analysis makes no presuppositions about the historical accuracy of the retrospective account, but simply draws from it the logic of any judgements about events and behaviour which are made, analytic induction must concern itself with historical veracity, for the causal sequence depends upon it. The tests of retrospective accounts which are prescribed by historians like Thompson (1978: 209ff) are thus required reading before collecting life history data with a view to causal interpretations depending on analytic induction. Amongst these tests, cross-checks and comparison with other, if possible independent, witnesses of the same events are amongst the most important matters to arrange before collecting the data. Logical analysis, on the other hand, which is not committed to strict retrospective realism, should ideally be tested by prospective practical inferences, and hence re-interviewing or observation over time is a desirable feature. Cross-referencing with other witnesses, in this latter perspective, is less a method of establishing the historical facts, than a substantive contribution to understanding processes of criticism, reinforcement and persuasion by others.

The analytical purpose, therefore, to which biographical data are to be put, affects in these ways the emphasis of the research design and the choice of data to be collected. In the remaining sections, key sets of choices are reviewed according to the way in which they emerged for investigators whose analytical aims varied along these lines.

8.7.4 Analytical issues in data collection

In drawing attention to the use of biographical material in normal comparative analysis, we have sought to erode the unsatisfactory boundary between "formal" life histories and other biographical material, and thus to emphasize the possibility, in comparative ethnography, of including the biographical interview (even one relatively limited and topical) along with customary observational or survey methods. At the same time, to make full use of such biographical material it is important that at some point it is also indexed and assembled under each individual, and thoroughly analysed in case studies.

However, even though examples of biographical data in general would be relevant in assessing methods of data collection, it is the "formal" life history which is, for most people, the distinctive instance of the method, and it is primarily with examples of these life histories that both the literature, and consequently this present discussion, is concerned.

8.7.4.1 *Relation of life histories to observation and survey method*

A "formal" life history collected on its own without any reference to other data has a very dubious status; collected with reference to other material, however, it can offer a distinctive contribution, at the same time as its weaknesses are complemented by the strength of other methods. Its distinctive usefulness is in the depth of its historical perspective, in the internal analytic connections which it reveals, and in the direct access it provides to the informant's own words. Its weaknesses lie in the informant's often variable ability for self-expression, and in uncertainties about how far the history can be generalized.

A concern with using "formal" life histories in relation with other methods is the particular strength of Oscar Lewis – whom Langness (1965: 13) described as "beyond question the greatest exponent of the life history". Although some reservations need to be made about the shortage of analysis in Lewis's life histories, his continuing involvement with the method over nearly three decades make his commentaries on the present issues of some importance.

The relation of observation to life histories concerned Lewis from the start. In Tepoztlan, however, although he collected life histories (1951: xix–xx), the material seems to have been wholly absorbed into the data gained in other ways; and in *Five Families* (1959) minute observation of a single day formed the dominant method, with life histories appearing only, somewhat awkwardly, in "flashbacks" (1959: 6). Lewis's recognition of the necessity for these "flashbacks", as well as his general desire for an authenticity as independent as possible of the observer, led him to the life history proper – a method which, he claimed, "tends to reduce the element of investigator bias because the accounts are not put through the sieve of a middle-class North American mind" (1961: xi).

However, although after 1961 the life history was Lewis's dominant method of presentation (see, for example, Lewis 1964) he came to see it as, on its own, one-sided; and in *La Vida* he reintroduced the observation of a single day, accompanied separately by verbatim biographical excerpts. And while he comments that "the two types of data supplement each other and set up a counterpoint which makes for a more balanced picture", he adds, interestingly, that "on the whole, the observed days give a greater sense of vividness and warmer glimpses of these people than do their own autobiographies" (1965: xxiii).

This re-evaluation sets in a fresh context the perhaps over-brief accounts given in Lewis's introductions of the questioning process by which the life histories were produced – the "hundreds of questions" asked, but eliminated (1961: xxi), the "patience and prodding" necessary to elicit certain types of information (1964: xxiv), the repeated questioning prior to the tape-recorded sessions which "served to sensitise the informant to the type of detailed information we wanted" (1965: xx), and the deletion of "uninformative material" or "obvious errors" (1977: xxvi). It seems clear that the autobiographical documents were a good deal more the product of Lewis's own preoccupations than would appear from a naive acceptance of his claim for their independence from the "middle-class North American mind". Now the preoccupations of the investigator can of course distort the result, and from this point of view it is a pity that we do not know more about Lewis's questioning; but they can also, in so far as they are based on observation and a full knowledge of the culture, serve as a corrective in eliciting relevant information. It is therefore still more the case that observational analysis, placed alongside the autobiography, can supply many details of the subjects' day-to-day behaviour which they themselves find no need to mention. However, in rediscovering the need for observation, Lewis does not seem to us to have gone far enough. If the aim of the observation is to reveal what natives take for granted, to describe a single day is to include much that is trivial, and to exclude still more that is significant; and to describe rather than analyse is to leave undone the essential task of interpreting unspoken cultural rules to the reader.

The problem of generalizability is a second issue which occupied Lewis, and again he argued for the use of survey and census material from the start (1951: xxi). In most of his life histories, and most notably in *La Vida* (1968), he attempts to present the autobiographies against a population frame established by survey methods and census records. Unfortunately he says too little about his sampling procedures – although it emerges that, probably because of the complex specification of the sample required, they are not always random (1968: 21). It is particularly important, in such cases, that a full account of the procedure is given; and here researchers can draw on discussions of random and non-random sampling such as that of Burgess (1982: 76ff) and Denzin (1970: 81ff), which recognize the peculiar problem of the anthropologist's interest in interactive groups. Both these discussions also recognize that sampling is only part of the problem – that generalizability is always in the end a theoretical issue, involving the identifiability of the unit of study (whether partially or, as often in anthropology, totally sampled) and its similarities and differences with other units.

Other important factors in data collection – obtaining multiple biographies, and re-interviewing over time — are exemplified in Lewis's material, but receive little comment from him; to examine these aspects, therefore, we turn to the work of other anthropologists.

8.7.4.2 *Multiple biographies and change over time as methodological resources*

For obvious reasons, the advantages of taking multiple biographies from the same social milieu are more often exploited than the advantages of having data recorded at various points in a long time sequence. What counts as the "same" milieu is, however, variable. Shaw (1980) contrasts the histories of two Aborigines from the same district, one history relatively picaresque, the other docile; and he links these differences with violent and peaceful stages of colonial expansion in the areas where each grew up, as revealed in the historical record. This analytic comparison, revealing the "merits of producing a series of life histories focused upon one geographic and culturally homogeneous region" (p. 320) is, Shaw claims, new. However the homogeneity of these men's background did not mean, apparently, that they knew each other's stories or could reflect independently on each other's experiences, and this would add a further dimension for analysis.

Multiple biographies of this richer kind, although available to Mandelbaum (1973), are little used by him; but he does make use of records at varying points of time. Also, like Lewis and Shaw, he makes use of the ethnographic background, and it is this which provides him with his baseline notion of the "cultural life plan" available to someone born at the time and in the social position of Ghandi (c.p. Jacobson 1978). Thereafter his notion of "turning points" depends heavily (perhaps for reason of space) on Ghandi's autobiography – too heavily indeed, since it is here that comparison with the versions of Ghandi's life by others who knew him could be important. However, the final "turning point" is of interest partly because it occurs 20 years after the publication of the autobiography. This point is crucial in the evolution of Ghandi's ideas, since it involved him not in united Indian opposition to the British, but in fighting between sectors of Indian society in the violent riots that followed partition. However, of the various courses open to him, his decision on a logical consistency with the ideas of his autobiography emerges in his explanation of his intervention: "My own doctrine was failing: I do not want to die a failure, but as a successful man". As subsequent events showed, the decision was taken at the cost of his life.

Finally, the evolution of ideas and events over time, and the interplay of biographical versions provided by mutually-acquainted informants about one another, come together in Loizos' chronicle of Cypriot refugees, *The Heart Grown Bitter* (1981). In 1968–1969, and on further short visits during 1969–1973, Loizos carried out the fieldwork for an ethnographic study of politics in a Cypriot village (Loizos 1975). These field notes then became an unexpected resource when in 1974, a Greek nationalist coup in the island precipitated a Turkish invasion which turned the villagers into refugees. Returning to the island, Loizos recorded the stories of the refugees, and the exceptional configuration of his data then enabled him to make two important kinds of statement. First, he was able to define truly antecedent conditions, independent of retrospective selection, such as those which made possible some of the most embittered responses to the refugee experience; and secondly, he was able to trace the evolution of political philosophies, and of the valuations which they introduced into villagers' own stories and into their stories of their neighbours, in response to the catastrophe.

Examples such as these suggest that life history taking, after a long period in which analysis was supposed to take second place to authenticity, may now be moving towards a point where considerable analytical gains are possible; but these gains are most likely where multiple biographies are obtained in an ethnographic setting – each, if possible, reflecting in part on another – and where attention is paid to spacing these records over significant periods of time.

8.7.5 Residual issues in data collection

Once decisions have been made about the analytic purpose of the life history, many practical issues of data collection resolve themselves; but some practical problems remain, which relate to the way in which life histories bear on general choices of technique. Thus we conclude by noting a few well-known instances of these.

The honesty and fullness of a life history may vary as the status of the anthropologist in the culture changes. Langness (1965: 35) argues that the taking of life histories should be left until relatively late in the fieldwork period when factors affecting the informant's sincerity have been assessed, and when the anthropologist is no longer entirely a stranger but has acquired some courtesy status. Even then, cross-sex encounters are likely to present difficulties; and Langness indicates the help or hindrance which may be experienced by the use of native assistance in such situations. However it must also be said that, particularly in urban studies where interviewers have a known status, some sort of limited life history at the start can offer a way into more extensive data collection.

A special relevance can also reside in the language used for a life history. Besides the usual problems of the anthropologist's understanding or translation of the language, some cultures present the problem of different languages or dialects being used for different ranges of discourse: for example, Gaelic for domestic relationships, English for public transactions (Mackinnon 1976); or demotic Greek for spontaneous, school Greek for formal speech (Loizos 1981: 190). For life histories, the domestic or spontaneous is usually more appropriate.

As regards the method of recording, biographical material is of course especially appropriate for tape-recording, particularly if extensive quotation is intended, or if the informant's exact words are needed for the analysis. But when it is not so much the exact words but the gist which matters, field notes are often used (Langness 1965: 46) – though in this case it is worth training and testing one's oral memory by the experimental use of both tape-recorder and field notes, with subsequent checking of the one against the other. The tape-recorder, it may be noted, has its problems too, which need sensitive handling (Loizos 1981: 190).

Finally, we have indicated that the part of the anthropologist as questioner should not be underestimated in any life history. Even if, unusually, literate subjects write their own autobiographies (Mandelbaum 1973), there still remain important questions to be asked about the context and purpose of the document (Burgess 1982; Denzin 1970). To have, instead, explicit questions guiding the story, as in oral autobiographies, is in some way more illuminating; but these questions need to be noted so that their influence can be considered in analysis. As regards the style of questioning, various choices are available: we have seen that Lewis had a definite range of topics to cover, and often prompted extensively; Loizos (1981: 190), however, used non-directive questions as far as possible. A further tactic, used by e.g. Shaw (1980), is to go over the completed life history with the informant, noting revisions.

The options which we have touched on in these four areas do not, of course, exhaust the range of practical issues which may trouble an investigator working with life histories in the field; and some further help may be obtained from the detailed comments of Langness (1965) and Denzin (1970) in particular. Our own emphasis in this brief survey has been mainly on the way in which data collection is affected by analytical issues. This course has seemed appropriate not only because of the analytical weakness long noted in many life histories, but also because the use to which the data is to be put will crucially determine the way in which it is collected.

Juliet du Boulay and Rory Williams

8.8 Survey premises and procedures

8.8.1 Introduction

Surveys are not a defining feature of "the anthropological method", but it is a mistake to think they are not or should not be included in the anthropologist's repertoire of practices. There must be few ethnographic field enquiries that have not included a numerical survey of some kind at some stage – whether to map the factual universe prior to analysing uses and interpretations of that universe, or to verify typicality and correlations of items reported or observed in the course of participant observation. The usefulness of these surveys depends of course on what they are done *for*, and on the various dimensions of the research context: no research method works equally well for every purpose in every setting. For some general discussions of survey research in anthropology see Bennett and Thais (1967), Leach (1967), Edgerton and Langness (1974), and Pelto and Pelto (1978: 77–82).

8.8.2 Non-numerical functions

The formal and explicit reason for doing any survey is the need to count and classify some set of items or events or opinions in the field. We will review a number of variations on this basic theme in a moment. Before doing so, however, we may note that the actual business of counting and recording can advance the anthropologist's general purposes in ways that are independent of the numerical data collected.

The fieldworker new to a research area, for example, tends to be shy, anxious and without routines. A survey of house types or other externals at that stage can have a three-fold non-numerical function: it can provide a purpose for early reconnaissance trips, an inoffensive topic for "small talk" exchanges with passers by and, perhaps most important, the beginnings of a structure to working time. Even at later stages, asking for simple answers to explicit survey questions can open the discussion of more delicate topics, and may also provide an acceptable reason for access to people's homes.

The latter effect is particularly useful in urban industrial settings where so little of life is enacted in public spaces that the anthropologist needs to do some part of his "hanging around" inside people's houses. But hanging around indoors without an agenda can discomfit the researcher and irritate the householder – neither of whom will be quite sure what is to be observed. It is no coincidence that surveys are most regularly used in urban settings since it is urban fieldworkers who most need something specific to ask of informants to ease their early exchanges. On the other hand, as we shall see later (§8.8.5.1), people living in this same cultural ambience may also find the presence of a "quasi-official", clip-board touting researcher intimidating or impertinent.

8.8.3 Scientific usefulness

At one level, the element of quantification in surveying permits increasing reliability, comparability and precision in testing theoretical propositions, as well as providing a rapid and systematic means of acquiring large amounts of information. However, the usefulness of a survey can only be measured against clearly recognized objectives – "What is it I want to know?" – and a sensible assessment of research options – "Are there better ways of finding it out?". These questions together depend on recognizing that not all information can be directly *asked*, let alone counted; some must be observed or inferred. Numbers of cattle or children, journeys away from home, time spent tending the garden and the like can be subjects of survey; the quality, extent or interconnectedness of social relationships cannot. This is a most important limitation given the general purposes of anthropology.

Leach (1967), in an article titled "An Anthropologist's Reflections on a Social Survey" spells it out quite cogently: it is the purpose of numerical survey to count units and of numerical *social* survey to count people *as units*.

It is the purpose of social anthropology by contrast to understand people not as units but as integral parts of systems of relationships. And these we know cannot be counted. Nevertheless, quantitative survey combined with more qualitative research strategies can provide dimensions of typicality for case material and will anyway enhance or verify the total ethnographic picture. Mitchell (1966) points out that in settings about which there is no pre-existing record of demographic or economic parameters it may be impossible to proceed to ethnographic work without it. Indeed, in many countries, even where official statistics are at hand, the categories employed may be unsuitable for proper research purposes (§8.8.4), and the accuracy of the figures may be highly questionable, particularly where political interference is suspected. On many aspects of gathering survey data in developing countries, see Hsin-Pao Yang (1955), Mitchell (1965) and Casley and Lury (1981), although these works are mainly concerned with the different types of agricultural and household census and are not specifically anthropological in character.

8.8.4 Types of survey

Just as the usefulness of a survey can only be assessed in terms of its purpose, so it is essential not to confuse the kinds of material being collected. A survey of "facts" that you count yourself is not the same as a survey of facts which are reported by respondents and (only) tallied by the researcher: the first is a count of items, the second a count of statements about items. Facts and figures important to the anthropologist might not be seen as such by informants, with the result that in such areas their memories may be particularly fallible. The counting of opinions is even further out of the anthropological perspective because of the assumption of meaning and context involved. But opinions too are social facts and we should resist the inclination to reject opinion and attitude surveys out of hand. It is more sensible to consider whether and how they can be put into social context and used to fill out the ethnographic picture. Notice that questions of opinion are invariably asked in traditional field studies – "What do you think about X?", "Do you prefer Y over Z?" etc. The difference is that the answers given are not usually tallied and tabulated, and that they tend to be interpreted non-specifically and in the round. It is the unit form rather than the "soft" content of opinion surveys that goes against the anthropological grain. In combination with other kinds of data there is no reason why opinions collected in this way should not be properly interpreted.

Census materials and government surveys of "hard fact" present a different problem. They cannot be achieved by field anthropologists as such, but many refer to them. Being both official and comprehensive they strike the eye as a reliable source of empirical data. For some purposes of course they are. But the population categories used and the assumptions on which they rest pertain to the official cognitive system. This is not likely to match the non-official system and will certainly not reflect variations within it. Thus, for example, the British census boxes the population into fixed "ethnic" groups and assumes the "head of household" to be main wage-earner, principle decision maker – and male. Because no study concerned with political or social process could proceed sensibly from these premises, the anthropologist cannot "use" the data which they structure as though he had collected it himself. In other cultural contexts different census categories may prove problematic, for example "occupation" and "residence" (see Hsin-Pao Yang 1955; Mitchell 1965).

Finally, we should consider the questions of a survey's statistical validity and significance. One is a matter of the way data are collected, the other of interpreting what they mean. This subject will be dealt with at some length in a later volume in this series, but a number of general issues should be introduced briefly at this point. It is, of course, useful (even essential) for anthropologists to have some training in statistics. Even where they have access to numerate members of other disciplines, it is always vitally necessary that they know the difference between numbers and statistics, and between statistical and sociological inference. If the researcher is thinking in terms of inferential statistics – either generalizing about whole populations or testing hypotheses – it will be necessary to think about appropriate forms of sampling. Any basic text will set the scene (see, e.g. Moser and Kalton 1971; Hoinville, Jowell and associates 1978; Marsh 1982). Computers are now widely used to handle large bodies of numerical data and can among other things, put it through any of the standard statistical tests (§9.4). Researchers wedded to qualitative methods may feel themselves threatened by even the idea of data processing, but anthropology as a whole will lose out unless its practitioners understand the possibilities for quantification by machine.

8.8.5 Survey design and administration

This section explores a number of steps necessary for a successful application of the survey method. Specific reference is made to two inner-city neighbourhood surveys in which Wallman and Dhooge collaborated as members of an interdisciplinary research team under social anthropological direction (Wallman and associates 1982).

8.8.5.1 *Getting to know the Field*

Survey research is often equated with opinion-polling and market research. In the popular image it involves little more than entering the field, knocking on people's doors and asking a number of questions. But the most useful surveys follow an intensive fieldwork period through which the researcher(s) have become familiar with the organizations, local areas or groups to be surveyed. It is a crucial phase in designing and conducting a survey. Exploratory interviews and observations help in the selection of units of study and contribute to refinement of any research hypotheses (e.g. Wallman 1983). Furthermore, by "hanging in" and talking to various people the survey researcher acquires an insight into the culture and issues of the setting or group of study. This qualitative information can lead to improvements in the questionnaire, in sampling, and in analysing and interpreting the survey data.

The importance of the fieldwork period extends beyond the design of the survey. The success of any social research – but anthropological research in particular – depends partly on the willingness of informants to cooperate with it. It has become increasingly difficult to gain entry and credibility even to do research which demands minimal time and emotional involvement of the respondents. Academic researchers are not the only professionals collecting information. As a result people have become very wary of answering questions. This is particularly true of organizations, areas and population groups which have been labelled as a "problem", and which are therefore the very areas that social scientists of all sorts are now most likely to be funded to study. The low response rate to surveys undertaken "cold" confirms the importance of a preceding fieldwork period. During that period some of the barriers may be broken down through key-persons who come to vouch for the credentials and credibility of the research, and whose personal networks play an important part in publicizing the survey and getting informants to cooperate.

8.8.5.2 *Questionnaires*

See Oppenheim (1966) and Belson (1981) in addition to general works cited in §8.8.4. Casley and Lury (1981) is useful for developing countries but see also comments and references on asking questions in §8.4.1.

In survey research where the emphasis is on systematization and standardization of the data-collection process, structured questionnaires should be used so that all respondents are asked the same questions in the same sequence; in most cases the possible answer categories are already given. This enhances reliability, allows efficient use of time and labour, and simplifies the coding, computing and tabulation processes.

Structured questionnaires vary a lot in complexity. Surveys concerned with attitudes or perceptions use complicated measuring techniques and scales often derived from psychology. Most social researchers lack the specialized training and knowledge required to construct these measuring devices. But even where technical difficulties are avoided by using more simple techniques, structured questionnaires which include questions of "meaning" still pose the problems of validity. How are we to interpret the answers without a thorough knowledge of the respondents' frame of reference? Without a whole series of closely related questions, "yes" answers to a question, like "Do you find this neighbourhood a nice place to live in?", are almost impossible to interpret. Such questions can only be asked usefully if the research integrates fieldwork and survey methods (see, e.g. Sieber 1982). In those situations where it is necessary to translate a questionnaire into a vernacular(s) the problems are compounded. One way of minimizing these is, having designed a culturally sensitive questionnaire in your own language, (a) translate into the vernacular, (b) get a second person to translate it back into your own language, and (c) identify the areas of difference and hammer out the precise wording in discussion with translators after a short pilot study.

The design of a good questionnaire is always difficult. It involves a number of stages all of which need careful attention. First, there is the problem of what questions to ask. How can the theoretical interests of the study be operationalized? In a study concerned with the resources necessary to managing everyday life in an inner-city area for example, several steps were involved. First it was necessary to postulate a set of six resources required for livelihood, then to decide which ones could usefully be translated into structured interview questions. For this purpose a distinction was made between the material and non-material (symbolic and affective) dimensions. The questionnaire was to deal only with the material dimension and to collect only factual information.

When questions have been selected, answer categories have to be drawn up. Ideally these should be mutually exclusive and should cover every possible type of respondent. In practical terms, however, it is important to resist the temptation to include more questions and answer categories than is feasible or even necessary. One should remember that surveys give respondents little freedom to talk about what they feel is relevant or important. If the questionnaire is too long, they are likely to lose interest and stop cooperating.

Secondly, the wording of questions it always vital. It should be so precise as to leave little scope for interpretation, whether by the interviewer or interviewee, and so neutral as to avoid pushing the respondent towards positive or negative answers. Sensitive issues are probably best approached by indirect questions, but in every case a simple, straightforward vocabulary and style tends to achieve the most reliable result.

Thirdly, there is the problem of where to place the questions. On this point, psychological rather than systematic considerations should prevail. One widely accepted rule is to start and end with neutral questions, putting the more controversial or potentially sensitive ones in the middle. We also have to try and keep the respondent interested. Another common strategy is a spread of open questions throughout the questionnaire which serve to keep respondents interested, by giving them the opportunity to express their views more freely.

It is in a *pilot study* in which the survey questionnaire is pretested that we can assess whether it has the right structure and the questions are properly worded. In pretesting the questionnaire for the inner-city survey referred to above, it was found that some questions were over-ambitious and the order of topics was "wrong". To get an idea how people felt about the density of their own networks and so an indication of their sense of involvement or isolation in the city, the questions "Roughly how many people do you know by sight?, by name?, as friends?" were included. Some respondents could not understand the question, others tried so hard to be accurate that they spent long minutes calculating actual numbers. Both reactions showed the question to be useless and made us aware of a "middle-class" bias in the order of the answer categories in this and other questions. The question was thrown out and it was possible to correct the more obvious biases in the questionnaire before the main survey was launched. The same pilot study showed that the sequence of the sections was wrong. Following the rule of starting with pleasant, neutral questions, the first section focused on leisure activities. People reacted very negatively. They had understood that this was a "serious" enquiry and in their view leisure activities were not serious. So the order was changed and the main survey started instead with questions about the neighbourhood. The reaction was much more positive as people could readily identify with the place they lived in and, apparently by extension, with the questionnaire as a whole.

8.8.5.3 *Interviewers*

Some surveys can or must be carried out by the lone researcher without the assistance of other interviewers. The following section, however, pre-supposes that a number of people are hired to ask the questions which the researcher wants answered.[13]

In every case the reliability and validity of the data collected depend not merely upon the design of the questionnaire – they also depend on the way the questions are asked and on the kind of rapport interviewers are able to develop with the respondents.

To take up the first point: questions must be asked and answers recorded accurately and consistently, and interviewers must be encouraged to follow any instructions they are given. Deviation from the interview schedule, par-ticularly from the wording of the questions, reduces the reliability of the data. It helps if interviewers have some understanding of the project's aims and of the reasons for posing the specific questions. Even simple, straightforward and pretested questions will not be clear to every respondent and the interviewers have then to be able to explain them and to give the right directions. A number of respondents are also likely to perceive some of the questions as threatening or totally irrelevant, and will want to know why they are asked. Without some knowledge of the research objectives, interviewers will be unable to answer such queries satisfactorily and will find it hard to convince respondents that complete information is necessary for the success of the survey. Difficulties in handling the questionnaire can be overcome by proper training sessions and administrative procedures (such as those des-cribed below).

Ability to be accurate and to pay attention to detail are only one set of criteria on which to base the selection of interviewers. Interviewing for social survey is an interactive process in which the personal characteristics of interviewers significantly affect the data collected. In this respect the selection of interviewers is much more complicated. Studies which have examined the factors influencing the rapport between interviewers and interviewees suggest that there is a close relationship between degree of response and the extent to which the social characteristics of the interviewer match those of the inter-viewee.[14] In practice the possibilities for matching even a limited number of characteristics are very limited. Factors such as age, sex, level of education,

[13] You may, of course, hire and train your own field-assistants (§7.11.3). On contracting out surveys to research organizations, see Pelto and Pelto (1978: 218). One British firm with considerable experience of undertaking survey research for academic purposes is Social and Community Planning Research, 35 Northampton Square, London EC1V 0AX (01-250-1866).
[14] This is a common assertion, but one which needs to be examined very critically. Compare, e.g. Hyman *et al.* (1954) and Denney *et al.* (1956–57) with the now extensive literature on the significance of the social characteristics of anthropologists in the field. On the question of sex, see e.g. Wallman (1978).

religion and political affiliation may all be important to some extent. The most appropriate selection of characteristics will depend on the kind of population the survey is aimed at and will not always be a matter of social resemblance: when carrying out a survey of the elderly, for example, women of any age seem to obtain a better response than men. But how are we to select the right interviewers when the population of study is in every sense mixed? The inner-city neighbourhood surveys posed this problem. It was known that the populations were mixed by age, socio-economic status and ethnic origin, but, as is usually the case, we had no more than a vague idea of who lived where within the total survey area. In such situations the limitations on matching by social characteristics are insuperable. But in local studies, all potential informants have one common feature: they are all local residents. On this basis *local* people were recruited to do the interviewing. The decision was taken against the background of the entry problems faced by researchers in inner-city settings. In these areas suspicion of officials and researchers is high and local interviewers, being less threatening, are more likely to succeed in securing trust and cooperation than outsiders.

But no one selection characteristic has the same meaning in every research setting – even if the research problem remains the same. The significance of "localness" for a good rapport between interviewers and interviewees proved to be right in the first survey area: for most residents the fact that the interviewer lived in the same neighbourhood was sufficient reason for their cooperation. In the second area interviewers seemed to need other qualifications in addition to living locally. Early in the survey the analysis of the refusals indicated that ethnic origin was also an important variable. Because the interviewers were of different ethnic origin the problem was solved to some extent by redistributing the local field force. But in order to obtain cooperation from ethnic categories not represented in our field force it was necessary to rely on "ethnic" interviewers from outside the area. In the second area therefore it would appear that the interviewers' ethnic origin was more important than their local status. The reasons for the contrast are discussed elsewhere. Here it is important only that the two areas had different notions of a good match.

8.8.5.4 *Administration and Debriefing*

Surveys are generally a costly and time-consuming business, particularly when it comes to coding and processing. Even when he or she is not directly involved in the data collection, a survey demands the researcher's continuous presence and participation. To administer a survey properly, detailed records have to be kept of the calls made and the time they were made, of successful interviews, and of refusals or misunderstandings. These records facilitate a better use of the field force and provide early warning signs of the possible bias or error in the survey results. The researcher can then decide to change the composition of the field force or give the interviewers new instructions.

Debriefing is equally necessary and time-consuming and is another task best done by the survey researcher. During debriefing sessions thorough checks can be made to see whether the interviewers are handling the question-naire correctly, and to monitor the various factors which may be influencing response rates. Even the most simple and highly structured questionnaires are open to mistakes and the chances of error increase with the size of the field force. A detailed examination of the completed interview schedule will also bring any conscious or unconscious cheating to light. At its best, however, debriefing is not just an effective way of enhancing the reliability of the survey data. It may also uncover weaknesses in the questionnaire which were not revealed in the pilot study. It is important when analysing the data to know how people reacted to the various questions. Because some of these reactions are situation specific, not all questions which might give problems can be detected while pretesting the questionnaire. For example, in seeking information about household tasks and the division of labour a long list of ordinary household machinery in the questionnaire was included, and in-formants were asked whether each item was present in the household. One street in the first area had suffered a spate of burglaries not long before the survey started. Consequently, the people living in that street (and only that street) were reluctant to tell the interviewer what was in the house. Being aware of this weakness before the survey of the second neighbourhood, the form of this question was altered so that it asked only about access to items. This time the question was readily answered except for one item: "com-mercial sewing-machine". In a number of households where respondents said the item was not present, interviewers reported seeing one in the house. In this case reluctance to admit ownership may be a reaction to official efforts to control or to tax money earned for outwork in unenumerated corners of the urban "rag trade".

This leads to a third and an often underused feature of debriefing. Like the non-numerical functions of the survey discussed at the beginning of this piece, debriefing interviewers is not only useful in allowing the researcher to make technical and quantitative corrections. The debriefing procedure is also an invaluable source of qualitative information. Interviewers tend to collect more and different information than is strictly asked for in the questionnaire. No anthropologist will be surprised to discover how often structured interview sessions extend into conversations in which the respondents talk in more depth about issues raised in the questionnaire. Moreover, encouraging interviewers to notice other things happening and keeping notes of all "extra" information discussed during the debriefing brings the standard social survey method very much closer to the central stream of anthropological research practice.

Sandra Wallman and Yvonne Dhooge

8.9 Social network data

8.9.1 Introduction

Social network analysis aims to make statements about the structure of social relationships among a set of actors through establishing *patterns* of linkages or relationships among these actors. In broad terms, therefore, social network analysis is no different from other types of social anthropological analysis: the distinction resides in which characteristics of the linkages the analyst finds convenient to emphasize for the task in hand. Whereas in general social anthropological analysis the *content* of social relationships is likely to be of paramount interest, in social network analysis the aim is to establish the pattern of linkages among actors *given* the specific content of the relationships. For example, in Kapferer's study of a strike in a Zambian clothing factory, he distinguishes two types of basic relationship viz. sociational or convivial and instrumental or exchange relationships. He then sought to show how the pattern of these relationships among the protagonists in the strike changed in a significant way during the six months leading up to the strike (Kapferer 1972).

The analyst may establish the pattern of linkages descriptively by using ethnographic evidence in the disciplined way characteristic of any social anthropological analysis. The notion of "social network" is used here in a metaphorical way to denote the complex set of interlinkages among the actors being described (see, e.g. Lomnitz 1977). Alternatively, the pattern of linkages may be established by using some formal procedure such as clique detection, or "blocking" procedures (for a non-technical account of some of these procedures, see Mitchell (1979) – a somewhat more technical account is provided in Burt (1980)). Both uses of the idea of the social network, of course, demand a systematic and complete assembly of the data which the analyst requires for exposition.

8.9.2 Network characteristics

There can be no simple general rule specifying which aspects of social networks need to be recorded either for formal or less formal analyses of social network data. The morphological and interactional characteristics of social networks which may possibly be of use in different analyses have been listed elsewhere (Mitchell 1969: 10ff) but clearly which of these will be relevant will depend on the particular problem being examined and what sort of statement the analyst wishes to make about it (see Mitchell 1974). In so far as formal analysis is concerned there appear to be two basic necessities. The first is that every effort must be made to provide information about every link of every actor with every other actor (see Holland and Leinhardt (1973) for a discussion of the distortions which may arise from incomplete recording of network data). Setting up a matrix in which each actor is allocated both a row and a column and ensuring each cell of the matrix is filled will provide a useful *aide-mémoire* to achieve this (see matrices in Kapferer (1972: 176–179)). Secondly, it will usually be found helpful to keep linkages relating to different social contexts separate, that is to establish the range and extent of multiplexity empirically. In addition, some indication of the intensity of each link will enable more subtle analyses to be performed than if the link is recorded as either existing or not existing (see the procedure advocated by Bernard and Killworth below).

8.9.3 Ego-centred and set-centred networks

Networks may be set-centred, that is they may refer to links among a set of actors that the observer assumes may potentially have contact with one another. Alternatively, networks may be ego-centred that they may be related to the links among a set of actors selected because they are linked directly or indirectly to some central actor. The network which pertains to a set of actors may be looked upon as made up of elements of the separately recorded ego-centred networks of each member of the set. The links between and among the members of the set may be abstracted from these ego-centred networks to constitute the set-centred networks. In so far as the formal analysis of network data is concerned, the distinction betwen ego- and set-centred networks is of little importance, but substantively it is very important (see Bernard and Killworth (1980) for discrepancy between network data collected by observational and interview methods). The analyst determines the set by some prior consideration and then systematically records the linkages among the members of the set. The *dramatis personae* in the rate-busting incident in the electrolytic cell room as described in Kapferer's account (1969) were defined by the physical lay-out of the work processs and Kapferer systematically recorded the significant interactions of all the members of the cell room. Although the dispute arose between two members of the cell room, the interactions recorded were not limited to these two only. The networks described by Boissevain (1974), however, were centred on two actors with whom Boissevain was friendly. The data used in the analysis were supplied by the two respondents though they were supplemented by Boissevain's own direct observations. Similarly, Epstein's discussion of Chanda's network (1969) was of necessity ego-centred, since the *dramatis personae* who featured in the incidents described were defined by their relationship to Chanda. The resource networks mobilized by people in bereavement as described by Boswell (1969) were also of necessity ego-centred, as was the network he describes of acquaintances surrounding a member of the élite in an African town (1975). The crucial distinction between set-centred and ego-centred networks turns on the extent to which the varying perceptions and interpretations of actions can be incorporated in the analysis. Typically, data related to ego-centred networks tend to be cognitive since the analyst works through a key actor. It is true that an effective analysis should take into account features of the relationships of the members of the network linked to ego, but usually these relationships are those described by ego and not by the actors themselves. It is in this sense that the data tend to be cognitive. There is no reason why the fieldworker should not record the perceptions of the relationships among the other members in ego's immediate or remote network, but this tends to be the exception rather than the rule. Equally there is no reason why the fieldworker should not record features of the behaviour of the members of ego's network towards one another when the opportunity arises, but if this does occur the data tend to become more like the richer set-centred network data than ego-centred network data.

8.9.4 Interviews

It follows that data on ego-centred networks tend to be assembled through contact with the ego upon whom the network is centred and are likely to be collected mainly by interview procedures. This was how Bott collected the data in her pioneering study of social networks. She reports that she and her co-worker conducted an average of 13 interviews of two hours each of the 20 families studied (Bott 1971: 17–24). Boissevain persuaded the two persons in his Maltese study to record the names of all the people they knew (1751 in one case, 648 in the other). Each of these two persons at the centres of the network then recorded some social data about each of the contacts. Boissevain remarks that "this task involved considerable work" (1974: 245). This is surely an understatement. In order to compute the density of the larger network which Boissevain does, he would have needed to enquire about 1.5 million separate relationships, and 200,000 for the smaller.

8.9.5 Observation

Data on set-centred networks are usually collected by procedures similar to those of normal ethnographic fieldwork. The studies by Kapferer of the electrolytic cell room (1969) and the tailor shop (1972) are examples. Kapferer was fluent in the *lingua franca* used in the two work situations. His own observations were supplemented by those of an African research assistant. Kapferer kept full notes of the various interactions going on in the work place. The details for subsequent network analysis were abstracted *after* the event. There is no reason, therefore, why network analysis should not be conducted on standard ethnographic records except that for formal analysis to be feasible the relevant characteristics of every link with every member of the set should be recorded. Given the 54 members of the tailor shop at time 2, and given that the relation of I to J need not be the same as that of J to I, Kapferer needed to have checked $54*53 = 2862$ potential instrumental relationships. In fact what seems to have happened is that only the obvious and ethnographically significant interactions were noted so that of the potential 2862 linkages, only 173 were in fact reported as existing. If the interactions culled from fieldnotes can be coded in machine-readable form a computer program may be used to abstract relevant data in a form suitable for formal analysis.

Bernard and Killworth (1973) used a more systematic procedure to collect network data in their study of the relationships of the crew and the scientists on an oceanographic research ship. This was a typical set-centred study (what they call a "closed network"). Bernard, the anthropologist, prepared a set of cards with the names of the crew and scientists on them. Each of the respondents who took part in the study was asked to arrange the cards into four piles (the respondents' own card having of course been removed): (i) those with whom he has "close" interaction, (ii) "some" interaction, (iii) "little" interaction, and (iv) "no" interaction. Interaction was defined as "working with", "bull-shitting" with, "saying hi to" and "not encounter-ing". The respondents were required then to sort each pile into rank order of intensity of interaction. Bernard and Killworth then used these rank orders to derive distances of each member of the ship's company to every other member, distances which they were able to use in a formal numerical analysis of their data (Bernard and Killworth 1973: 167 ff). I am not sure that rank ordering which, as Bernard and Killworth report, repsondents sometimes find difficult is strictly necessary. Reasonable representations of the structure of relations may be derived from using the pile membership as an ordered category.

Reliance on a formal procedure such as this ensures that the data are collected systematically, but it is not clear how practicable it would be in non-literate field situations. Anwar (1979: 231–233) described his use of this technique among Pakistani immigrants to Britain, some of whom were illiterate. He concludes that card-sorting is difficult for illiterate people, but it works well for educated respondents (1979: 233) and makes the interesting suggestion that photographs might be used instead of cards for illiterate respondents.

8.9.6 Diaries

A technique used in literate societies where close observation of interaction is not possible is to persuade respondents to record their interactions in diaries. This technique, supplemented by interviewing and observation, was used by Cubitt (1973), on a sample of couples for a period of one week. The couples were asked to record the names of those with whom they had been in contact each day and what the contact had entailed, i.e. the "contact" of the relationships. Cubitt remarks that while this technique provides useful in-formation, not many respondents are prepared to endure the chore for more than a week. Clearly the network data are ego-centred but Cubitt provides data on 35 urban families ranging in size from 18 to 53. She estimated densities from focal person's assessment of the links between members of the personal network (1973: 75). Cubitt appreciates that it would have been better to interview all those in the network, but constraints on time and resources prevented this.

8.9.7 Social surveys

Data about ego-centred networks have been collected in some social surveys. Notable examples are studies conducted by the Centre for Urban and

Community Studies of the University of Toronto (Wellman and Crump 1977) and the Institute of Urban and Regional Development, University of California, Berkeley (Jones and Fischer 1978; Fischer 1982: Appendix A). The questions relating to social networks which are feasible to use in social surveys must be fairly simple and direct, should not involve complex and time consuming questioning, and must relate to social characteristics which are amenable to sufficiently unequivocal specification so as to enable responses from different respondents to be compared with one another. These requirements contrast sharply with those of direct observation or deep interviewing in which nuances of relationships can be explored, and detail is an advantage rather than an impediment.

The demand for "objectivity" in social survey methods has led to an emphasis on exchange rather than affective or normative relationships. The standard practice is for the interviewer to elicit a list of names from a respondent in response to questions of the type: "If you have to go away for several weeks, who would you ask to come in to water the plants or keep an eye on things?". Typical issues are – "discuss work with", "take advice from", "discuss personal problems with", "help in the home", "could borrow money from". In the pilot surveys for the Northern California Community Studies, questions of this sort elicited about 12 names per respondent (Jones and Fischer 1978: Table 1). Once a list had been established, personal attributes of these persons could be established and more detailed enquiries (including an estimate of network density) could be made for a sub-sample of the set.

While the East York Social Networks Project (Toronto) also covered the content of exchange links (Wellman and Crump 1977: 22), the emphasis in this survey, unlike that of the Northern California Community Study, was much more on the "intensity" of network links, that is on the closeness and durability of links (Wellman and Crump 1977: 4ff).

Obviously, the constraints imposed upon data collection in normal survey work mean that the sort of information collected will be of a different quality from that collected by observation and by extended inverviewing. Survey data are likely to provide estimates of the variability and extent of different patterns of association but they naturally do not lend themselves to following out leads or to exploring the cognitive and affective implications to the respondent of the links being studied.

J. Clyde Mitchell

8.10 Some other interactionist methods

8.10.1 Introduction

Ethnographers like to think that most of their research involves the observation and description of "natural" events as they happen. Although we know that ethnographers may have a very dramatic and measurable affect on the social situations which they describe (§5.3, §8.1), and while the information they assemble is constantly solicited, prompted and obtained through interaction, these preferred techniques are for the most part indirect or non-interactionist. While these terms may hardly seem justified (at least when compared with other specific techniques which are sometimes used, most noticeably "unobtrusive measures" (§8.1, n.1)), they indicate approaches which are in sharp contrast to the major forms of obviously interactionist, obtrusive, artificial, controlled, or direct procedures. Among these we must include the more structured kind of interview (§8.8), the collecting of verbal generalizations, historical reports, simulation and formal field tests. The last four of these are dealt with here.

8.10.2 Verbal generalizations

All ethnographers rely on these kinds of data to some degree, and for many they are their main source of information. Such generalizations may form a focus of interest in themselves, or they may be a means of gaining access to social processes, events and cultural practices which, while continuing, are inaccessible to the ethnographer directly. This may be because of the insufcient duration of a fieldwork period, because the ethnographer is not permitted to be present, or because the events are physically inaccessible. The information solicited in questionnaires, interview schedules, and other similar formal techniques is frequently of this kind (§8.4, §8.8).

8.10.3 Historical reports

Information on particular past events, or on customary events no longer practised (whether of the immediate or distant past) can only be obtained through the soliciting of oral biographical accounts (§8.7) or through consulting written records (§1.1). It is always important to ensure that data on beliefs and practices acquired through historical reports are not accepted as generalizations of continuing practices.

8.10.4 Simulation

This may involve the special performance of ceremonies, dances, rituals, technical procedures or other kinds of event which are still current but inaccessible (§8.10.2), or the re-enactment or reconstruction of the same which are no longer practised. The latter, in particular, is a hazardous exercise, but legitimate so long as the artificiality of the event is taken into account in its interpretation. It should be especially remembered that ceremonies no longer regularly practised as part of a co-ordinated ritual repertoire and set of beliefs, and brought out occasionally as "folk art", become highly stereotyped, depleted and, of course, acquire new meanings and elements.

The historical reconstruction approach of §8.10.3 and §8.10.4 has been part of anthropology since the early days (Gruber 1970; Mulvaney 1970) and much of the continuing stimulus for investigation is what might loosely be termed "ethnographic salvage". See also Phillips (1971) and §4.4.4.

8.10.5 Tests and field experiments

Ignoring a vaguer non-technical sense in which the term "experiment" has been used in ethnographic research (e.g. Freilich 1963), these are of two kinds: those administered to individuals in series, and those involving groups. Such experiments come closest to laboratory techniques in the context of fieldwork, and it is perhaps understandable that for the most part these have consisted of psychological tests. Such tests have been employed intermittently for decades by ethnographers (e.g. Nadel 1937a, b, 1939a; §3.4, §3.5). Many of them are essentially Freudian-based projective techniques originating mainly in clinical psychology. They include thematic apperception (picture) tests, Rorschach ink-blots, sentence completion exercises, various drawing tests and doll-play techniques. Other psychological research instruments have included optical illusions, acuity of perception tests, judgements of aesthetic quality, pursuit-rotar tests and games. For a survey of tests, and examples of their application, see Jules and Spiro (1953), Phillips (1971), Pelto and Pelto (1978: 81–102); see also Johnson (1978).

Increasingly, formal tests are being employed in other areas of ethnographic enquiry, especially in the analysis of cognitive and semantic aspects of culture. The techniques used have included card-sorting (e.g. Lieberman and Dressler 1977), sorting and arranging objects (e.g. Berlin and Berlin 1975), colour chips (e.g. Heider 1972), drawings of human representations (Harris 1970), painting tests (Forge 1970; Korn 1978), and semantic differential techniques (Osgood 1964). Experiments in the field of art, subsistence and technology overlap with physical simulation techniques (§8.10.4), and have been used extensively by both anthropologists and archaeologists (e.g. Coles 1973). Some of the tests listed may be administered in a questionnaire form and may involve formal techniques of elicitation discussed elsewhere in this volume (§8.4, §8.8).

Experiments involving groups of interacting individuals are rare in ethnography. Some sociologists have relied extensively on laboratory studies of group dynamics and interpersonal interaction (these may involve the use of role-plays, and the construction of hypothetical situations in order to test cultural, linguistic and social responses; see, e.g. Cicourel 1964; ch. 2; also Madge 1957; Denzin 1970). Experiments involving groups have also been used in ethno-archaeology.

Various testing and experimental techniques have been listed here but not discussed. Their various applications are all too specific to warrant extended treatment in a general volume. It is hoped, however, that they will be the subject of more detailed discussion in the various specialized volumes to follow.

8.11 Checklists and *aides-mémoires*

It is rare for any researcher to embark on fieldwork without at least a mental checklist of the kinds of data which are to be collected. However, there is much to be said in favour of a systematic listing of the particular topics, practices and beliefs to be enquired about, and looked for. Such lists may take three forms:

(a) checklists encoded in questionnaire forms;
(b) checklists employed in the context of semi-formal interviews: schedules, interview guides (Lofland 1971: 75ff), lists of question frames (Frake 1964); and
(c) background checklists for occasional reference and to provide guidelines for research in general.

Questionnaires have already been dealt with in some detail (§8.8), while something has also been said about schedules and question frames (§8.4).

The point about schedules is that they are designed to cope with particular interview situations; there can be no general rules and they must continually be updated as new information becomes available.

Background checklists may begin with published lists, such as those contained in *Notes and Queries*, in Mauss's (1947) *Manuel d'Ethnographie*, the *Tozzer Library Index to Anthropological Subject Headings* (1981), in Arensburg (1954), or in the *Outline of Cultural Materials* (Murdock *et al.* 1967). These may serve as useful starting points, act as reminders about information all too often excluded in descriptive accounts, and as occasional reference works. However, they are quite inappropriate for mechanical use in the field. For a start, they lack the detail required. There are a few checklists available designed for work in particular fields of enquiry: for example, on socialization (Whiting 1970; Hilger 1960) and language (Societe pour l'Etude des langues Africaines 1971).[15] We hope that more checklists on particular subjects will become available in the course of this series; while references to the existing field literature on particular subjects will be collected in the relevant volumes. Williams (1967: 25) recommends that in the absence of such guides it is useful to retain several articles or published texts on a subject. These should cover the domain in detail and generally be regarded as authoritative (e.g. Conklin (1961) on swidden cultivation). A good standard ethnographic account may serve the same purpose.

Secondly, a checklist will vary depending on the particular interests of the research project and the hypotheses to be tested. For this reason, it makes sense to compile project-specific checklists, perhaps using available checklists for reference, but sensitive to the needs of a particular project. Hubert *et al.* (1963) have published a lengthy *aide-mémoire* used in their work on middle class kinship in London.[16]

Thirdly, a checklist (like an interview schedule) must be under constant revision in the field. Some ethnographers have found it useful to keep a

[15] This publication consists of volumes on (1) general enquiries in linguistic fieldwork and grammatical description, (2) questions for use in the establishment of grammars, (3) a phrase questionnaire, (4) a checklist of types of tools, techniques and their parts, and (5) checklists for the study of ethnobiology, kinship terms, social organization, proper names, oral traditions, emotions and conceptions. It should perhaps be stressed that in each case the questions are designed for the elicitation of linguistic data.

[16] The Highgate study *aide-mémoire* ran to 34 pages and was organized under 20 broad headings. It is discussed in detail in Hubert *et al.* (1968; see also Firth *et al.* 1969: 34–35). A portion of it is reproduced here as Table 8.2. Note that the section reproduced consists of introductory general guidance for the "enumerator", followed by grouped lists of queries about the information to be elicited. These are accompanied by further explanations and directions. Note also that this was rewritten at the end of about one year to meet the theoretical implications of new material and discussion (see Hubert *et al.* (1968: 7, 24–25) on the use of this *aide-mémoire*).

Table 8.2 *Fragment of* aide-mémoire *used in Highgate kinship study (reproduced from Hubert* et al. *1968: 113–114).*

*5. CHILDREN (See also SERVICES)

The emphasis in this section is on *services*,* i.e. what alternative sources of help are (i) available, and (ii) of available sources, which are utilized, and why. All questions should be asked with these alternatives in mind:

> No help
> Bought services
> Wife's mother
> Wife's mother-in-law
> Other relative
> Friend (not neighbour)
> Neighbour (i.e. on the Estate)

or any combination of any of them. In some families most or all of these alternatives may be available, and in others not, but in all cases it is important to know not only who actually did help, but who the informant would have liked to use if she could have, who she would expect to be willing to help and what sanctions are involved.

All information in these sub-sections must be got as fully as possible, and in such a form as to be easily quantifiable in terms of the *Services Form*, which must be filled in.

Material regarding birth and infancy should be collected for all informants' children up to the age of 15 years old. (For older children, born during the war, details of birth under difficult conditions may be revealing.)

   (a) *Pre- and post-natal care*

> *Who helped in the house if necessary?
> *Who advised, and gave moral support?
> *Any clinics attended? If so, where?
> If not local, where and why?
> Were any new contacts made among neighbours,
>   or others?
> Use of family doctor?

   (b) *Confinement*

This is a time when a woman, if she has any other children, has to rely (almost without exception) on someone apart from her husband to take over at least some of the duties in the home. Unlike other times when she may be ill, or have an accident, etc., there is some time beforehand to plan, and to a certain extent choose, who will take over the care of the child(ren) and possibly the house. Thus, it should be possible to get in some detail the preferences, expectations and processes of decision-making involved.

> For each child:

> *Were the confinements at home or in hospital?
> If hospital, where?
> If not local, why?
> What attitudes of family? . . . [and so on]

separate notebook series containing lists of new queries as they arise, as well as ideas and other notes relating to the conduct of fieldwork, but which should not be confused or incorporated with general fieldwork notes (§8.12.6).

Field guides can act as reminders and stimulus material when making enquiries about events not physically witnessed, or about conceptions not generally mentioned in conversation. They may also provide useful prompts for detailed and systematic observation. However, they may prevent or hamper certain enquiries if used too mechanically and may also create artifacts of their own.

It is hoped that checklists for individual substantive fields will be covered in separate volumes in this series, and that in time lists will appear which update and replace those contained in Sections II and III of the sixth edition of *Notes and Queries*. However, there are basic kinds of data which are required (to varying degrees, depending on the study) in most ethnographic research which cannot be confined to a specialized volume. These include certain aspects of the environmental context, fundamental demography, settlement patterns in their widest sense, age, gender, social inequality, and habitual individual and group routines and customs which fit only awkwardly into the conventional institutional categories. For some indicators see *Notes and Queries* (1951: 35, 58–70) and Murdock *et al.* (1967: sections 13, 18, 36, 54 and 56). On the ascertainment of age, see also Weiner and Lourie (1969: xx–xxii).

R. F. Ellen

8.12 Notes and records[17]

8.12.1 Introduction

The compilation of notes and other written records is arguably the most important part of the fieldwork enterprise, and something which must begin at the very outset of research. Bailey (n.d.: 11) estimated that in the course of 12 months' fieldwork ethnographers are likely to write more than 500,000 words, on average between 1980 and 2375 words a day. This means that

[17] See also Malinowski (1933), Herskovits (1948: ch. 6), Paul (1953), Madge (1957), Mills (1959), Junker (1969: ch. 2), Wolff (1960), Geer (1964), McCall and Simmons (1969: 144–162), Boissevain (1970), Norbeck (1970), Lofland (1971: 101–109), Schatzman and Strauss (1973), Agar (1980: 111–113) and Burgess (1982).

about one-third of all working time in the field is spent in writing. Whether or not these calculations are entirely accurate is difficult to say, but writing is a major occupation and preoccupation. For these reasons, the amount of attention paid to note-taking in the professional literature is amazingly (almost scandalously) slight.

In all cases it is important to remember that the note or item of written information has been abstracted from a complicated context. It is often an extremely partial summary. It may sometimes include extraneous comments, and will invariably employ categories which have no local cultural significance, and arrange data in ways which bear little relation to indigenous concepts (Goody 1977).[18] In this sense, data have been *created*, and although we must guard against the gratuitous mixing of "data" with analysis (Radin 1965: 115–119), it is difficult to see how we can entirely avoid it. Accurate, reliable and "valid" ethnographic description is, in fact, extremely difficult to accomplish and can never be perfect (see den Hollander 1967). The words we use are never entirely free of ambiguity, and should be chosen with care. A good ethnography, which (incidentally) is only as good as the records upon which it is based, is that which is able to distinguish between those words which require explanation and those which do not.

It is also false to imagine that once notes have been written-up, they and they alone constitute the data which are subsequently analysed. Each note, at the point at which it is incorporated into an intermediate or final written text, actually consists of two things: the information written down (as this would be available to any subsequent reader) and a host of unwritten memories and associations accessible only to the original fieldworker. The note, therefore, acts as a trigger for the release of remembered information; the problem being that such information is often incorrect or, at least, has been cerebrally reconstructed since the original event.

8.12.2 Temporary notes

In many situations it is impossible, inconvenient or inappropriate to write a permanent record directly. It is usual to make temporary notes first. Although this may be unavoidable it is important to recognize that it adds a further

---

[18] On the various ways in which our own categories and constructs affect ethnographic descriptions in general, and written field notes in particular, see Bennett (1948: 673), Guest (1960), Ennew (1976) and Cohen (1978: 3). Halfpenny (1979) tries to show the different sources for the categories which we use (e.g. professional, cultural, experimental, analytically-induced, common-sense and "grounded" in the data).

complication to the process by which data are included in a finished text. Thus (referring back to Fig. 8.1), if DATA 1 constitutes what we observe and select with our senses then DATA 2 must be separated into DATA 2a (temporary records) and DATA 2b (permanent records). Of couse, DATA 2a is sometimes avoidable (either by design or by default), but where it is present it provides possibilities for futher misinterpretation and reconstruction.

Ideally, temporary notes are made in pencil or ballpoint pen in a stenographer's notebook or bound tablet. These usually have a stiff back and are conveniently kept in a large pocket or small satchel. It is very important to have fool-proof writing implements. Remember that fountain pens are quite unsuitable and that ballpoint pens are often unreliable and may leak; on wet paper they may be useless. Chagnon (1974: 104) found expensive Parker pens best, but kept a quantity of cheap ballpoint pens in reserve. Soft lead pencils are probably preferable. Paper quality (for both temporary and permanent records) is also an important consideration: it should not absorb water, and ink should not run. Chagnon recommends that, when working under rough conditions, paper should be top-grade heavy bond with a high rag content. Some anthropologists have gone so far as having notebooks specially manufactured to meet their specific requirements, or have had them modified (for example, by the addition of resilient plastic covers) before going into the field. To avoid insect and rodent damage, as well as damp, all stationery should be carefully stored in tins, trunks or other metal containers, preferably together with an insecticide (paradichlorobenzine crystals or commercial mothballs) and silica-gel. Of course, this will not always be necessary.

Temporary notes are characteristically scruffy, sometimes no more than memory triggers. They may be written in shorthand. Although Pitman's or some other form of commercial shorthand may prove useful, it is generally impossible to avoid inventing some personal abbreviations, jargon and other short cuts to information storage. Certain key words may become useful ways of stating complex information (say, on social relationships) concisely. As fieldwork proceeds such notes will be increasingly written in local vernaculars and registers, or other languages through which the research is being conducted. Such notes may be taken while things are happening, say during a conversation or while watching or participating in an event, where active observation or even participation may prevent the immediate writing of fuller accounts. Indeed, they may be quite unintellibigle to anyone else, and after several days perhaps unintelligible to the person who wrote them.

The making of temporary notes usually requires quick decisions about what to write down, what to observe, what to listen to, what to commit to memory and what to leave out or forget about. However, it may be well to remember that what you record and how you record it will very much depend on the speed of events and their structure. You will necessarily impose a structure on the events you perceive and write about. Also you will necessarily be selective and your construction of what is going on will be determined very much by culturally biased preconceptions and prior knowledge. If folk explanations are difficult to understand at the time, then the notes and comments should convey this. It is better for a note to appear on first sight obscure than to suggest exactness or accuracy where this is neither intended nor expressed. Each temporary note should record date, time and place, report on those persons present (including those who supplied the information, but not excluding others). It is also often helpful if the note is accompanied by a tentative suggestion as to how the note might be classified, perhaps connecting it with some earlier sequence.

Problems sometimes arise over the appropriateness of being seen to be taking notes. If as soon as you arrive in the field you are seen taking extensive notes on matters of all kinds this might occasion some consternation. Clearly note-taking must begin early, but it is best to begin in a limited and private fashion, judging people's responses and attitudes. Once well into fieldwork, when people understand your purposes, a more public and comprehensive approach may be perfectly acceptable. Similarly, conversations with individuals may, to begin with, be affected by constant interruptions to write everything down; once people understand your needs it may be quite possible to sit down at a table to write down what is being said, ask informants to go slowly or to repeat, or to read back what you have written.

8.12.3 Permanent records

Figure 8.2 indicates the various kinds of written record found in ethnographic work. These are, naturally, "permanent" only in a relative sense; that is when compared with temporary notes. It is usual for "permanent" records to be altered, added to, corrected and updated, and as data they may be subject to differing interpretations depending on the occasion. However, they often attain an almost artefactual permanence; that is they become reified as "documentation". The dangers of such attitudes and assumptions have already been discussed in §8.1.

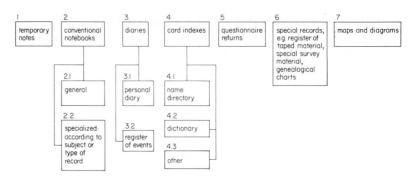

Fig. 8.2 *Types of written record produced during fieldwork.*

8.12.4 Conventional notebooks

Conventional notebooks are treated first as they are the commonest means of assembling field data. All other kinds of written record are, however desirable, optional and highly variable.

Temporary notes, because they are bound to become increasingly indecipherable and unintelligible, should be transcribed into permanent notebooks at the earliest convenient moment. If this is not done, there is the additional hazard of new experiences and sensations either eliminating previous memories, reconstructing inaccurately, or both. Transcription should, ideally, occur every day, preferably at a fixed time. It is necessary to be organized and systematic, otherwise the contents of notes will be lost and therefore represent wasted effort. Routines are essential.

Notes may be transcribed by hand into duplicate or triplicate notebooks. This is still the most common form of making a permanent record. Punched loose-leaf paper in standard ring-binders is also to be recommended: it can be easily manipulated, and also enables some notes to be extracted and carried around and others to be stored safely (Chagnon 1974: 103). Some anthropologists have argued strongly that consolidated notes should be typewritten, on paper or into a portable computer. Whether or not notes are handwritten or typed will depend in part on a personal assessment of the adequacy of handwriting, and on the ease of bringing hardware into a particular field situation. In either case, wide margins should be left for later additions, indexes, annotations and (in the case of typescripts) for binding. Write or type on one side of a piece of paper only. If you use a portable typewriter choose your model with care, ensure that you have tools for making routine repairs, plenty of spare ribbons, and, where necessary, added keys for locally-appropriate orthographies and diacritical marks.

The importance of making multiple copies of notes and other field documents cannot be over-stressed. Not only does this make subsequent analysis easier, it may also prevent the horrendous consequences of the loss of a unique set of data. The number of anthropologists who have lost part or all of the field notes at some time or another is surprisingly large: one thinks of Edmund Leach and at least one distinguished contributor to this volume. However, it is not wise to seek to emulate the masters all of the time.

The top copy in a duplicate notebook is bound into the book, and so remains a chronological and serial sequence of notes. The top copy of typed notes should also be retained as a chronological record, accumulated in a loose-leaf hardback binder. The chronological sequence is a record of how research time is spent and of the order in which information accumulates. Although the chronological set is necessarily retained, marginal cross-references are possible. A cumulative contents page is also useful.

The second (and sometimes third) page in a triplicate notebook is removed as each notebook is completed, and at least one carbon copy should periodically be mailed or sent to a safe address. On return from the field, copies over and above the chronological sequence can be used to roughly divide the field of enquiry topically. Clearly, the advantage of a third copy is that a topical record can be kept in the field in addition to a safe copy elsewhere. If it is inconvenient to make copies mechanically using carbon paper, it may be possible to photograph your notes using fine-grained film (§7.9.1, n. 2).

The *consolidation* of notes is an important part of fieldwork and it is necessary to understand its consequences. It is a key part of the processing of information and thus of the emergence of data. It involves translation, degrees of summarization and expansion, simplification and generalization, and inevitably, therefore, also analysis. It organizes notes in a way that enables back-referral, but at the same time facilitates a constant and cumulative dialogue with your material. It shows up obvious gaps, generates new ideas, permits the re-interpretation of earlier ideas, frames plans for the next steps in research, and may help in the development of long-term strategies of enquiry and tactics for use in particular cases. In other words, it forces you to be creative. It provides an intellectual and imaginative uplift which can be very satisfying and practically rewarding, but which also has its dangers. All this, of course, is a council of perfection, and all field notes will probably fall short of the various recommendations made here.

The writing and consolidation of notes should begin as soon as you enter the field. To begin with it is advisable to spend time simply writing detailed descriptive accounts of things which require a minimum of linguistic competence: physical scenes, material objects, technical procedures and descriptions of behaviour. Note-taking must be practised, and the more practise the better; it will also be helpful to potential informants who will thereby get used to your writing things down. Much of what you write may at first seem obvious and trivial; but it may not be so obvious a few years later, and you may regret not having written more copiously at the time. Remember to write as much down as possible, and include in what you write your own social interactions and their consequences. Note how people behave towards you. As your knowledge of the language improves so you will be able to ask questions and initiate conversations. Begin with innocuous and acceptable subjects, such as local history, technical operations or plant classification; subjects which it is legitimate for a curious outsider to ask about. What is appropriate will naturally vary from one cultural setting to the next.

If not typed, permanent notes must be written carefully in legible handwriting. Legibility can make all the difference in subsequent interpretation, by yourself or by others, especially when it comes to the meaning and construction of indigenous words, and the original intention of the note-taker. It may save time and space to employ a range of abbreviations, but if so these should either be standard abbreviations which can be checked in the relevant reference works or if invented should be used consistently and the range of conventions listed at some convenient place in a notebook. It is also worth being familiar with the full range of technical terms of ethnographic description. When native words appear in notes these should be written according to standard orthographic practice for the language concerned, and if none exists then a transcription which is consistent with the recommendations of the International Phonetic Association should be used. This, however, is not to suggest that a transcription is necessarily mechanical or that it need be narrow; simply that it should be clear and consistent.

In transcribing temporary notes it is necessary to write in sufficient fullness to ensure that the note is going to make sense at some indefinable point in the future, and will be (up to a point) intelligible to another person. Do not worry about stating the obvious; it may not seem so obvious to someone else, or to you ten years later. Also, do not abhor repetition if in doubt. Keep gratuitous interpretation down to a minimum and include everything which might be relevant.

It should be remembered that the completed note forms a record of some permanence which should be treated critically as a text. If we look at a note it will contain certain different kinds of information:

(a) Verbatim texts: words and phrases taken from informants. These may be altered in the sense that they are partly decontextualized. This is so because the complexity of the context of utterance is necessarily reduced, and because rendering them orthographically can never completely capture the full character of utterance.
(b) Translations of verbatim statements.
(c) Generalizations about behaviour in particular situations.
(d) Crude summary translations from informants and actors statements; these are usually highly selective.
(e) Descriptions of actions (or activities) perceived (for example, rituals). These are affected by our own cultural construction of what we regard to be "events".
(f) Interpretation. Without it concise and full description is impossible, and yet it is necessarily highly selective.
(g) Diagrams and illustrations in support of (a) to (e).
(h) Quantified statements (e.g. lists of numbers), relating to (e).

8.12.5 Classifying, coding and indexing (§9.3)

Classifying is important since it may have a profound effect on the efficiency and quality of subsequent information retrieval and, therefore, on the course of analysis. Indeed, the processes of theory construction, hypothesis formulation, categorization, data production, coding and interpretation are inextricably bound-up with one another (Bulmer 1979; Halfpenny 1979).

Classifications facilitate rapid, accurate and comprehensive searches of stored field material, but a poor classification or careless retrieval may be worse than having none at all. In connection with this, particular attention should be paid to classifications which separate data which are otherwise related. For example, if "name-giving ceremonies" are indexed only under RITUAL, a search intended to assemble all data on KINSHIP may fall short of the mark. Similarly, a note on who borrows a cup of sugar might be classified under "friendship", "obligation", "neighbourliness" or "debt" (Cohen 1978: 4). Consequently, some thought to *cross-referencing* may in the long term be time well spent. Many publishing houses produce their own notes for authors which contain useful guidance on indexing. One of the more detailed guides to the subject is Cutler (1970).

Notes must, in the first instance, be coded so that they can be subsequently located in a mass of material. You will probably wish to refer back to earlier notes quite frequently in the field, to check up on certain matters and test informal hypotheses. At the very least, all sheets should be numbered sequentially. In prebound duplicate notebooks, page numbering is generally provided, and if these are being used it also makes sense to number each book sequentially. Since each standard duplicate notebook usually consists of 100 numbered pages, notes may be referred to using a combination of book and page number.

> 1-20, 120 or 0120
> 6-43, 643 or 0643
> 11-12 or 1112

It may sometimes additionally be helpful to provide each note with a number, particularly where more than one note (and on different subjects) appears on the same page. Thus the code then becomes:

> 1-20-2, 1202 or 01202
> 6-43-6, 6436 or 06436
> 11-12-3 or 11123

The other basic information which should be available for each note is as follows:

(a) Date and time of an observation.
(b) Place: village, neighbourhood, house, room . . .
(c) Name, sex, age and status of informant.

Information recorded on different days should not generally appear on the same page, but if information appearing on a single page was obtained under identical circumstances it is clearly unnecessary to repeat the basic information for each note.

In addition to using codes to identify material sequentially, they are also necessary to divide up the field of enquiry. It has been the custom to use basic "institutional" categories to indicate content: for example, *kinship*, *religion*, *economics* and *law*. However, it may sometimes be useful to classify by type of record: thus you may decide to use such categories as *life-histories* or *dispute cases*. Most permanent notes will generally be part of a single sequence, although it may sometimes be useful to run several series of notebooks, either to divide up the field of enquiry exhaustively or to provide additional specialist notebooks on a particular subject (e.g. ethnobotany), or for a particular type of research document (e.g. genealogies or interview files). Some ethnographers (surely unsatisfactorily) have dismembered a chronological sequence of notes and stored them in different envelopes according to subject (Wolff 1960).

For a full classification of a field of enquiry it is possible to utilize or modify existing standard published codes, such as that used for the Human Relations Area Files (Murdock *et al.* 1967). The advantage here lies in the codes being ready-made, each category often being numerically indicated. The sequence of numbers possesses its own internal classificatory logic linking related types of subject-matter. For example:

52 RECREATION
53 FINE ART
531 decorative art
532 representative art
533 music
534 musical instruments
535 dancing
536 drama
537 oratory
538 literature
539 literary texts
54 ENTERTAINMENT

The disadvantage of such codes is the obvious one: no culture will fit precisely into the framework. Under no circumstances should published reports be written in slavish accordance with such schemes.

It will often be necessary to allocate notes to more than one category. In such circumstances repetition can be avoided by making some instant judgement as to the most appropriate category in which to file the note, and then cross-referencing to other relevant categories. Generally speaking, every study will require the development of a more specialized classification to cope with special needs.

It is sometimes difficult to allocate a note to a particular category at the time of writing a permanent record, and certainly the issue should not be prejudged by allocating a note to a doubtful category. If in doubt place the note in the category *general* or *miscellaneous*. It can then be reclassified more appropriately later on. If a note is wrongly classified in an early stage of research, it may prove extremely difficult to retrieve at a later date.

Useful notes on certain subjects cannot be recorded each day as separate events, but are more appropriately written up as cumulative reflections on repeated events over a period of time: e.g. notes on the daily cycle. This, however, should not suggest that they are the result of a synchronic survey.

How material is indexed will depend on individual field situations. It may be possible to keep a running index on cards. This will enable you to make rapid checks when the occasion arises. Sometimes indexing may only really be possible and practicable on returning from the field (§9.3). But in making categories you have selected some criteria as being more important than others. These will reflect matters which your informants regard as important, and which you think will be important for the purposes of subsequent analysis.

It is impossible to work out in any detail a satisfactory classification before embarking on fieldwork, or indeed during the early days of fieldwork, since you will not know what the important criteria are. Furthermore, once you have adopted a classification it will become exceedingly difficult to substitute another, for both practical and cognitive reasons. If you begin with a, b, c, you cannot then switch to d, e, f. What you may be able to do is add further lateral categories or sub-divide existing ones (e.g. a (a_1, a_2, a_3,), b, c, d). But in this case you must understand the consequences of so doing for the character of the data which emerge. The classificatory arrangement used must be able to grow and change organically in response to newly discovered needs. For this reason it may prove best to settle for a crude and conventional scheme which will be adequate for the purposes of retrieval, but not terribly sophisticated. A cross-cutting classification, say *type of record* as opposed to *content*, can be generated by using a colour code retrospectively. Rapid reference can be ensured in this way by underlining or writing with different coloured ballpoints, by using transparent marker pens, or capitalizing key words.

In the end, no doubt, you will be forced to adopt a classification which takes account of both type of record and content. Thus you may employ basic institutional categories, adapted to special interests and in response to the situation; but you will also have categories of types of record, such as pages of notes devoted to a photographic register, a separate series of cards on persons, and so on. Classifications must, in the end, be pragmatic rather than ideal.

Although it is impossible to separate ideas and interpretation entirely from notes, it is probably wise to keep a separate notebook for more general interpretative observations. This provides an opportunity for experimental analysis without feeling that the integrity of the field notes themselves is being violated. Here also can be registered future plans and personal reminders.

Notes of any kind should only be destroyed under the most extreme conditions. Much of what you write may subsequently be rejected as having no value: it may contain repetitions, obvious errors, or may simply be embarrassingly naive. However, it does record the process of establishing insights, and may often contain information, the significance of which only becomes recognized at a much later date.[19]

[19] There is a general, non-anthropological, literature on the art of making notes: see, for example, Berry (1966).

8.12.6 Diaries

Private diaries in the field are not only useful as a form of emotional catharsis, as a means of separating professional from personal observations, but they may also contain ideas, information and comments, that may later become very useful as an additional source of information and as an aid to interpretation. Like field notes in general, they may also be an important means of gauging your reactions and adaptations to an alien world (Larson 1964). As a rule, however, it is important to keep such records separate, whatever subsequent use might be made of them. The reasons may be professional, personal or intellectual. You may, for example, wish to say how unpleasant you find a particular person. Of course, this is not to suggest that notes are not also a personal record. Complementary information of a similar kind to that committed to private diaries may also be extracted from personal letters (§5.3).

Other forms of diary which can be useful in the field are simple daily registers of events, as well as appointment and planning calendars.

8.12.7 Card indexes

The most common use for which card indexes may be put in ethnographic fieldwork is for the compilation of *dictionaries* (not dealt with in this volume; but see Samarin 1967) and *personal and census records*. In the latter case, cards are filed for each individual in a total selected universe (say, a village), or in a sample of that universe. The information recorded on such a card will vary depending on the enquiry, but the basic information is likely to include:

(a) name,
(b) sex,
(c) age (date of birth),
(d) residence/domicile,
(e) name of father and group affiliation,
(f) name of mother and group affiliation,
(g) marital status, and if appropriate name of spouse(s), living and dead,
(h) names of children. Ideally, names of additional persons mentioned on such a record card will also have their own separate card.

In addition to this, depending on the enquiry, it might be useful to provide additional genealogical information, ethnic or religious affiliation, language(s)

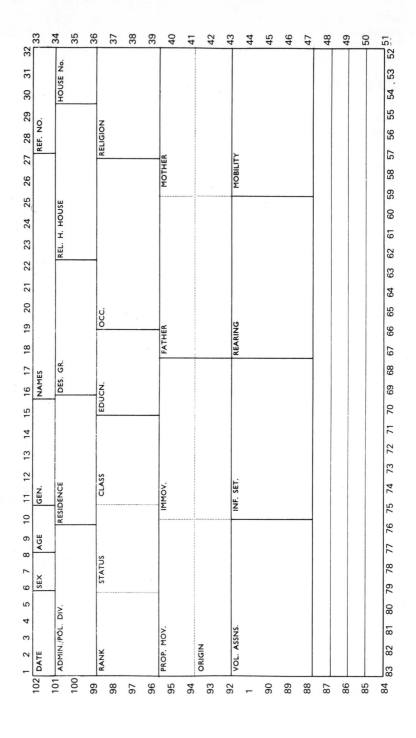

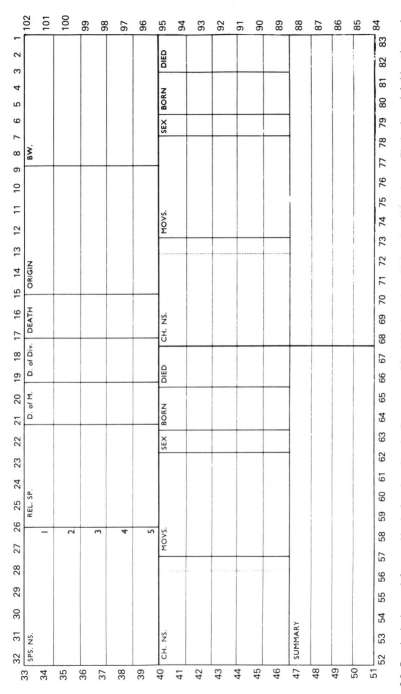

Fig. 8.3 Punched edge card: front and back of card produced at Department of Social Anthropology, University of Manchester. The basic cards (without internal divisions) can be obtained in various sizes from Copeland-Chatterson, Dudbridge, Stroud GL5 3EU, U.K.

spoken, land holdings, wealth, and so on. Such systematic records can be endlessly modified to include the most esoteric information, for example, on the meaning of personal names.

Such basic records can be adapted in various other ways. For example, although it is usually a record of the living, it may also prove useful to extend such a card system to include personal records of the dead. At the same time, it may be useful to code cards in such a way that makes subsequent use of computers easier; that is in a machine-readable form which enbles smooth transference to IBM cards or magnetic tape (Chagnon 1974: 118–119). Chagnon also found it useful to prepare print-outs of personal data in such a way that they could be taken back to the field and used routinely to check findings and update records.

The cards themselves may be of the basic office variety ($5 \times 3, 6 \times 4, 8 \times 5$ cm) which can be fitted into standard cardboard or plastic desk boxes. Alternatively the punched edge card can be used. These have the distinct advantage of permitting coding to be conducted in the field in a way which makes retrieval and preliminary hypothesis testing easy. However, such card systems are expensive. Some institutions have produced their own punched cards based on this model, for example the Manchester card reproduced in Fig. 8.3. Although it is usual for one card to be used for each unit of enquiry (e.g. person, household, village), some punched card systems (such as the B-V card) use cards for each attribute. This may have certain advantages when working under field conditions (Garbett 1965).

A basic problem in keeping useful records of persons, especially where there are large numbers of individuals and where it is necessary to be continually updating and cross-checking information, is the identification of individuals. Photographs may prove helpful, although for immediate use a polaroid camera would be necessary. This photographic technique was pioneered by Rose (1960) and has been widely used since. Chagnon (1974: 11) has suggested developing this one stage further by writing identification numbers on informant's arms and then photographing them. This avoids errors in written inventories or films, but has obvious ethical drawbacks under normal fieldwork conditions.

8.12.8 Audio-tape records (§7.9)

Certain kinds of data lend themselves to audio-tapes. Clearly, they are indispensable in the study of music, and may prove vital in the study of oratory or story-telling. But they are also useful for collecting large amounts of systematic data where the oral quality is itself secondary; for example, genealogies and the texts of myths. Some ethnographers (e.g. Chagnon 1974: 115) are now using audio cassettes for making personal records. Others are recording entire interviews and conversations on tape (Schatzman and Strauss 1973). In addition to the obvious advantages of increased verisimilitude, tape-recorders can be used in situations where writing is difficult. The major disadvantages are the enormous amounts of tape required and the need for an efficient back-up transcription service. Audio cassettes are obviously preferable for collecting this kind of data, where quality is less important than quantity; large reels are generally too bulky. Whether interviews are taped or written must in the end depend on what you are trying to get out of them.

R. F. Ellen

9 *Data into text*

9.1 Introduction[1]

"Writing up" is a transformation of data from the category "what you know" into a new category: "what you communicate". The two are related: at least in an ideal world, what you know sets nearly all the limits of what you can write. And because you create both categories they may be linked also by your intentions from a very early stage: people often plan their research to settle some controversy once and for all, or to modify the practice of a development agency, to secure protection for a threatened people against the destruction of their environment. In each of these cases what the researcher knows is shaped by his earlier aspirations about whom he would persuade of what. "The Ethnographer", Malinowski remarks, ". . . must be an active huntsman". You may not have been so well-organized, so inflexible perhaps, as the anthropologist who caught the boat-train to his field with empty notebooks in his pack, each marked with the title of a chapter of his thesis. Nevertheless you will be a poor researcher if you went to the field without any ideas: "Preconceived ideas are pernicious in any scientific work, but fore-shadowed problems are the main endowment of the scientific thinker" (Malinowski 1922: 8–9). No anthropologist is simply a sponge. This implies, among other things, that since you do plan your research with ideas about what you will eventually write (§7.2), it is worth your while to be aware of the foreshadowed problems about how you will write it. Permit a banal example:

[1] Numerous sociological works deal with matters covered in this chapter: many of them of indifferent quality, others restricted to the discussion of quantitative data. It does not seem sensible to review them here. Galtung (1967) deals with all aspects of data analysis; Selltiz *et al.* (1966) contains a section on the writing of research reports. Mills (1959) remains a classic statement on the conduct of social inquiries and the presentation of results.

ETHNOGRAPHIC RESEARCH
ISBN 0 12 237180 1

when you read the literature on a subject you intend to write about, you will be tempted (because you come to know it so thoroughly, like the back of your hand) to keep only a rough shorthand bibliography. That is foolish because later, when your article or book is finished, you will have to append a list of references: it is inexpressibly tedious to have to go back to the library to copy the full bibliographic details. Foreshadowing the problem of presentation saves time and good temper.

However well-organized you are, transformation of knowledge into communication implies another shifting of material, perhaps equally crucial. Events happen one after another, at any rate are written down one after another. That is as unlikely to be the order in which you wish to communicate them as it is that you read books in alphabetical order of authors. A bibliography can be kept on index cards and each item entered in its proper order, but that is not possible for ethnographic data which has no order until you decide what it will be, and does not consist of discrete units. To transform the order of events into an order of argument you will need to do two things – to create an argument and to devise an information retrieval system, usually an index. So far as the first of these is concerned it seems safe to say that most readers will not welcome guidance of any substantive kind. Nevertheless, it may be worth saying that, for ordinary people, arguments and theories do not spring toothed and armed into the mind, nor are they created by pure cerebral processes, abstract and reproducible: for most of us they are the product of discussion, reading and the contemplation of data – of argument. In short, if you find yourself in the position of having little idea what some smaller or greater part of your data could possibly bear on you should talk to people and argue with them. Often you will find that your companions are less impressed with the intractability of your problem than you are. "Oh, yes," they say, "that's clearly a case of . . .", as if once the class to which something belongs is established, problems of analysis disappear. You can profitably resist such airy categorizations, working out why it is not a case of whatever. That is to say, although the evolution of argument and explanation is indeed social, it is not often a matter of assent – rather, one of dogged and pernickety contestation of convention and easy assumption.

Ideally the class of activities called indexing creates an intermediate stage in which data is held ready for redeployment in a wide variety of future arguments, some of them unforeseen at the time you create the index. It has to be a flexible system, and it is better not to rush into indexing, but to spend a week or two contemplating your data and its kinds, considering what sorts of arguments it will support: you will then have some fairly clear idea of what it is that you will index.

9.2 Kinds of data

Most anthropologists complete their fieldwork with data of the following kinds:

(i)   systematic and exhaustive,
(ii) systematic but not exhaustive,
(iii) descriptive,
(iv) intuitive, unwritten.

Some social scientists think their work is more scientific if the data is homogeneous, is collected in standardized and controlled conditions. That is not necessarily so and is contingent on particular opinions about what science may be; but in any case anthropologists usually insist that their kind of work (tending towards a rounded account of society more than to a survey of a particular institution or set of behaviour) is exempt from this requirement. Anthropologists should translate as well as explain, and that entails additional sorts of thoroughness and discipline. So to list kinds of information does not imply that one kind is better than another. Indeed, in an ideal book, statistics and intuition are marshalled to bear on the same problem, and description is handmaiden to analysis: the kinds are not ranked, even if they support some arguments better than they do others.

9.2.1 Systematic and exhaustive data

These are usually simple in form, however complex their implications may be. A complete or carefully sampled survey of land ownership or of households comes into this category. So too do ordinary lists – membership of associations, say, or of occupational categories. Maps, which list items (fields, houses, wells, shrines) by their spatial relations, or genealogies which list people by their kinship relations, are examples of structured lists, different from simple ones whose order (as it might be, alphabetical) is not *prima facie* of sociological significance.

This information is likely to appear in your text as diagrams and tables: although it is sensible to use prose for simple figures ("fifty-two percent of votes were cast for the socialist party"; "eight adjoining houses are occupied by the descendants of the sisters Latifa and Naziha"), anything more complex is best represented graphically. If you do decide to present lists in graphic form, you should be aware of the conventions and ground rules. Tables, maps and genealogies are highly schematic representations, not very suited to complex data; they are capable of suggesting relations which do not in fact exist. A good discussion of table-making can be found in Goody (1977). Cast as a dissection of Left and Right symbolism, its more general implications are suggestive to any thoughtful person who needs to reduce complex reality to graphic simplicity. More specific rules of presentation are discussed, illustrated by examples, by Moore (1979): readers have expectations of graphs and histograms and, unless you wish to mislead them, it is sensible to conform to the conventions. Figure 9.1 may suggest how expectations, for example that time intervals on a graph should be evenly spaced and on the horizontal axis, can colour perception of statistical data.

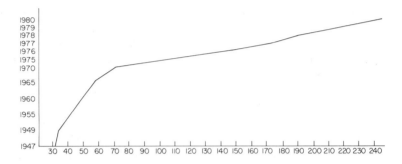

Fig. 9.1 *United Kingdom retail price index, 1974– 1980 (1974 = 100).*

Anthropologists, not always easy with figures, usually aware of complexities in their data, are perhaps predisposed to overload tables and diagrams with information. It is not uncommon to read theses in which every case has its own rank in a magnificent table stretching across several columns of only remotely interconnected information, with a final column of remarks to squeeze in every last drop of scruple: the next 30 or so pages of the thesis scattered with references to "Table *n*". The reader's heart sinks. The most notorious case of table-induced confusion is E. R. Leach's *Pul Eliya* (1961), where schematic representation of data in minute detail does little to persuade his readers of the (famous) argument. Consider, for example, the list which continues over the page-turn 121–122, which refers to two other tables and which concludes: "This again makes it clear that while there is some connection between compound group membership and *pavula* allegiance, the correlation is nowhere very close". In all, it is an inept presentation in a book devoted to the proposition that statistical patterns of behaviour are more significant than the rules of conduct in which they may be dressed.

You will be wise to have good reasons to ignore the precept that schematic representations should contain as little information as necessary: two tables, two genealogies, two maps are nearly always better than one; and three than two. It is the nature of schemata to violate subtlety and nuance, and you cannot overcome that by making them confusing.

The use of analytical statistics in anthropology requires careful collection of information in standard forms, hence careful planning right from the very early stages of research. The classic guidance, and justification, is given by J. A. Barnes (1960, 1967), and will be elaborated by J. S. Eades in a forthcoming volume in this series. It is a forlorn hope that your data will meet the stringent requirements of formal statistical analysis by accident: if you have not planned in advance it is too late to bother now.

Statistics in anthropological argument are usually descriptive: difficult to collect, covering short time-spans, small populations, it is sensible not to derive arguments from figures but to trust your knowledge of the society, to use figures to illustrate arguments derived from understanding of the culture and institutions. You might take as a model P. Loizos' discussion of the relative wealth of 82 households in Kalo, Cyprus. No simple counting of land owned at a particular time would reflect local perceptions of wealth; and Loizos therefore counts the amounts of land held by each household at its peak in the domestic property cycle. But even that is misleading: the villagers are "less concerned with or impressed by what a family has, than [by] what its children will have at marriage" (1975: 311), and a graphic representation should take that into account. So Loizos' table divides maximum landholding of families by the number of children they have, thus combining subtlety with simplicity, producing a table which approximates local perceptions.

The temptation to overload maps is perhaps not so current, partly because anthropologists often get professionals to draw their maps for them. If that highly recommended course is not open to you, you will find Monkhouse and Wilkinson (1963) a thorough guide, especially if, as may be the case, you are able to construct your maps from other ones.

Genealogies (which are discussed at length in volume 2 of this series) share some characteristics with tables and maps. They are highly conventional ways of representing kinship, descent and alliance. They can lend a reality to data which may be in some respects illusory, and the conventions by which complexity is reduced to simplicity may distort the data. Clearly, the conventional sign "=" represents a minimal part of the local and particular meaning of "is married to": but the simplification is generally acceptable. Similarly, you may note, Freeman was forced by "the technical difficulty of depicting a kindred in a single compact diagram" to draw his paradigmatic kindred limiting the degree of cousinage, eliminating most spouses and the marriage of kin, restricting the number of children born to a couple – all for the ease of drawing (1960: 205). Diagrams of unilineal descent are easier to draw but, as Peters' exquisite account of Bedouin shows, may be thoroughly misleading: people who say "Marry your paternal cousins and your paternal and maternal relatives will be one", and who do so marry, do not have a system of actual relationships which can be drawn easily within the framework of the conventional patrilineal genealogy. For such marriages create knots of reduplicated cognatic kinship, and a lineage "united by the bond of agnation will also have within it smaller groups of kin having different sets of cognatic links" (1970: 388–389). Such technical difficulties of representation could result in an untoward convergence between the diagram of actual relationships and those highly schematic charts of patrilineages found in introductory texts on kinship and marriage – some accounts of Middle Eastern societies do in fact seem to have fallen into that trap.

Genealogies, like tables and maps, should be drawn *ad hoc*, to illustrate specific points in your argument. Cluttered ones seem often to result in a certain diffuseness of discussion, as if, having included information in a schema, the author feels bound to qualify and to explain it in words. You may have managed to collect complete or very large genealogies, and that effort, together with the additional effort of having drawn them, can lead you to think that the whole genealogy should be reproduced in the text: you will rightly earn admiration, but you will not necessarily create instant comprehension of your argument.

So much for the preparation of tables, diagrams, maps and genealogies. The general precepts are to keep them simple and specific, to recognize their conventionality and (unless you have compelling reasons for deceiving your less acute readers) to keep to the conventions, recognizing that they may suggest, perhaps even entice you into a distortion of your data.

9.2.2 Systematic but not exhaustive data

These data are the stock-in-trade of social anthropologists. It is generally qualitative information gathered by careful and meticulous inquiry, equally carefully recorded in notes or on tape. In that sense it it systematic; but it is often not complete, either because the topic is inexhaustible, or because the ethnographer became ill, thought he had enough information, or was warned off by officials.

Data in this category requires equally scrupulous exercise of judgement as it is transformed from notes into argument. You should be self-conscious about the procedures you use to reduce scattered and at least initially naive notes to general statements of the kind, "*x* (a category of people) say – or do – or think – *y*". You should also be open about the undoubtedly different degrees of certainty you offer to your readers. This will depend in part on the ways you collected the information: you may find it useful to distinguish among at least the following kinds.

> The results of deep investigations, the matters you had planned to inquire into, had worked out a programme of inquiry for, are likely to be among your most reliable data.

> The results of contingent investigations of matters which arose during your research may not be so reliable.

> Texts, usually formal speech recorded or reduced to writing by you from people you selected or who volunteered information may be biased: careful assessment is required, and the information presented with other confirmatory or qualifying data.

> Recordings of informal speech, collected in a systematic way, are often illuminating: the editor of this volume, for instance, has recordings of Nuaulu as they identify the categories of particular fauna. That kind of discussion is difficult to write down; the discussion is nevertheless important for an account of taxonomic activity.

> Documents may be collected systematically; they are again formal language, written by someone who had a purpose and a relation to what is written (and what is left out). It is important to recognize that tax-collectors, tax-farmers, court clerks, local historians and colonial officers are not dispassionate objective observers as anthropologists are.

> If you are a blue-eyed ethnographer you will also have a considerable amount of latent information: you will, for example, have written down the names of the people who were fellow guests or participants at most of the events you were present at. That is exceedingly valuable information, although you will need an index to extract it.

Clearly, to use this information in argument you have to make judgements about its status, about the weight of argument it will bear. In essence you will make that judgement on the basis of its representativeness and of the consistency of the various bits among themselves.

It is considered vulgar to say so, but for the most part representativeness is a question of quantities. True, ethnography which consists of detailed analysis of cases is sometimes contrasted with generalized accounts of structures, for which quantitative assessment is certainly required: but the differences between structural and case studies are theoretical, not epistemological – it is not as if case studies imply arbitrariness, structural studies merely a plodding accumulation of instances. For the selection of which cases to analyse in detail must be governed by their typicality, their representativeness. In some ethnography, you may feel, the case study method has been selected because the ethnographer had not got much information, but that is not intrinsic to the case study – it is just a characteristic of bad ethnography. Mitchell's *The Yao Village* (1956) is exciting and credible because the reader is sure that the cases are not an arbitrary list of what fell into the ethnographer's net. The same point holds good for other kinds of analysis. If you wish to give an account of the semantic range of the word *bread* you do not accept the casual usage of one speaker, but look for a variety of speakers in a variety of contexts – some of which you will have created yourself. Similarly, part of the weight of structuralist elucidations should come from the recurrence of those oppositions in different contexts – in for example, different versions of the same myth, and in different myths.

So judgements about representativeness imply a weight of evidence which is theoretically quantifiable – so many cases, so many exceptions, such and such a range. You will not produce detailed statistical accounts of your data, and you should not be expected to do so. On the other hand it is well to be self-conscious about how you decide whether some data is significant or not, rather than to rely on intuition. That is not a recommendation to become a statistician, to replace the richness of your understanding with a thin gruel of meaning-free figures; it is a suggestion that your arguments should be constructed with some regard to the amount of evidence which supports them. Some anthropologists will tell you of the temptation to use just one case, perhaps even a discrepant one, as the basis for an argument because the argument so constructed is particularly pleasing and elegant. You should resist it. Remember that you can always say "Although there is only one case in my notes . . ."; and "I have no evidence at all whether x is the case or y, but given a and b, I am very inclined to think I can assume x". Remember too that if you do demolish the widely admired analysis of an eminent scholar, one of his very first moves against you, if he notices at all, will be to say that he really cannot believe your case is typical.

Consistency is not less important. Arguments of the kind sketched above ("I have no evidence at all whether x is the case or y . . .") are not only common in anthropology, but are based on an assumption of consistency which may be more appropriate to the seminar than to the field (Gellner, 1962). Consistency can also refer however to the way different kinds of data relate to each other. The ideal text brings different kinds of data to bear on the one argument, integrates statistics and description and deep analysis. Of course, you will try to avoid apparent contradictions. You should also try to avoid disjunction: a chapter on population growth and rainfall and mean annual temperatures, followed by a chapter of sophisticated analysis of kinship is an unnecessarily disjointed procedure.

9.2.3 Descriptive data

These are those notes which you make on particular events, processes, things. They may be recurrent, as weddings, sacrifices; they may be unique events, at any rate in your experience. Some of the information you have in your descriptions – how much bridewealth was paid, who was present, who tethered the goat – is of course latent, usable with data of the "systematic but incomplete" kind. A further use of descriptive notes, as of photographs, is to recall and to convey the feel, the existential quality of the events described. That has two functions. Your readers, however dryly and abstractly you write, will read with imagination: they will try to get some sense of what it is to live in such company with such kinds of institutions and cosmology. That is not mere sensationalism: in seminars, for example, a person may be suspicious of your argument because he senses an incoherence between descriptions and analysis – his comments can help you to sharpen an argument, augment your descriptions and produce better ethnography. Then again some description is argumentative. Evans-Pritchard's discussion of the nature of Nuer religious life is reinforced by his description of the casualness of those present at some sacrifices, leading to his conclusion that "the many attempts that have been made to explain primitive religions in terms of supposed psychological states – awe, religious thrill, and so forth – are, as far as the Nuer are concerned, inept . . ." (1956, 207).

Mention of Evans-Pritchard's descriptions raises another matter. If you have several accounts of recurrent events in your notes you have the options of presenting a generalized account or an annotated account of a particular case (or of several cases in succession). The choice depends mostly on your preferred method of coping with variation – no two procedures are ever carried out in exactly the same way, but if you are sure that the essentials of the procedure are common to all instances it is reasonable to consider distilling a general account:

> The first act is . . . the driving into the ground of a tethering peg and the tethering of the animal to it. . . . Sometimes, after the victim has been staked, a libation of milk, beer or water is poured over, or at the foot of, the peg (Evans-Pritchard 1956, 208).

One disadvantage of this option is that it forces an indirect language: you will note the present tense and the verbal nouns; the distance introduced by the necessary "sometimes", the frequent use of parenthesizing "or". It is more direct and more immediate to write:

> In the late afternoon another ceremony was performed – the *kava* of the canoe. Food from a large oven was brought into the chief's house, a series of libations poured, and offerings made to the gods of the vessel and of the chief. About a dozen men were present inside, but the expert and some of the workers refused an invitation to come in (Firth 1939, 123).

If you choose this direct style – in the past tense, describing the actions of particular men on particular days in March 1929 – you achieve directness, but encounter problems when you wish to note variations. Should you note every point of variation in the description as it arises? If you do, you may interrupt the flow of your account. Should you save them to the end? If you do, it might cause the reader to ignore them, to make frequent references backwards, or may leave him more impressed with the annotations than with your main story. In Firth's description of canoe repair his narrative is interspersed with comments, generalizations and variants: that is not distracting in a long description (14 pages), but may be so in a shorter one. Finally, note that while particularized narratives give an impression of work done at a lower level of abstraction, that is not necessarily the case, for the narrative is only one part of your text, the rest of it written at whatever levels you wish and are capable of. Moreover, specificity may be considered a virtue if your readers are able to see for themselves what you build your abstractions from. Some people indeed value texts which allow for reinterpretation (not excluded by generalized descriptions), and tend to be understimulated by texts created at a high level of abstraction.

9.2.4 Intuitive knowledge

This last category is distinct from the others because it lies in your memory and is not written down, externalized. In some cases perhaps you could have made notes, in others you may have decided not to – considerations of tact, even of policy render some kinds of information sensitive. In still other cases, the knowledge you have of a culture creeps into you unawares so that it is not until you write a paper or a chapter that you realize that you know, say, the range of greetings and how they are generally used. It is also likely that you retain in your memory a better sense of the feel of some events than you were able to capture in hurried notes made late at night. It is false purism, utterly bogus, to deny validity to this kind of knowledge. Someone who argues that what is not on paper is therefore not objective mistakes externality for objectivity. The one is no guarantee of the other: things you wrote when you were tired, cross, ill, lazy are unreliable, for example, and the fact that they are in ink and on paper, does not confer any special status on them. For the most part you can discount the bias, recognize that because you were tired you left out a "not" in a crucial sentence. An analogous sifting is required, on perhaps rather different principles, of your memories. You will be aware of the weaknesses of human nature, at any rate in your colleagues; and you will know that the exigencies of argument may stimulate wholly honest men, whom you would trust with your handbag, to create facts and to colour events retrospectively; you will be on your guard to be sure that you are not as other people are. In general, it is fair to say, memory is not a good base for rigorous or systematic treatment unless you can demonstrate to yourself in some other area of your inquiry that you have total recall: if you can show by comparison with your notes, that you do recall all the fishermen, you may have some trust that you recall all the secret policemen whose names you did not write down. Otherwise you should be very careful before you base elaborated or crucial arguments on intuitive data and on memory; and you may agree that you should trust memories which run counter to your general line of argument perhaps rather more than ones which seem to confirm a preferred line otherwise weakly supported with evidence. It is a matter of judgement, about yourself and about the weight of memory in argument. Most anthropologists feel most confident using their memories to convey the flavour of events, the character of people; but that should not exclude you from using them, from time to time, in argument.

9.3 Indexing and sorting (§8.12.5)

If you cannot remember exactly everything you have in your notes, transformation from secular to argumentative order usually requires some form of data retrieval system. It is just about possible to avoid this by making multiple copies of your notes, cutting them into pieces and distributing them among files. But that is bulky and expensive; it is also extremely inflexible – once you have sorted the slips into files it is very difficult to re-sort them into new files. If you think it possible that the categories you devise at the beginning of your writing-up will continue to reflect your needs, then you might consider this method. Otherwise, and unless you have access to a computer, indexes are better.

An index consists of three basic elements: mnemomics, references, structure.

Mnemomics are short names for information; references indicate where the information is to be found; the structure is the arrangement of mnenomics.

9.3.1 Mnemomics

Mnemomics in an index serve as category names, e.g. "marriage", "John Smith", "canoe building" are categories into which you can sort information. Because they are categories they are essentially imprecise, bound to overlap, and bound to have gaps between them. They are arbitrary, and are therefore likely to be useful for a year or a decade or so, but to lose their utility as you change your interests. The absurd but not unknown categories "Chapter 1", "Chapter 2" last only for one piece of work: your categories should be more flexible. Most people begin to categorize their data as they write it down, by writing paragraph headings or marginal notes. Precocious categories often die an early death, however, and as you progress through your notes, either in the field or later when you come to index them, you will almost certainly want to create new categories, and you will realize that information indexed earlier should also be noted under the new category. That is inevitable; it is infuriating, but you can minimize the number of times it happens to you by thinking carefully beforehand, and by contemplating your list of kinds of information. Bulmer (1979) has a helpful and illuminating analysis of the different kinds of explicit and implicit categories social researchers employ while they collect data and while they analyse it.

Since categories indicated by mnemomics inevitably overlap, and inevitably have gaps between them, you will necessarily find it difficult to classify some pieces of information. The rule is simple: if you are not sure which of two categories something should go in, put it in both.

9.3.2 References

References should be unique, simple and precise. Many researchers number the pages or sections of their notes as they write them. These numbers can be used later as references; and if they are in one sequence you can avoid complex references (as "vol. 2, p. 34"). If you kept specialist notebooks or have files of documents you will need complex references, but that should be kept to a minimum. The degree of precision with which you identify the location of information is largely a matter of common sense: pages or paragraphs usually serve, but if you have a list of, say, members of an association which is not in any special order, you may want to reference lines. Some computer-based indexes will give a unique number to every word (Read 1980).

9.3.3 Structure

Structure is the order in which you keep your life of mnenomics. Indexes to books are mainly in alphabetical order, with substructures:

"Marriage . . .; age of . . .; ceremonies . . .; sororate . . .".

and may include cross references ("see *divorce*; *fornication*; *widowhood*"). These are designed to help strangers to a book to find their way around it and it may be that a person who is familiar with the data can use a more meaningful structure. You could collect mnenomics into broad groups – "Family and household", "Kinship", "Administration and politics", and so on. You will have to duplicate many of these references: administration overlaps with kinship in many societies, symbolism with politics as well as with religion. Aim to do the minimum necessary to be able to find your way around your notes: you do not have to provide a guide to a complete stranger; but do try to imagine yourself returning in ten years' time to find the data for a new issue that will crop up.

You can choose among a variety of methods of making an index. Some people prepare a number of sheets of paper with the mnenomics at the head and as they read through their field notes they make a reference on the sheet. That costs time as you sift through the sheets each time you wish to make an entry – and you may want to make several entries for each paragraph of your notes. An alternative is to make a slip for each separate reference and to sort them every now and then. So for John Smith's wedding you would fill in various slips:

Weddings: John and Mary Smith's; long description 101–105.
Smith: John's wedding: 101–105
Smith: Mary (*née* don't know: her F said he works for local council): wedding, 101–105

And you might need, in a long account, to make others:

Ritual weeping: at John Smith's wedding, 103
Social embarrassment: John Smith's boss sick at wedding, 104

and so on. That generates an enormous amount of paper; sorting the slips is tedious, but can be a relief from the almost equal tedium of making them. You will find sophisticated advice on making indexes from Cutler (1970), and at a less technical level from Anderson (1971). Both authors are concerned chiefly with public indexes, not with private guides to familiar material.

Should you have more than one index? Some people distinguish routine or "mundane" from "analytical" indexes – the terms are Lofland's (1971: 118); Anderson (1971: 4) uses analytic for Lofland's mundane, synthetic for his analytic. A mundane index could, for example, include the names of all individuals or of places, referring to mentions of them in the notes. An analytic index groups information on related topics: the distinction, not absolute, may nevertheless help you to decide whether or not you might need to keep a separate, differently structured index for all the references to John Smith, in addition to the more analytic categories "Birth – Marriage – Death" in which he and others will no doubt appear. If you have got structured lists of systematic and complete data (such as genealogies, maps) it is sometimes worthwhile using these as the basis for an index. There may be half a dozen people called John Smith and it is in some circumstances convenient to identify them and everyone else in a meaningful way, rather than just giving them arbitrary numbers. Certainly if you plan to use a computer and to do any matching of data you must provide unique identifiers.

9.4 The use of computers

Computers are useful for manipulating data, for testing arguments and for writing text. Forward planning is essential. In some circumstances it is sensible to buy your own micro-computer.

The full power of a computer is most evident in manipulating data. A task which takes a mind several days or weeks can be done in seconds by a machine. For example, the Zuwaya, a tribe in Libya, class their 11,500 people into 164 named minimal lineages, live in about 40 different places and distinguish about 98 different occupations. To prepare a table of how many men of which lineages, living in which places, work at which jobs takes a machine less than 15 seconds, is more accurate than a mind is likely to be, and requires no extra physical space. MacFarlane (1977: 207–214) makes the point even more forcefully, contrasting the use of manual and machine indexes to his historical data in terms of years of work (1977: 207–214). Even so, it is often sensible to do such work by hand. The striking savings in time and accuracy are achieved only after a high initial cost in preparation: data on the Zuwaya took about six months to code and to punch on cards. The job took so long partly because it was not anticipated, and could have been cut to four months if various makeshift intermediate stages had been omitted, the path to the machine mapped out in advance, and more willingly. In short, if you wish to do only one count or major operation with the machine the costs of preparation are high. They diminish as you plan and as you use the machine for more jobs.

Most statistical jobs are catered for by programs within the machine, standard "packages" which will count, tabulate and perform quite complex analyses on any data which is presented in proper form. You do not need to learn a language to use these, but you do need to learn how to send the data to the package and how to instruct it to do the routine jobs you want done. That requires some effort, but is not beyond average intelligence. The most well-known and most used package of this kind is Statistical Package for the Social Sciences (Nie *et al.* 1979) which suits most needs to a sophisticated level. More recent packages do not do more complex tasks, but accept a wider variety of data – the gain is in flexibility. Jobs which are not statistical are less well catered for, partly because people began rather late to use machines for what is somewhat misleadingly called "qualitative" analysis (in contrast to the wholly appropriate "quantitative" uses); partly because literary critics, historians, anthropologists and their like do not make standardized demands of the machines, tend to be more uncertain about what operations they want to use the machine for. Nevertheless, most university-based machines carry some programs for textual analysis; and some centres (Cambridge, Cardiff) develop programs for other purposes. In these cases it is worth writing to them to find out what the state of the science is. For record matching, for example, MacFarlane's Cambridge group is the leading source. Similarly, to find out what is available in the field of literary analysis you should read the *Bulletin of the Association for literary and linguistic computing*, published from Swansea.

Most ordinary demands you might make are catered for, or advice is available from people working on analogous problems. You may however need to carry out jobs which are not catered for, and in this case you will need tailor-made programs. Should you write them yourself? Commercial programmers are expensive. University computing centres usually have friendly programmers; it seems to be their nature to be interested in new problems, and you should in all cases consult one. What they will ask you for is a precise specification, an algorithm, for the tasks you want done. It is not enough to say "the job is to trace all kinship connexions between spouses"; you have to say how the data is arranged, what the steps are for identifying kinship, for identifying spouses, how you want the results presented. All this has to be done in unambiguous terms presenting either—or choices to the machine at each stage. You will find it quite illuminating to prepare algorithms for even standard anthropological tasks, especially if you are under the illusion that minds also work in an either–or way.

If you go so far as to prepare the algorithm, you might as well learn one of the languages which the machine accepts. They are not complex, although they are unambiguous and some people find that difficult. Remember too that you have to learn only to write the language – you are never called on to speak it or to listen to it, you need only to read your own messages. The advantage in learning the language is that you can – always with help and advice at first – write your own programs so that you know that the results are indeed results of operations you specified. Also, you are then able to adjust the program yourself. Most programs fail the first time you use them, they do not produce results in the form you find most apt for your further purposes: tinkering with programs is a common activity, and is best done by the person who uses them (Pelto and Pelto 1978).

What tasks might you use the computer for? Some have been mentioned. The following list is undoutedly incomplete, but may suggest some of the more time-saving possibilities.

9.4.1 Indexing

Various guides to computer indexing exist: they are of variable utility. Read (1980) and Walter (1975) discuss the techniques for research data. Other writers are chiefly concerned with the problems of librarians and publishers. It can be said again that indexes for strangers are different from those for people who are familiar with the material. The problems chiefly arise, as they do with card-indexes, from the natural inadequacy of the categories used, and with the need to follow the references once the machine has listed them for you, and then to copy the text into your working papers. One way to avoid these inadequacies is to enter all the data into the computer and *not to index it at all*. You can then give commands of the kind "Print a file (on the video screen or on paper) containing all the paragraphs which include the word 'marriage'". You will of course have to be sure that paragraphs are clearly demarcated, and that all the notes on marriage do in fact contain that word – although you can also demand the paragraphs on "marriages", "nemakai", "zawaj", "nozze" and any other fancy words you are likely to have used. This method is analogous to photocopying and cutting and filing notes, except that it is *ad hoc*, flexible and less bulky. Foreign languages do not present a problem, provided the orthography can be accepted by the machine. This method is particularly suited to notes, not very useful for lists of either kind. Here the problem is not so much the inadequacy of indexes as the need for cross-referencing. The command "Print a file with all the references to John Smith" produces all the texts and all references to him in lists – but the lists are useful only when they help you to see how he relates to other people in the same list, whether or not he appears on different lists, and how his distribution of appearances compares to other people's. For this you may need to match J. Smith on list A with John Smith on list B. If you have few lists and few people, matching is best done by mind. Structured lists are easily incorporated. A map can be coded with grid references and the codes used to show the spatial relationship of house A to house B. Genealogies can be coded to show the kinship between individuals. Such coding schemes should be open so that new houses or people can be added as they are noted. The advantage of such coding schemes is that they provide each indexed unit (house, field, person) with a unique identity code which incorporates his relation to other units. Other simple information can be added to the identity code to form a record. While such an index is "mundane" in formal terms, considerable information can be extracted from it.

9.4.2 Text editing

The main advantage of text-processing by computer is that the text is stored in memory and needs to be typed only once. Successive modifications can be made at any stage without extensive retyping. This is a substantial saving in time even if you do not compose directly at the keyboard.

9.4.3 Other operations

With ingenuity and patience machines can be instructed to do many tasks which involve routine operations. An early list is provided by articles in Hymes (1965: especially pp. 507ff). More recent examples, using more powerful and cheaper machines, include Chagnon's use of a machine to check the consistency and completeness of his field data (Chagnon 1982), and a fascinating re-analysis of Firth's data on marriage patterns in Tikopia (Kasakoff and Adams 1977). Both of these are suggestive and stimulating. In addition you might consider the use of machines for controlled comparison. The Human Relations Area Files are now computerized (although not available in the UK), and can be used to discover broad patterns of cross-cultural association. It is true that a certain amount of snobbism exists in Britain about the HRAF, but that often seems to depend on a misunderstanding of what they are for. Goody has used them to make preliminary hypotheses about the institutional correlates of bridewealth and dowry in an argument which has certainly stimulated discussion (Goody 1973). Machines can also be used to simulate social processes. If you create an argument that a phenomenon A is produced by the interaction of institutions X and Y over a period of time, you can test your reasoning by making the machine simulate the interaction of X and Y, comparing its results with yours. For example, arguments about the factors which produce the known switches in southern Europe between some form of bridewealth and some form of dowry, and back again, are obvious candidates for simulation (Owen Hughes 1978; Loizos 1975; Ifeka-Moller 1976). That has not yet been done. But some anthropologists have used machines to simulate workings of marriage preferences. The procedure is to estimate how many marriages of a particular kind might be the result of demography, residence and other factors (such as a propensity to marry the nearest eligible spouse), and to compare the results of that simulation with observed rates (Gilbert and Hammel 1966; Goldberg 1967; Hammel and Goldberg 1971). Clearly this basic procedure is repeatable for other effects of alleged human intention. Simulation is also used by Hammel to calculate the effects of incest rules, and is widely used in medical anthropology – for a review of recent work see Johnson (1978: 186–192) and Dyke (1981).

9.4.4 Forward planning for computers

Part of the purpose of this chapter is to suggest that the conventional distinction between research and writing up corresponds to a real bureaucratic distinction ("I need a year's money for research in the field, and a year's money for writing up at home"), but does not correspond very precisely to any likely divisions in the intellectual process of creating a text. This chapter necessarily trespasses on the territory mapped out for Chapter 8; most markedly in the case of computers where, if you plan to use them in the final stages of your research, you should include learning about the machine's capabilities and limitations among your preparations for the field.

Machines are powerful, quick and accurate; but they need their grist in an easily digestible form. To prepare data for entry to a machine is a time-consuming and tedious task, bearing in mind also that the machines are particularly suitable for dealing with large amounts of data. You can minimize the task by recording data in ways which are machine-compatible, or nearly so. It seems improbable that recording data with a view to feeding it to a machine necessarily affects the kind of data you collect: it is possible to imagine people so obsessed with the machines that they become subservient to them; but that is a contingent not a necessary defect. So if you even contemplate vaguely the remote possibility that you might perhaps want to use a computer at a later stage, the golden rule is: talk to an expert before you go to the field; learn something about the ways of presenting data to machines; work out how your data can be recorded in ways compatible with mind and machine; check back with the expert before you implement them.

If you plan to use a computer to index your notes (or to non-index them, as suggested above) you might consider taking a machine to the field. If that seems too expensive or impractical (no computers run on kerosene, but some do run on car-batteries), you can still take steps to minimize later work: your references to your notes, for example, should be kept in a single and numerical sequence as far as possible. When you draw maps, draw them on squared paper, as near to scale as you can manage, and use the grid-reference as part of the identity code for the items ("Francesco's farm: 0108"). When you collect genealogies give each individual a unique code number, ideally one which shows their relation to other individuals – you can then easily attach other indexing information or substantive data to the code number (Chagnon 1974; Davis forthcoming).

9.4.5 Should you buy a computer?

If you can get access to a large machine it is sensible to use that rather than the smaller micro-computers which are those normally bought by private persons. The issue is partly one of the power of the machines, partly of the sizes of their memories. Micro-computers store information on small disks which you put into a disk-drive: you can store the equivalent of about one learned article on a disk, but if you want to store more information it has to go onto another disk. That is not very different from (say) writing notes in a series of notebooks, and is acceptable if you wish to use a micro-computer in the field to record your field notes. It is not very satisfactory if you wish the machine to consider a lot of data all at once: it needs a larger memory. You can buy larger memories for such machines, but they are often at least as expensive as the machines themselves. Similarly, the powerful program packages such as S.P.S.S. are too big to fit into a micro-computer, and in any case are expensive to buy. A computer of your own can be used as a notebook, will do some useful preliminary jobs for you in the field, but will not really serve to do the big jobs you are likely to want it for. The other purpose for which it really can be worthwhile buying your own machine is writing your text. It is still the case that a big computer will do more things for you than a micro-computer – it will store more text in a continuous sequence; its text editing facilities are likely to be more sophisticated, and the printing machines do have a wider range of type-faces. Nevertheless you may not have permanent access to a terminal; university terminals are often in demand during term-time; you may not have the possibility of leaving your papers around you as you write, if that is how you like to do it. There are obvious inconveniences in using shared facilities. If you can afford it, therefore, it may be sensible to buy your own machine. Do not get anything too elaborate, or too specialized. Word-processors, for example, are very specialized – they have their own printers, and do not in all cases have much computing power. Moreover their memories tend to be relatively small. The sensible thing to do, if you can arrange it, is to type your text as you might type your field-notes, into your machine; to store the text on disks, and to transfer them for processing and printing to a larger machine.

If you plan something of this kind you will need to make choices about what kind of machine to buy. You will find that books of advice, usually impartial, can give you a general view of the possibilities; but they have the defects that they are not often specific enough about the kinds of facility an anthropologist is most interested in; that they come out with great frequency, and are rapidly outdated. It is in any case essential that you consult with the managers of the big machine you plan to use: you must be sure that what you record on disk can be transferred easily to the mainframe machine. Among several similar machines that you might buy, some are more easily compatible with some big machines, some with others: that should determine your choice. If you plan to take a machine to the field you should also make sure that the machine can take the local electricity supply.

9.5 Presentation and style

If you are writing for a publishing firm they will provide you with a guide to their house-style which will tell you how they would like you to make references, where they habitually put apostrophes in "the 1930s", "Jones' essay", and (if you are lucky) in the plural of "brother-in-law's". It does no harm to standardize these things: all evidence suggests readers notice inconsistency of usage much more than writers do. If you are not working for a publisher it is probably sufficient to produce your text in conformity with B.S.4821 (B.S.I. 1972). More elaborate guidance is given by Gibaldi and Achtert (1980), Turabian (1982) and by Money and Smallwood (1978).[2]

Style in a more literary sense is a delicate matter, can lead to spectacular rows and wounded feelings because people feel that their particular voice is questioned. That is indeed sometimes the case; and although not all critics of other people's style are reasonable, it is also true that a confused or indistinct voice may not communicate very well. Another reason people are sensitive about style is that they can think that a special voice is required for particular kinds of work: they sometimes mistake pomposity for precision, leading them to resist attempts to make their language plain. If you are charged with this defect in a polite way it is sensible to consider as calmly as you can the proposition that you can achieve high epistemological status by writing sentences which have abstract nouns for subjects and people for objects. If you are convinced of its truth you may resist all attempts to change your style with a quiet mind.

Here are some general principles, cast in a particularly prescriptive form. You will recognize that they are not fixed rules, but that you might need to have good reasons for regularly breaking them.

    (i) Prefer the active to the passive voice.

    (ii) Prefer verbs to nominalized verbs.

    (iii) Prefer people or things to abstractions for the subjects of your sentences.

    (iv) Prefer adjectives to nouns as qualifiers.

    (v) Prefer "I", "you", to "we", "one".

    (vi) Vary the length of your sentences.

    (vii) Connect sentences with "for", "since", "nevertheless".

    (viii) Try to begin rather few of them with "The", "It is", "There is".

    (ix) Try to avoid negative sentences (but do not take this principle to extremes as in the preceding sentence).

These are guidelines: texts which regularly ignore them tend to be less intelligible than ones which keep to them. Incidentally, if you do use a big computer to prepare your text you will quite likely find that the standard

[2] These discuss many technical aspects of presentation: abbreviations, contractions, numerals, spelling, punctuation, capitalization, underlining, quotations, footnotes, references, tables, illustrations, transliteration, spacing, pagination, corrections, insertions, proof-corrections, and bibliographies. On this last matter, see Bowers (1962) for a thoroughly technical discussion; a simple method, widely used by anthropologists, is suggested by the Editor of *Man* (see the *Notes to contributors* in any issue). Publishers' guides which can be bought include Hart (1979), Butcher (1980), the University of Chicago Press (1969) and Leggett *et al.* (1974).

programs include a text analyser. UNIX, a widely diffused system, supports a program STYLE which will count your sentences and sentence structures, your active and passive verbs, your nouns and adjectives and so on. An associated program, DICTION, calls your attention to your use of any of about 400 pleonasms and other infelicities. Table 9.1 is the result of running this chapter through the programme STYLE. These programs are not wholly accurate, do not help you to achieve a beautiful style: but if your text scores "unreadable" on each of the four scales by STYLE you might consider revision.

Table 9.1 *Computer analysis of the style of Chapter 9*

readability grades:[a]
   (Kincaid) 14.4 (auto) 13.0 (Coleman-Liau) 8.3 (Flesch) 14.1 (42.9)
sentence info:
   no. sent 407 no. wds 11654
   av sent leng 28.6 av word length 4.27
   no. questions 7 no. imperatives 1
   no. nonfunc wds[b] 6923 59.4% av leng 5.21
   short sent (<24) 45% (184) long sent (>39) 22% (91)
   longest sent 126 wds at sent 208; shortest sent 4 wds at sent 2
sentence types:
   simple 34% (138) complex 32% (132)
   compound 9% (38) compound-complex 24% (99)
word usage:
   verb types as % of total verbs
   tobe 36% (442) aux 23% (289) inf 22% (269)
   passives as % of non-inf verbs 13% (122)
   types as % of total
   prep 10.0% (1171) conj 3.1% (363) adv 4.7% (543)
   noun 24.2% (2825) adj 22.4% (2608) pron 6.9% (799)
   nominalizations 2% (229)
sentence beginnings:
   subject opener: noun (75) pron (66) pos (3) adj (91) art (45) tot 69%
   prep 6% (26) adv 8% (32)
   verb 2% (9) sub conj 9% (36) conj 1% (6)
   expletives[c] 4% (18)

[a] *Readability grades* measure the reading skills required to understand a given text. In each case the higher the number, the more difficult the text. Kincaid measures on a scale 5.5–16; Automated Readability Index measures on a scale from 0 to 7; Coleman-Liaeu on a scale from 0.3 to 16.3 and the Flesch Reading Ease Score runs from 0 to 17 in the Unix version – it is converted from the original index which runs from 100 to 0, the most difficult text scores 0.
[b] *Non-functional words* are those which are not conjunctions, prepositions, auxiliary verbs, pronouns.
[c] *Expletives* designates sentences which begin with "There is ..." or "It is ...".

If you sense that your writing reads uneasily you could consider reading good prose and working out how the writer achieves clarity and simplicity. You might also consider using a style or usage dictionary: they are good for looking things up in when you are not sure of conventional usage, and many people browse in them for pleasure. Gowers' edition of Fowler (Gowers 1965) or his *Complete plain words* (1967) are useful for most purposes. Strunk (1972) is sensible, and Nicolson (1957) discusses American usage. Each of these is prescriptive, and you may care to consult also the more descriptive work by Crystal and Davy (1969).

9.6 Anonymity and garbling (§6.5.4)

There are solutions to dilemmas which occupy an indistinct middle ground between the ethical and the aesthetic. Just as you may have kept your notes discreet, you may think it is wise to preserve the anonymity of your informants: you may have information which most people would agree is discreditable; you may have information which you and they consider creditable but which police or tourists might take a different view of; it is not unknown for anthropologists to be sued for libel. Anthropologists have used several techniques to resolve these difficulties – none of them seem especially desirable.

They have used pseudonyms either for whole communities or for individuals. You should consider whether that really does protect informants: a false name for a place may misguide tourists, but if the police are as literate and efficient as you fear they are likely to be able to pierce the pseudonym, for example to have records of which foreigners have lived for how long where. You could lessen the risk of discovery by giving yourself a pseudonym as well: nobody seems to have adopted this solution however, perhaps because even those colleagues who conceal their people seek recognition for themselves; though, to be fair, even the most self-effacing anthropologist might reasonably hesitate before offering an anonymous seminar or conference paper. False names for individuals may protect them from the police, be a weaker shield against spouses, children, neighbours. Disguising only the wicked or corrupt may help identification by elimination, and for that reason it may be sensible to invent a complete set of false names for everyone: but that takes a long time, and can confuse you. Peters' work on a Lebanese village (1963, 1972) never gives the village a name at all: that makes the village anonymous, requires all commentary to be cast in the frame "In Peters' Lebanese village . . .", which some writers might consider a disadvantage. A discussion of some aspects of anonymity is conducted between Pitt-Rivers (1978, 1979), Davis (1978, 1979) and Jenkins (1979).

Garbling is another technique used by some anthropologists, and has little to recommend it. You construct fictitious individuals to whom you attribute adultery, political subversion, lunacy, wife-beating and so on. Plausibility requires that scapegoats be not overloaded with sins. You should be careful to construct likely combinations of wickedness; you should be on your guard against readers who will take your construct for real: "Would you please give information about the socio-economic status of the transvestite you described, in particular mentioning whether or not he married uxorilocally?". Ingenious listeners who have constructed a demolition of your analysis on the basis of garbled ethnography may resist your assertion that it is all an invention. If invented characters have a tendency to get out of hand the problem is worse when you garble events or situations: it becomes very difficult to keep them under control, however important it may be to hide the inessential parts of the truth.

Finally, one solution is absolutely confusing to your readers, and should be avoided. Abu Zahra (1982) has met the delicate task of protecting the identity of informants by giving no names to individuals but describing them by their relations to others, by their activities and roles. "An educated man"; "the circumcised's father"; "One of the deceased's father's brother's wives who is also a distant relative of the Qawwala's wife (and whose daughter is married to the deceased's brother) was in a delicate position . . ." (p. 110). Apart from the difficulty of the prose, and the difficulty of distinguishing between action and generalized description of roles, it becomes very difficult to follow people through the text: a person is father's brother in one context, sister's husband in another. It is often important to be able to recognize the ethnologized as individuals.

Probably no solution is perfect; each anthropologist with fears for his people will camouflage them as best he can. It is of course always sensible to consider whether or not it is essential to publish difficult information at all.

J. Davis

Bibliography

Note: *In most cases, dates provided are to the first edition of a volume. Where this differs from that of the edition consulted or referred to, date of first publication is placed in brackets.*

Aberle, David F. (1966). Peyote religion among the Navaho. *Viking Fund Publications in Anthropology* **14**, 227–243.

Abu-Zahra, N. (1982). *Sidi Ameur. A Tunisian Village*. St. Anthony's Middle East Monographs **15**. London: Ithaca.

Ackroyd, S. and Hughes, J. A. (1981). *Data Collection in Context*. London: Longman.

Adair, J. (1960). A Pueblo G.I. In *In the Company of Man: Twenty Portraits by Anthropologists* (ed. J. B. Casagrande). New York: Harper and Row.

Adam, J. M. (1966). *A Traveller's Guide to Health*. London: Hodder and Stoughton, for the Royal Geographical Society.

Adams, R. N. (1971). Responsibilities of the foreign scholar to the local scholarly community. *Current Anthropology* **12**(3), 335–339. In Joseph G. Jorgensen and others, Toward an ethics for anthropologists. *Current Anthropology* **12**(3), 321–356.

Adams, R. N. (1981). Ethical principles in anthropological research: one or many? *Human Organization* **40**(2), 155–160.

Adams, R. N. and Preiss, J. J. (eds) (1960). *Human Organization Research: Field Relations and Techniques*. Homewood, Ill.: The Dorsey Press, Inc., for the Society for Applied Anthropology.

Agar, M. H. (1980). *The Professional Stranger: An Informal Introduction to Ethnography*. Orlando, New York and London: Academic Press.

Aguilar, J. L. (1981). Insider research: an ethnography of a debate. In *Anthropologists at Home in North America: Methods and Issues in the Study of One's Own Society* (ed. D. A. Messerschmidt). Cambridge: Cambridge University Press.

Albert, E. M. (1960). My 'boy', Muntu (Ruanda-Urundi). In *In the Company of Man: Twenty Portraits by Anthropologists* (ed. J. B. Casagrande). New York: Harper and Row.

Allerton, D. J. (1979). *Essentials of Grammatical Theory: A Consensus View of Syntax and Morphology.* London: Routledge and Kegan Paul.

Almy, S. W. (1977). Anthropologists and development agencies. *American Anthropologist* **79**(2), 280–292.

American Anthropological Association (1971). *Principles of Professional Responsibility.* Washington, D.C.: American Anthropological Association. [Reprinted partially or wholly in Reynolds (1982); Diener and Crandall (1978); Rynkiewich and Spradley (1976) q.v.].

American Anthropological Association (1973). *Professional Ethics: Statements and Procedures of the American Anthropological Association.* Washington, D.C.: American Anthropological Association.

American Anthropological Association (1981). *Guide to Departments of Anthropology 1981–82* (September), 20th edn.

American Journal of Sociology (1956). **62**(2), September.

The American Sociologist (1978). Exchange [three papers by Richard M. Stephenson (q.v.), Joan A. Cassell (q.v.) and Kathleen Bond, with comments and replies on ethical problems in social research]. *The American Sociologist* **13**(3), 128–177.

Anderson, M. D. (1971). *Book Indexing* (Cambridge authors' and printers' guides). Cambridge: Cambridge University Press.

Anscombe, G. E. M. (1957). *Intention.* Oxford: Blackwell.

Anthropological Forum (1977). Anthropological research in British colonies: some personal accounts. *Anthropological Forum* **IV**(2), 137–248.

Anwar, M. (1979). *The Myth of Return: Pakistanis in Britain.* London: Heinemann.

Appell, G. N. (1974). *Basic Issues in the Dilemmas and Ethical Conflicts in Anthropological Inquiry.* New York: MSS Modular Publication, Inc., Module 19 (1974), pp. 1–28. [*N.B.* This publication is sometimes dated 1973 in citations; here the publisher's suggested citation is followed.]

Appell, G. N. (1976). Teaching anthropological ethics: developing skills in ethical decision-making and the nature of moral education. *Anthropological Quarterly* **49**(2), 81–88.

Appell, G. N. (1978). *Ethical Dilemmas in Anthropological Inquiry: A Case Book.* Waltham, Mass.: Crossroads Press.

Appell, G. N. (1980). Talking ethics: the uses of moral rhetoric and the function of ethical principles. *Social Problems* **27**(3), 350–357.

Ardener, E. (ed.) (1971). *Social Anthropology and Language* (ASA Monogr. 10). London: Tavistock.

Ardener, E. (1971a). Social anthropology and the historicity of historical linguistics. In *Social Anthropology and Language* (ed. E. Ardener) (ASA Monogr. 10). London: Tavistock.

Ardener, E. (1972). Belief and the problem of women. In *The Interpretation of Ritual* (ed. J. S. La Fontaine). London: Tavistock. Reprinted (1975) together with "The Problem Revisited" in *Perceiving Women* (ed. S. Ardener). London: Dent; New York: Wiley.

Ardener, E. W. and Ardener, S. G. (1965). A directory study of social anthropologists. *British Journal of Sociology* **16**(4), 295–314.

Ardener, S. (ed.) (1975). *Perceiving Women*. London: Dent; New York: Wiley.

Ardener, S. (ed.) (1978). *Defining Females: The Nature of Women in Society*. London: Croom Helm; New York: Wiley.

Ardener, S. (ed.) (1981). *Women and Space: Ground Rules and Social Maps*. London: Croom Helm; New York: St. Martin's Press.

Ardener, S. (1983). The presentation of women in scientific discourse. In *Visibility and Power: Anthropological Essays in Society and Development* (ed. L. Dube). Delhi: Oxford University Press (in press).

Arensberg, C. M. (1954). The community-study method. *American Journal of Sociology* **60**, 109–124.

Arensberg, C. M. and Kimball, S. T. (1940). *Family and Community in Ireland*. Cambridge, Mass.: Harvard University Press.

Asad, T. (ed.) (1973). *Anthropology and the Colonial Encounter*. London: The Ithaca Press; New York: Humanities Press.

Asad, T. (1973). Introduction. In *Anthropology and the Colonial Encounter* (ed. T. Asad). London: The Ithaca Press; New York: Humanities Press.

Babchuch, N. (1962). The role of researcher as participant observer and participant-as-observer in the field situation. *Human Organization* **21**(7), 225–228.

Back, K. W. (1956). The well-informed informant. *Human Organization* **14**(4), 30–33. Reprinted (1960) in *Human Organization Research: Field Relations and Techniques* (eds R. N. Adams and J. J. Preiss). Homewood, Ill.: The Dorsey Press Inc., for the Society for Applied Anthropology.

Bailey, F. G. (1977). *Morality and Expediency: The Folklore of Academic Politics*. Oxford: Blackwell.

Bailey, F. G. *et al.* (1967–68). *Social Anthropology: Handbook*. The University of Sussex, School of African and Asian Studies. [Unpublished cyclostyled documents.]

Barnard, A. (1980). Sex roles among the Nharo Bushmen of Botswana. *Africa* **50**(2), 116–124.

Barnes, J. A. (1947). The collection of genealogies. *Human Problems in British Central Africa. (Rhodes-Livingstone Journal* **5**, 48–55).

Barnes, J. A. (1949). Measures of divorce frequency in simple societies. *Journal of the Royal Anthropological Institute* **79**, 37–62.

Barnes, J. A. (1960). Marriage and residential continuity. *American Anthropologist* **62**, 850–866.

Barnes, J. A. (1967). The frequency of divorce. In *The Craft of Social Anthropology* (ed. A. L. Epstein). London: Tavistock.

Barnes, J. A. (1967a). Some ethical problems in modern fieldwork. In *Anthropologists in the Field* (eds D. G. Jongmans and P. C. W. Gutkind). Assen: Van Gorcum. An earlier version was published (1963) in the *British Journal of Sociology* **14**, 118–134.

Barnes, J. A. (1969). Politics, permits, and professional interests: the Rose case. *The Australian Quarterly* **41**(1), 17–31. Reprinted in Appell (1978) q.v.

Barnes, J. A. (1977). *The Ethics of Inquiry in Social Science: Three Lectures*. Delhi: Oxford University Press.

Barnes, J. A. (1979). *Who Should Know What? Social Science, Privacy and Ethics.*

Harmondsworth: Penguin Books.

Barnes, J. A. (1981). Ethical and political compromise in social research. Mss., F/1/322/2. June 1981. Quoted with permission. A Dutch version is available: "Ethische en politieke compromissen in sociaal onderzoek". In *Hoe weet je dat? Wegen van sociaal onderzoek* (eds C. Bouw, F. Bovenkerk, K. Bruin and L. Brunt), pp. 27–37. Amsterdam: Uitgeverij De Arbeidspers/Wetenschappelijke Uitgeverij (1982).

Barnes, J. A. (1982). Social science in India: colonial import, indigenous product, or universal truth? In *Indigenous Anthropology in Non-Western Countries* (ed. H. Fahim). Durham, North Carolina: Carolina Academic Press.

Barnett, H. G. (1970). Palauan Journal. In *Being an Anthropologist: Fieldwork in Eleven Cultures* (ed. G. D. Spindler). New York: Holt, Rinehart and Winston.

Barth, F. (1974). On responsibility and humanity: calling a colleague to account. (With a reply by Colin Turnbull). *Current Anthropology* 15, 99–103.

Bartlett, F. C. (1937). Psychological methods and anthropological problems. *Africa* 10, 401–420.

Basso, K. H. (1979). *Portraits of "the Whiteman"*. Cambridge: Cambridge University Press.

Bastide, R. (1973). *Applied Anthropology*. New York: Harper and Row.

Bateson, G. and Mead, M. (1942). *Balinese Character: A Photographic Analysis*. New York: Academy of Sciences.

Baviskar, B. S. (1979). Walking on the edge of factionalism: an industrial co-operative in rural Maharashtra. In *The Fieldworker and the Field: Problems and Challenges in Sociological Investigation* (eds. M. N. Srinivas, A. M. Shah and E. A. Ramaswamy). Delhi: Oxford University Press.

Beals, A. R. (1970). Gopalpur, 1958–1960. In *Being an Anthropologist: Fieldwork in Eleven Cultures*. (ed. G. D. Spindler). New York: Holt, Rinehart and Winston.

Beals, A. R. (1978 [1972]). Ethical dilemmas in anthropological fieldwork. In *Ethical Dilemmas in Anthropological Inquiry: A Case Book* (ed. G. N. Appell). Waltham, Mass.: Crossroads Press. (Originally published in *The Piltdown Newsletter* 3(3), 1972: 1–3).

Beals, R. L. (1969). *Politics of Social Research: An Inquiry into the Ethics and Responsibilities of Social Scientists*. Chicago: Aldine Publishing Company.

Beattie, J. (1964). *Other Cultures*. London: Cohen and West.

Beattie, J. (1965). *Understanding an African Kingdom: Bunyoro*. New York: Holt, Rinehart and Winston.

Beauchamp, T. L., Faden, R. R., Wallace, R. J., Jr. and Walters, L. (eds) (1982). *Ethical Issues in Social Science Research*. Baltimore: The Johns Hopkins University Press.

Becker, H. S. (1954). A note on interviewing tactics. *Human Organization* 12(4), 31–32.

Becker, H. S. (1958). Problems of inference and proof in participant observation. *American Sociological Review* 23, 652–660.

Becker, H. S. (1964). Problems in the publication of field studies. In *Reflections on Community Studies* (eds A. J. Vidich, J. Bensman and M. R. Stein). New York: John

Wiley and Sons.

Becker, H. S. (1967). Whose side are we on? *Social Problems* **14**(3), 239–247.

Becker, H. S. (1971). *Sociological Work: Method and Substance.* London: Allen Lane (The Penguin Press).

Becker, H. S. and Geer, B. (1957). Participant observation and interviewing: a comparison. *Human Organization* **3**, 28–32.

Beckham, R. S. (1967). Anthropology. *Library Trends* **15**(4), 685–703 (April).

Beidelman, T. O. (1974). Sir Edward Evan Evans-Pritchard (1902–1973): an appreciation. *Anthropos* **69**, 554–567.

Bell, C. and Newby, H. (eds) (1977). *Doing Sociological Research.* London: George Allen and Unwin.

Bell, C. and Newby, H. (1977a). Epilogue. In *Doing Sociological Research* (eds C. Bell and H. Newby). London: George Allen and Unwin.

Bellwinkel, M. (1979). Objective appreciation through subjective involvement: a slum in Kanpur. In *The Fieldworker and the Field: Problems and Challenges in Sociological Investigation* (eds M. N. Srinivas, A. M. Shah and E. A. Ramaswamy). Delhi: Oxford University Press.

Belshaw, C. S. (1976). *The Sorcerer's Apprentice: An Anthropology of Public Policy.* New York: Pergamon Press.

Belson, W. A. (1981). *The Design and Understanding of Survey Questions.* Farnborough: Gower.

Bennett, J. W. (1948). The study of cultures: a survey of techniques and methodology in fieldwork. *American Sociological Review* **13**, 672–687.

Bennett, J. W. (1960). Individual perspectives in fieldwork. In *Human Organization Research: Field Relations and Techniques* (eds R. N. Adams and J. J. Preiss). Homewood, Illinois: The Dorsey Press Inc., for the Society for Applied Anthropology.

Bennett, J. W. and Thais, G. (1967). Sociocultural anthropology and survey research. In *Survey Research in the Social Sciences* (ed. C. Y. Glock). New York: Russell Sage Foundation.

Bensman, J. and Vidich, A. (1965). Social theory in field research. In *Sociology on Trial* (eds M. Stein and A. Vidich). Englewood Cliffs, N.J.: Prentice-Hall.

Berk, R. A. and Adams, J. M. (1970). Establishing rapport with deviant groups. *Social Problems* **18**(1), 102–117.

Berlin, B. and Berlin, E. A. (1975). Aguaruna color categories. *American Enthnologist* **2**, 61–87.

Berlin, B., Breedlove, D. and Raven, P. (1974). *Principles of Tzeltal Plant Classification: An Introduction to the Botanical Ethnography of a Mayan-speaking People of the Highland Chiapas.* Orlando, New York and London: Academic Press.

Bernard, H. R. and Killworth, P. D. (1973). On the social structure of an ocean-going research vessel and other important things. *Social Science Research* **2**, 145–184.

Bernard, H. R. and Killworth, P. D. (1980). Informant accuracy in social network data IV: A comparison of clique-level structure in behavioral and cognitive network data. *Social Networks* **2**(3), 191–218.

Berreman, G. D. (1962). *Behind Many Masks: Ethnography and Impression Management in a Himalayan Hill Village.* Ithaca, New York: Society for Applied Anthropology, Monograph No **4**. (Reprinted as the Prologue to his *Hindu of the Himalayas,* Berkeley, University of California Press, 1972, pp. xvii–lvii).

Berreman, G. D. (1968). Ethnography: method and product. In *Introduction to Cultural Anthropology: Essays in the Scope and Methods of the Science of Man* (ed. J. A. Clifton). Boston, Mass.: Houghton-Mifflin.

Berreman, G. D. (ed.) (1973a). Anthropology and the Third World. In *To See Ourselves: Anthropology and Modern Social Issues* (ed. T. Weaver). Glenview, Ill.: Scott Foresman and Company.

Berreman, G. D. (ed.) (1973b). The social responsibility of the social anthropologist. In *To See Ourselves: Anthropology and Modern Social Issues* (ed. T. Weaver). Glenview, Ill.: Scott, Foresman and Company.

Berreman, G. D. (1978). (Comment on "Ethics and responsibility in social science research"). *The American Sociologist* **13** (August), 153–155.

Berreman, G. D., Gjessing, G., Gough, K. (and others) (1968). Social Responsibilities Symposium. *Current Anthropology* **9**(5), 391–435 (with comments).

Berry, R. (1966). *How to Write a Research Paper.* Oxford: Pergamon Press.

Béteille, A. and Madan, T. N. (1975). *Encounter and Experience: Personal Accounts of Fieldwork.* Delhi: Vikas Publishing House.

Blalock, H. M. and Blalock, A. B. (eds) (1968). *Methodology in Social Research.* New York: McGraw Hill.

Blanchard, K. (1977). The expanded responsibilities of long-term informant relationships. *Human Organization* **36**(1), 66–69.

Blaxter, M. and Paterson, E. (1982). *Mothers and Daughters: A Three-generational Study of Health Attitudes and Behaviour.* London: Heinemann.

Bloch, M. (ed.) (1975). *Political Language and Oratory in Traditional Society.* London, Orlando and New York: Academic Press.

Blumer, H. (1969). *Symbolic Interactionism.* Englewood Cliffs, N.J.: Prentice-Hall.

Boas, F. (1906). Some philosophical aspects of anthropological research. *Science* **23**(591), 641–645.

Boas, F. (1919). Scientists as spies (Letter). *The Nation* **109**, 797. Reprinted (1973) in *To See Ourselves: Anthropology and Modern Social Issues* (ed. Thomas Weaver) Glenview, Ill.: Scott, Foresman and Company.

Boas, F. (1920). The methods of ethnology. *American Anthropologist* **22**(4), 311–321.

Bohannan, L. (pseud. Elenore Smith Bowen) (1954). *Return to Laughter: An Anthropological Novel.* London: Gollancz; New York: Harper and Row. 1964, New York: Doubleday.

Bohannan, L. (1960). The frightened witch (Nigeria). In *In the Company of Man: Twenty Portraits by Anthropologists* (ed. J. B. Casagrande). New York: Harper and Row.

Bohannan, L. (1967). Miching Mallecho, that means witchcraft. In *Magic, Witchcraft and Curing* (ed. J. Middleton). New York: Natural History Press.

Bohannan, P. (1972). Dago'om: a man apart. In *Crossing Cultural Boundaries: The Anthropological Experience* (eds S. T. Kimball and J. B. Watson). San Francisco: Chandler Publishing Company.

Bohannan, P. (1981). Unseen community: the natural history of a research project. In *Anthropologists at Home in North America: Methods and Issues in the Study of One's Own Society* (ed. D. A. Messerschmidt). Cambridge: Cambridge University Press.

Boissevain, J. F. (1970). Fieldwork in Malta. In *Being an Anthropologist* (ed. G. D. Spindler). New York: Holt, Rinehart and Winston.

Boissevain, J. F. (1974). *Friends of Friends: Networks, Manipulators and Coalitions.* Oxford: Basil Blackwell.

Bollinger, D. (1968). *Aspects of Language.* New York: Harcourt, Brace and World.

Boruch, R. F. (1982). Methods for resolving privacy problems in social research. In *Ethical Issues in Social Science Research* (eds T. L. Beauchamp, R. R. Faden, R. J. Wallace, Jr. and L. Walters). Baltimore: The Johns Hopkins University Press.

Boruch, R. F. and Cecil, J. S. (1979). *Assuring Confidentiality in Social Science Research.* Philadelphia: University of Pennsylvania Press.

Boserup, E. (1970). *Woman's Role in Economic Development.* New York: St Martin's Press.

Boswell, D. M. (1969). Personal crises and the mobilization of the social network. In *Social Networks in Urban Situations: Analyses of Personal Relationships in Central African Towns* (ed. J. C. Mitchell). Manchester: Manchester University Press, for Institute for African Studies.

Boswell, D. M. (1975). Kinship, friendship and the concept of a social network. In *Urban Man in Southern Africa* (eds C. Kileff and W. C. Pendleton). Gwelo: Mambo Press.

Bott, E. (1971). *Family and Social Network* (2nd edn). London: Tavistock Publications.

du Boulay, J. (1979). "Some elusive dimensions of participation". Unpublished paper presented to S.S.R.C. Workshop on Participant Observation, University of Birmingham, September 24/25.

Bourdieu, P. (1977). *Outline of a Theory of Practice* (trans. R. Nice). Cambridge: Cambridge University Press.

Bourque, S. C. and Warren, K. B. (1981). *Women of the Andes: Patriarchy and Social Change in Two Peruvian Towns.* Ann Arbor: University of Michigan Press.

Bovin, M. (1966). The significance of the sex of the fieldworker for insights into the male and female worlds. *Ethnos* **31** (suppl.), 24–27.

Bowen, E. S.: *see* Bohannan, L.

Bower, R. T. and de Gasparis, P. (1978). *Ethics in Social Research: Protecting the Rights of Human Subjects.* New York: Praeger.

Bowers, F. (1962 [1949]). *Principles of Bibliographical Description.* Princeton, New Jersey: Princeton University Press.

Brabrook, E. W. (1893). On the organisation of local anthropological research. *Journal of the Anthropological Institute* **22**, 262–274.

Bradley, R. T. (1982). Ethical problems in team research: a structural analysis and an agenda for resolution. *The American Sociologist* **17** (May), 87–94.

Brass, T. (1982). The sabotage of anthropology: and the anthropologist as saboteur. *Journal of the Anthropological Society of Oxford* **13**(2), 180–186.

Brazil, D., Coulthard, M. and Johns, C. (1980). *Discourse Intonation and Language Teaching*. London: Longman.

Brew, J. O. (ed.) (1968). *One Hundred Years of Anthropology*. Cambridge, Mass.: Harvard University Press.

Briggs, J. L. (1970). *Never in Anger: Portrait of an Eskimo Family*. Cambridge, Mass.: Harvard University Press.

Brim, J. A. and Spain, D. A. (1974). *Research Design in Anthropology: Paradigms and Pragmatics in the Testing of Hypotheses*. New York: Holt, Rinehart and Winston.

Brislin, R. W. and Holwill, F. (1979). Indigenous views of the writings of behavioural/social scientists: towards increased cross-cultural understanding. In *Bonds without Bondage: Explorations in Transcultural Interaction* (ed. K. Kumar). Hawaii: University Press of Hawaii, for the East–West Centre.

Britan, G. M. (1979). Some problems of fieldwork in the Federal bureaucracy. *Anthropological Quarterly* **52**(4), 211–220.

British Sociological Association (1973). *Statement of Ethical Principles and Their Application to Sociological Practice*. London: B.S.A.

British Standards Institute (B.S.I.) (1972). *Recommendations for the Presentation of Theses* (B.S. 4821). London: B.S.I.

Broadhead, R. S. and Rist, R. C. (1975–76). Gatekeepers and the social control of social research. *Social Problems* **23**, 325–336.

Brokensha, D. (1963). Problems in fieldwork: a study of Larteh, Ghana. *Current Anthropology* **4**, 533–534.

Brookfield, H. C. (1973). Full circle in Chimbu: a study of trends and cycles. In *The Pacific in Transition* (ed. H. C. Brookfield). Canberra: Australian National University Press.

Brookfield, H. C. and Brown, P. (1963). *Struggle for Land: Agriculture and Group Territories among the Chimbu of the New Guinea Highlands*. Melbourne: Oxford University Press.

Brower, R. A. (ed.) (1959). *On Translation*. Cambridge. Mass.: Harvard University Press.

Brown, G. W. (1973). Some thoughts on grounded theory. *Sociology* **7**, 1–16.

Brown, P. (1974). Mediators in social change: new roles for big men. *Mankind* **9**, 224–230.

Brown, P. (1979). Change and the boundaries of systems in highland New Guinea: the Chimbu. In *Social and Ecological Systems* (eds P. Burnham and R. Ellen). London, Orlando and New York: Academic Press.

Brown, P. and Brookfield, H. C. (1959). Chimbu land and society. *Oceania* **30**, 1–75.

Brown, P. J. (1981). Field-site duplication: case studies and comments on the "my-tribe" syndrome. *Current Anthropology* **22**(4), 413–414.

Brown, Robert (1963). *Explanation in Social Sciences*. London: Routledge and Kegan Paul.

Brown, Richard (1973). Anthropology and colonial rule: Godfrey Wilson and the Rhodes-Livingstone Institute, Northern Rhodesia. In *Anthropology and the Colonial Encounter* (ed. T. Asad). London: Ithaca Press.

Brunt, E. and Brunt, L. (1978). Achter de scherman van stadsetnografisch onderzoek. (Behind the scene of urban ethnographic research.) *Sociologische Gids.* **25**(5), 373–385.

Brunt, L. (1974). Anthropological fieldwork in the Netherlands. *Current Anthropology* **15**(3), 311–314.

Bruyn, S. T. (1966). *The Human Perspective in Sociology: The Methodology of Participant Observation.* Englewood Cliffs, N.J.: Prentice-Hall.

Bruyn, S. T. (1970). The methodology of participant observation. In *Qualitative Methodology: Firsthand Involvement with the Social World* (ed. W. J. Filstead). Chicago: Markham.

Bucher, R., Fritz, C. E. and Quarantelli, E. L. (1956). Tape recorded interviews in social research. *American Sociological Review* **21**(3), 359–364.

Buechler, H. C. (1969). The social position of the ethnographer in the field. In *Stress and Response in Fieldwork* (eds. F. Henry and S. Saberwal). New York: Holt, Rinehart and Winston.

Bulmer, M. (ed.) (1977). *Sociological Research Methods: An Introduction.* London: Macmillan.

Bulmer, M. (ed.) (1979). *Censuses, Surveys and Privacy.* London: Macmillan.

Bulmer, M. (1979a). Concepts in the analysis of qualitative data. *Sociological Review* **27**(4), 651–677.

Bulmer, M. (ed.) (1982). The merits and demerits of covert observation. In *Social Research Ethics.* London: Macmillan.

Bulmer, M. (ed.) (1982a). *Social Research Ethics: An Examination of the Merits of Covert Participant Observation.* London: Macmillan.

Bulmer, R. N. H. and Tyler, M. J. (1968). Karam classification of frogs. *Journal of the Polynesian Society* **81**, 472–499; **82**, 86–107.

Bureau of Anthropology (1908). *Journal of the Royal Anthropological Institute* **38**, 489–492.

Burgess, R. G. (ed.) (1982). *Field Research: A Sourcebook and Field Manual.* (Contemporary Social Research **4**). London: George Allen and Unwin.

Burgess, R. G. (1982a). Approaches to field research. In *Field Research: A Sourcebook and Field Manual* (ed. R. G. Burgess). (Contemporary Social Research **4**). London: George Allen and Unwin.

Burgess, R. G. (1982b). The unstructured interview as conversation. In *Field Research: A Sourcebook and Field Manual* (ed. R. G. Burgess). (Contemporary Social Research **4**). London: George Allen and Unwin.

Burgess, R. G. (1982c). Keeping field notes. In *Field Research: A Sourcebook and Field Manual* (ed. R. G. Burgess). (Contemporary Social Research **4**). London: George Allen and Unwin.

Burguière, A. (1977). *Bretons de Plozévet.* Paris: Flammarion.

Burling, R. (1969). Linguistics and ethnographic description. *American Anthropologist* **71**, 817–827.

Burling, R. (1970). *Man's Many Voices: Language in its Cultural Context.* New York: Holt, Rinehart and Winston.

Burridge, K. O. L. (1960). *Mambu: A Melanesian Millenium.* London: Methuen.

Burridge, K. O. L. (1975). Other peoples' religions are absurd. In *Explorations in the Anthropology of Religion* (eds W. E. A. Van Beek and J. H. Scherer). The Hague: M. Nijhoff.

Burt, R. S. (1980). Models of network structure. *Annual Review of Sociology* **6**, 79–141.

Butcher, J. (1980). *Typescripts, Proofs and Indexes* (Cambridge authors' and publishers' guides). Cambridge: Cambridge University Press.

Bynner, J. and Stribley, K. M. (eds) (1979). *Social Research: Principles and Procedures.* London: Longman and The Open University Press.

Callaway, H. (1870). *The Religious System of the Amazulu.* London: Trubner.

Callaway, Helen (1981). Spatial domains and women's mobility in Yorubaland, Nigeria. In *Women and Space* (ed. S. Ardener). London: Croom Helm; New York: St. Martin's Press.

Callaway, Helen (1981a). Women's perspectives: research as re-vision. In *Human Inquiry: A Sourcebook of New Paradigm Research* (eds P. Reason and J. Rowan). Chichester: John Wiley.

Campbell, D. T. (1955). The informant in quantitative research. *American Journal of Sociology,* **60**, 339–342.

Campbell, D. T. and Levine, R. A. (1970). Field-manual anthropology. In *A Handbook of Method in Cultural Anthropology* (eds R. Naroll and R. Cohen). New York: Columbia University Press.

Canadian Sociological and Anthropological Association (1979). Code of professional ethics. *Society-Société* **3**(2), 3–8.

Caplan, P. (ed.) (1979). Cross-cultural perspectives on women. In *Women's Studies International Quarterly* **2**(4).

Carpenter, E. (1960). Ohnainewk, Eskimo hunter. In *In the Company of Man: Twenty Portraits by Anthropologists* (ed. J. B. Casagrande. New York: Harper and Row.

Carroll, J. D. and Knerr, C. R. (1976). The APSA Confidentiality in social sciences Research Project: a final report. *PS* **6**(3), 416–419.

Casagrande, J. B. (1954). The ends of translation. *International Journal of American Linguistics* **20**, 335–340.

Casagrande, J. B. (ed.) (1960). *In the Company of Man: Twenty Portraits by Anthropologists.* New York: Harper and Row.

Casagrande, J. B. (1960a). John Mink, Ojibwa informant. In *In the Company of Man: Twenty Portraits by Anthropologists* (ed. J. B. Casagrande). New York: Harper and Row.

Casley, D. J. and Lury, D. A. (1981). *Data Collection in Developing Countries.* Oxford: Clarendon Press.

Cassell, J. (1978). Risk and benefit to subjects of fieldwork. *The American Sociologist* **13**(3), 134–143.

Cassell, J. (1980). Ethical principles for conducting fieldwork. *American Anthropologist* **82**(1), 28–41.

Cassell, J. (1982). Does risk-benefit analysis apply to moral evaluation of social research? In *Ethical Issues in Social Science Research* (eds T. L. Beauchamp, R. R. Faden, R. J. Wallace, Jr. and L. Walters). Baltimore: The Johns Hopkins University Press.

Castaneda, C. (1970). *The Teachings of Don Juan; A Yaqui Way of Knowledge.* Harmondsworth: Penguin.

Castaneda, C. (1971). *A Separate Reality: Further Conversations with Don Juan.* London: The Bodley Head.

Castaneda, C. (1973). *Journey to Ixtlan: The Lessons of Don Juan.* London: The Bodley Head.

Caudill, W. and Roberts, B. H. (1951). Pitfalls in the organization of interdisciplinary research. *Human Organization* **10**(4), 12–15. Reprinted (1960) in *Human Organization Research: Field Relations and Techniques* (eds R. N. Adams and J. J. Preiss). Homewood, Ill.: The Dorsey Press Inc., for the Society for Applied Anthropology.

Chagnon, N. A. (1968). *Yanomamö: The Fierce People.* New York: Holt, Rinehart and Winston.

Chagnon, N. A. (1974). *Studying the Yanomamö.* (Studies in Anthropological Method.) New York: Holt, Rinehart and Winston.

Chakravarti, A. (1979). Experiences of an encapsulated observer: a village in Rajasthan. In *The Fieldworker and the Field: Problems and Challenges in Sociological Investigation* (eds M. N. Srinivas, A. M. Shah and E. A. Ramaswamy). Delhi: Oxford University Press.

Chambers, E. (1980). Fieldwork and the law: new contexts for ethical decision-making. *Social Problems* **27**(3), 330–341.

Chambers, E. and Trend, M. G. (1981). Fieldwork ethics in policy-oriented research (Letter). *American Anthropologist* **83**(3), 626–628.

Cherry, L. L. and Vesterman, W. (1980). *Writing Tools. The STYLE and DICTION Programmes.* New Jersey: (Vax/Unix papers) Murray Hill (Bell Laboratories).

Chilungu, S. W. (and others) (1976). Issues in the ethics of research methods: an interpretation of the Anglo-American perspective. *Current Anthropology* **17**(3), 457–481 [with comments and reply].

Chinas, B. (1971). Women as ethnographic subjects. In *Women in Cross-cultural Perspectives: A Preliminary Sourcebook* (ed. S.-E. Jacobs). Urbana: University of Illinois, Department of Urban and Regional Planning.

Chrisman, N. J. (1976). Secret societies and the ethics of urban fieldwork. In *Ethics and Anthropology: Dilemmas in Fieldwork* (eds M. A. Rynkiewich and J. P. Spradley). New York: John Wiley.

Christian Dental Fellowship (C.D.F.) (1965). *Emergency Dentistry.* Christian Dental Fellowship.

Cicourel, A. V. (1964). *Method and Measurement in Sociology.* New York: The Free Press; London: Collier-Macmillan.

Circular of inquiry (1888). In "3rd Report of the Committee . . . appointed for the purpose of investigating and publishing reports on . . . the North-western tribes of the Dominion of Canada". *Reports of the 57th Meeting of the BAAS 1887*. London: John Murray.

Clammer, J. R. (ed.) (1983). *The Année Sociologique and the Development of Modern Anthropology*. London: Macmillan.

Clarke, M. (1973). *Survival in the Field*. Faculty of Commerce and Social Science Discussion Papers, University of Birmingham. Series E, number 23. Reprinted (1975) as "Survival in the field; implications of personal experience in fieldwork" *Theory and Society* 2, 95–124.

Clifford, J. (1980). Fieldwork, reciprocity and the making of ethnographic texts: the example of Maurice Leenhardt. *Man* 15, 518–532.

Codere, H. (1959). The understanding of the Kwakiutl. In *The Anthropology of Franz Boas: Essays on the Centennial of his Birth* (ed. W. Goldschmidt). (Memoir of the American Anthropological Association 89.)

Codere, H. (1966). Introduction. In *Kwakiutl Ethnography* (F. Boas). Chicago: University of Chicago Press.

Codere, H. (1970). Field work in Rwanda, 1959–1960. In *Women in the Field* (ed. Peggy Golde). Chicago: Aldine.

Coffield, F., Robinson, P. and Sarsby, J. (1981). *A Cycle of Deprivation? A Case Study of Four Families*. London: Heinemann.

Cohen, A. P. (1978). Ethnographic method in the real community. *Sociologica Ruralis* 18(1), 1–22.

Cohen, F. G. (1976). The American Indian Movement and the anthropologist: issues and implications of consent. In *Ethics and Anthropology: Dilemmas in Fieldwork* (eds M. A. Rynkiewich and J. P. Spradley). New York: John Wiley.

Coles, J. (1973). *Archaeology by Experiment*. New York: Charles Scribner's Sons.

Colfer, C. J. P. (1976). Rights, responsibilities, and reports: an ethical dilemma in contract research. In *Ethics and Anthropology: Dilemmas in Fieldwork* (eds. M. A. Rynkiewich and J. P. Spradley). New York: John Wiley.

Collier, J. (1970). *Visual Anthropology: Photography as a Research Method*. New York: Holt, Rinehart and Winston.

Collins, H. M. (1979). "Concepts and methods of participatory fieldwork". Unpublished paper presented to S.S.R.C. Workshop on Participant Observation. University of Birmingham, September 24/25.

Colson, E. (1954). The intensive study of small sample communities. In *Method and Perspective in Anthropology* (ed. R. F. Spencer). Minneapolis: University of Minnesota Press. Reprinted (1967) in *The Craft of Social Anthropology* (ed. A. L. Epstein). London: Tavistock.

Colson, E. (1967). Competence and incompetence in the context of independence. *Current Anthropology* 8, 92–100, 109–111.

Colson, E., Scudder, T. and Kemper, R. V. (1976). Long-term field research in social anthropology. *Current Anthropology* 17(3), 494–496.

Colvard, R. (1969). Interaction and identification in reporting field research: a critical reconsideration of protective procedures. In *Ethics, Politics and Social Research* (ed. G. Sjoberg). London: Routledge and Kegan Paul.

Condominas, G. (1973). Ethics and comfort: an ethnographer's view of his profession. (Distinguished Lecture 1972). American Anthropological Association, *Annual Report 1972*, pp. 1–17. Washington D.C.

Conklin, H. C. (1960). Maling, A Hanunóo girl from the Philippines. In *In the Company of Man: Twenty Portraits by Anthropologists* (ed. J. B. Casagrande). New York: Harper and Row.

Conklin, H. C. (1960a). A day in Parina. In *In the Company of Man: Twenty Portraits by Anthropologists* (ed. J. B. Casagrande). New York: Harper and Row.

Conklin, H. C. (1961). The study of shifting cultivation. *Current Anthropology* 2, 27–61. Also published (1963), Pan American Union Studies and Monographs 6, Washington, D.C.

Conklin, H. C. (1968). Some aspects of ethnographic research in Ifugao. *Transactions of the New York Academy of Sciences* (Ser. II) 30, 99–121.

Coon, C. S. (1932). *Flesh of the Wild Ox: A Riffian Chronicle of High Valleys and Long Rifles*. London: Cape.

Crane, J. G. (1982). A Lun Dayeh engagement negotiation. *Contributions to Southeast Asian Ethnography* 1, 142–178.

Crane, J. G. and Angrosino, M. V. (1974). *Field Projects in Anthropology: A Student Handbook*. Morristown: General Learning Press.

Crapanzano, V. (1970). On the writing of ethnography. *Dialectical Anthropology* 2, 69–73.

Crick, M. (1982). Anthropological field research, meaning creation and knowledge construction. In *Semantic Anthropology* (ed. D. Parkin). (A.S.A. Monogr. 22). London, Orlando and New York: Academic Press.

Crystal, D. (1971). *Linguistics*. Harmondsworth: Penguin.

Crystal, D. and Davy, D. (1969). *Investigating English Style*. London: Longmans.

Cubitt, T. (1973). Network density among urban families. In *Network Analysis: Studies in Human Interaction* (eds J. Boissevain and J. C. Mitchell). The Hague: Mouton.

Cunnison, I. (1960). The Omda (Baggara Arabs, Sudan). In *In the Company of Man: Twenty Portraits by Anthropologists* (eds. J. B. Casagrande). New York: Harper and Row.

Curr, E. M. (ed.) (1886). Questions concerning the Aborigines of Australia. In *The Australian Race*, Vol. 2. Melbourne: J. Ferres.

Current Sociology (1982). Confidentiality: theory and practice. *Current Sociology* 30(2), 1–82.

Currier, M. (1976). Problems in anthropological bibliography. *Annual Review of Anthropology* 5, 15–34.

Cutler, A. G. (1970). *Indexing Methods and Theory*. Baltimore: Waverly Press and Williams and Wilkins.

Damon, F. (1980). The Kula and generalized exchange: considering some unconsidered aspects of 'The elementary structures of kinship'. *Man* (N.S.) 15, 267–292.

Darnell, R. (1976). The Sapir years at the National Museum, Ottawa. In *The History of Canadian Anthropology* (ed. J. Freedman). (Canadian Ethnol. Soc. Proc. **3**).

Davies, M. and Kelly, E. (1976). The social worker, the client and the social anthropologist. *British Journal of Social Work* **6**(2), 213–231.

Davis, F. (1961). Comment on "Initial Interactions". *Social Problems* **8**(4), 364–365.

Davis, J. (1978). The value of the evidence [letter]. *Man* (N.S.) **13**(3), 471–473.

Davis, J. (1983). Tracing kinship by computer. A coding scheme and algorithm (in press).

Dean, J. P. and Whyte, W. F. (1958). How do you know if your informant is telling the truth? *Human Organization* **17**(2), 34–38.

Debenham, F. (1955). *Map Making* (3rd edn.). London: Blackie.

Degérando, J.-M. (1969 [1800]). *The Observation of Savage Peoples*. London: Routledge and Kegan Paul.

Deitchman, S. J. (1976). *The Best Laid Schemes: A Tale of Social Research and Bureaucracy*. Cambridge, Mass.: MIT Press.

Deloria, V. Jr (1969). *Custer Died for Your Sins*. New York: Macmillan.

Denney, M., Riesman, D. and Star, S. A. (1956–7). Age and sex in the interview. *American Journal of Sociology* **62**, 143–152.

Dentan, R. K. (1970). Living and working with the Semai. In *Being an Anthropologist* (ed. G. D. Spindler). New York: Holt, Rinehart and Winston.

Denzin, N. K. (ed). (1970). *Sociological Methods: A Sourcebook*. London: Butterworths.

Denzin, N. K. (1978). *The Research Act: A Theoretical Introduction to Sociological Methods* (2nd edn.). New York: McGraw-Hill; London: Butterworths.

Devereux, G. (1967). *From Anxiety to Method in the Behavioral Sciences*. The Hague: Mouton.

Diamond, N. (1970). Fieldwork in a complex society: Taiwan. In *Being an Anthropologist* (ed. G. D. Spindler). New York: Holt, Rinehart and Winston.

Diamond, S. (1964). Nigerian discovery: the politics of fieldwork. In *Reflections on Community Studies* (eds. A. J. Vidich, J. Bensman and M. R. Stein), pp. 119–154. New York: Wiley.

Diamond, S. (ed.) (1980). *Anthropology, Ancestors and Heirs*. The Hague: Mouton.

Diamond, S. (1980a). Anthropological traditions: the participants observed. In *Anthropology: Ancestors and Heirs* (ed. S. Diamond). The Hague: Mouton.

Diener, E. and Crandall, R. (1978). *Ethics in Social and Behavioral Research*. Chicago: The University of Chicago Press.

Dillman, C. M. (1977). Ethical problems in social science research peculiar to participant observation. *Human Organization* **36**(4), 405–407.

Dingwall, R. (1980). Ethics and ethnography. *Sociological Review* **28**(4), 871–891.

Djamour, J. (1959). *Malay Kinship and Marriage in Singapore*. London: The Athlone Press.

Dollard, J. (1949). *Caste and Class in a Southern Town*. New York: Doubleday.

Douglas, J. D. (ed.) (1971). *Understanding Everyday Life: Toward the Reconstruction of Sociological Knowledge*. London: Routledge and Kegan Paul.

Douglas, J. D. (1976). *Investigative Social Research: Individual and Team Field Research.* Beverly Hills, London: Sage Publications.

Douglas, J. D. (1979). Living morality versus bureaucratic fiat. In *Deviance and Decency: the Ethics of Research with Human Subjects* (Eds C. B. Klockars and F. W. O'Connor). Beverly Hills, London: Sage Publications.

Douglas, M. (1973). Torn between two realities. *The Times Higher Education Supplement*, 15th June. Reprinted (1975) as "The authenticity of Castaneda" in *Implicit Meanings: Essays in Anthropology* (M. Douglas). London: Routledge and Kegan Paul.

Dua, V. (1979). A woman's encounter with Arya Samaj and untouchables: a slum in Jullundur. In *The Fieldworker and the Field: Problems and Challenges of Sociological Investigation* (eds M. N. Srinivas, A. M. Shah and E. A. Ramaswamy). Delhi: Oxford University Press.

Dube, L. (1975). Woman's worlds – three encounters. In *Encounter and Experience* (eds A. Béteille and T. N. Madan). Delhi: Vikas Publishing House.

Du Bois, C. (1960). The form and substance of status: A Javanese–American relationship. In *In the Company of Man: Twenty Portraits by Anthropologists* (ed. J. B. Casagrande). New York: Harper and Row.

Du Bois, C. (1970). Studies in an Indian town. In *Women in the Field* (ed. P. Golde). Chicago: Aldine.

Dumont, L. (1966). The village community from Munro to Maine. *Contributions to Indian Sociology* **8**, 67–89.

Durkheim, E. (1897). Review of A. Labriola: Essais sur la conception materialiste de l'histoire. *Revue Philosophique de la France et de l'Etranger* **44**, 645–651.

Durkheim, E. (1964). *The Rules of Sociological Method* (8th edn.). Translated by S. A. Solovay and J. H. Mueller (ed. G. E. G. Catlin). London: Collier-Macmillan.

Dwyer, K. (1977). On the dialogic of fieldwork. *Dialectical Anthropology* **2**, 143–151.

Dyke, B. (1981). Computer simulation in anthropology. *Annual Review of Anthropology* **10**, 193–207.

Dyke, B. and MacCluer, J. W. (1973). *Computer Simulation in Human Population Studies.* Orlando, New York and London: Academic Press.

Eames, E. and Goode, J. G. (1977). *Anthropology of the City.* Englewood Cliffs: Prentice Hall.

Easterday, L., Papademas, D., Schorr, L. and Valentine, C. (1977). The making of a female researcher: role problems in fieldwork. *Urban Life* **6**(3), 333–348. (Reprinted in R. G. Burgess (ed.) 1982 q.v.)

Easthope, G. (1974). *A History of Social Research Methods.* London: Longman.

Eddy, E. M. and Partridge, W. L. (eds) (1978). *Applied Anthropology in America.* New York: Columbia University Press.

Edgerton, R. B. and Langness, L. L. (1974). *Methods and Styles in the Study of Culture.* San Francisco: Chandler and Sharp.

Edholm, F., Harris, O. and Young, K. (eds) (1977). *Critique of Anthropology* (Women's Issue) **3**, (9–10).

Edholm, O. G. and Bacharach, A. L. (1965). *Exploration Medicine*. Bristol: John Wright.

Efrat, B. and Mitchell, M. (1974). The Indian and the social scientist: contemporary contractual arrangements on the Pacific Northwest Coast. *Human Organization* **33**(4), 405–407.

Eggan, F. (ed.) (1937). *Social Anthropology of North American Tribes*. Chicago: Chicago University Press.

Eggan, F. (1961). Ethnographic data in social anthropology in the United States. *The Sociological Review* **9**, 19–26.

Eggan, F. (1963). The graduate program. In *The Teaching of Anthropology* (eds D. G. Mandelbaum, G. W. Lasker and E. M. Albert). Berkeley: University of California Press.

Eggan, F. (1974). Applied anthropology in the mountain province, Philippines. In *Social Organization and the Applications of Anthropology: Essays in Honor of Lauriston Sharp* (ed. R. J. Smith). Ithaca: Cornell University Press.

Elkin, A. P. (1975/76). R. H. Mathews: his contribution to Aboriginal studies. *Oceania* **46**, 1–24, 126–152, 206–234.

Ellen, R. F. (1976). The development of anthropology and colonial policy in the Netherlands, 1800–1960. *Journal of The History of The Behavioural Sciences* **12**, 303–324.

Ellen, R. F. (1982). Review of *The Structure of Folk Models* (eds L. Holy and M. Stuchlik). *Man* (N.S.) **17**(2), 371–372.

Emerson, R. M. (1981). Observational field work. *Annual Review of Sociology* **7**, 351–378.

Ennew, J. (1976). Examining the facts in fieldwork. Considerations of method and data. *Critique of Anthropology* **7**, 43–66.

Epstein, A. L. (ed.) (1967). *The Craft of Social Anthropology*. London: Tavistock Publications.

Epstein, A. L. (1967a). Editor's Preface. In *The Craft of Social Anthropology* (ed. A. L. Epstein). London: Tavistock Publications.

Epstein, A. L. (1969). The network and urban social organization. In *Social Networks in Urban Situations: Analyses of Personal Relationships in Central African Towns* (ed. J. Clyde Mitchell). Manchester: Manchester University Press, for Institute for African Studies.

Epstein, T. S. (1979). Mysore villages revisited. In *Long-term Field Research in Social Anthropology* (eds G. M. Foster, T. Scudder, E. Colson and R. V. Kemper). Orlando, New York and London: Academic Press.

Evans-Pritchard, E. E. (1932). The Zande corporation of witchdoctors. *Journal of the Royal Anthropological Institute* **62**, 291–336.

Evans-Pritchard, E. E. (1940). *The Nuer: A Description of the Modes of Livelihood and Political Institutions of a Nilotic People*. Oxford: Clarendon Press.

Evans-Pritchard, E. E. (1956). *Nuer Religion*. Oxford: Clarendon Press.

Evans-Pritchard, E. E. (1969). The perils of translation. *New Blackfriars* **50** (December), 813–815.

Evans-Pritchard, E. E. (1973). Some reminiscences and reflections on fieldwork. *Journal of the Anthropological Society of Oxford* 4(1), 1–12.

Evans-Pritchard, E. E. (1976). *Witchcraft, Oracles and Magic Among the Azande* (Abridged, with an Introduction by Eva Gillies). Oxford: Clarendon Press.

Fabian, J. (1971). On professional ethics and epistemological foundations. *Current Anthropology* 12(2), 230–231.

Fabian, J. (1979). Rule and process: thoughts on ethnography as communication. *Philosophy of the Social Sciences* 9, 1–26.

Fahim, H. M. (1977). Foreign and indigenous anthropology: the perspectives of an Egyptian anthropologist. *Human Organization* 36(1), 80–86.

Fahim , H. (1979). Field research in a Nubian village: the experience of an Egyptian anthropologist. In *Long-term Field Research in Social Anthropology* (eds G. M. Foster, T. Scudder, E. Colson and R. V. Kemper). Orlando, New York and London: Academic Press.

Fahim, H. (ed.) (1982). *Indigenous Anthropology in Non-Western Countries*. Proceedings of a Burg Wartenstein Symposium. Durham, North Carolina: Carolina Academic Press.

Fahim, H., Helmer, K., Colson, E., Madan, T. N., Kelman, H. C. and Asad, T. (1980). Indigenous anthropology in non-Western countries: a further elaboration. *Current Anthropology* 21(5), 644–663.

Faraday, A. and Plummer, K. (1979). Doing life histories. *Sociological Review* 27(4), 773–798.

Favret-Saada, J. (1980). *Deadly Words: Witchcraft in the Bocage*. Cambridge: Cambridge University Press.

Feinberg, R. (1979). The role of the ethnographer in shaping attitudes towards anthropologists: a case in point. *Pacific Studies* 2(2), 156–165.

Festinger, L. and Katz, D. (eds) (1953). *Research Methods in the Behavioral Sciences*. New York: Holt, Rinehart and Winston.

Festinger, L., Riecken, H. W. and Schachter, S. (1956). *When Prophesy Fails*. Minneapolis: Minnesota University Press.

Fetterman, D. M. (1981). New perils for the contract ethnographer. *Anthropology & Education Quarterly* 12(1), 71–80.

Feuchtwang, S. (1973). The colonial formation of British social anthropology. In *Anthropology and the Colonial Encounter* (ed. T. Asad). London: Ithaca Press.

Feyerabend, P. (1975). *Against Method: Outline of an Anarchistic Theory of Knowledge*. London: New Left Books.

Fiedler, J. (1978). *Field Research: A Manual of Logistics and Management of Scientific Studies in Natural Settings*. San Francisco, London: Jossey-Bass.

Field, D., Clarke, B., with the assistance of Goldie, N. (1982). *Medical Sociology in Britain: A Register of Research and Teaching*. [London]: British Sociological Association, Medical Sociology Group.

Fielding, N. (1982). Observational research on the National Front. In *Social Research Ethics* (ed. Martin Bulmer). London: Macmillan.

Filmer, P., Phillipson, M., Silverman, D. and Walsh, D. (1972). *New Directions in Sociological Theory*. London: Collier-Macmillan.

Filstead, W. J. (ed.) (1970). *Qualitative Methodology: Firsthand Involvement with the Social World*. Chicago: Markham Publishing Company.

Finer, R. (1982). Reference and enquiry work. In *Handbook of Special Librarianship and Information Work* (ed. L. J. Anthony). London: Aslib.

Finnegan, D. E. (1978). Dealing with animosity arising from a previous study. In *Ethical Dilemmas in Anthropological Inquiry: A Case Book* (ed. G. N. Appell). Waltham, Mass.: Crossroads Press.

Firth, J. R. (1957). Ethnographic analysis and language with reference to Malinowski's views. In *Man and Culture: An Evaluation of the Work of Bronislaw Malinowski* (ed. R. Firth). London: Routledge and Kegan Paul.

Firth, Raymond (1936). *We, The Tikopia*. London: George Allen and Unwin.

Firth, Raymond (1939). *Primitive Polynesian Economy*. London: Routledge and Kegan Paul.

Firth, Raymond (1959). *Social Change in Tikopia*. London: George Allen and Unwin.

Firth, Raymond (1960). A Polynesian aristocrat (Tikopia). In *In the Company of Man: Twenty Portraits by Anthropologists* (ed. J. B. Casagrande). New York: Harper and Row.

Firth, Raymond (1967). Themes in economic anthropology: a general comment. In *Themes in Economic Anthropology* (ed. R. Firth). (A.S.A. Monogr. **6**). London: Tavistock.

Firth, Raymond (1971). Economic aspects of modernization of Tikopia. In *Anthropology in Oceania: Essays Presented to Ian Hogbin*, (eds L. R. Hiatt and C. Jayawardena). Sydney: Angus and Robertson.

Firth, Raymond (1975). Seligman's contributions to Oceanic anthropology. *Oceania* **45**, 272–282.

Firth, Raymond (1981). Bronislaw Malinowski. In *Totems and Teachers: Perspectives on the History of Anthropology* (ed. S. Silverman). New York: Columbia University Press.

Firth, Raymond, Hubert, J. and Forge, A. (1969). *Families and Their Relatives: Kinship in a Middle-class Sector of London*. London: Routledge and Kegan Paul.

Firth, Rosemary (1971). Anthropology within and without the ivory towers, *Journal of the Anthropological Society of Oxford* **2**(2), 74–82.

Firth, Rosemary (1972). From wife to anthropologist. In *Crossing Cultural Boundaries: The Anthropological Experience* (eds S. T. Kimball and J. B. Watson). San Francisco: Chandler.

Fischer, A. (1970). Field work in five cultures. In *Women in the Field* (ed. P. Golde). Chicago: Aldine.

Fischer, C. S. (1982). *To Dwell Among Friends: Personal Networks in Town and City*. Chicago: University of Chicago Press.

Fison, L. and Howitt, A. W. (1880). *Kamilaroi and Kurnai*. Melbourne: G. Robertson.

Flaherty, D. H. (1979). *Privacy and Government Data Banks: An International Perspective*. London: Mansell Publishing.

Flaherty, D. H., Hanis, E. H. and Mitchell, S. P. (1979). *Privacy and Access to Government Data for Research: An International Bibliography*. London: Mansell Publishing.

Forcese, D. and Richer, S. (1970). *Stages of Social Research: Contemporary Perspectives*. Englewood Cliffs, N.J.: Prentice-Hall.

Ford, J. (1975). *Paradigms and Fairy Tales: An Introduction to the Science of Meanings*. London: Routledge and Kegan Paul.

Forge, A. (1967). The lonely anthropologist. *New Society* (17 August) 9, 221–223.

Forge, A. (1970). Learning to see in New Guinea. In *Socialization: The Approach from Social Anthropology* (ed. Philip Mayer). London: Tavistock Publications.

Forrest, D. W. (1974). *Francis Galton: The Life and Work of a Victorian Genius*. London: Paul Elek.

Fortes, M. (1957a). Malinowski and the study of kinship. In *Man and Culture: An Evaluation of the Work of Bronislaw Malinowski* (ed. R. Firth). London: Routledge and Kegan Paul.

Fortes, M. (1957b). Siegfried Frederick Nadel 1903–1956. In *The Theory of Social Structure* (S. F. Nadel). Melbourne: Melbourne University Press.

Fortes, M. (1963). Graduate study and research. In *The Teaching of Anthropology* (eds. D. G. Mandelbaum, G. W. Lasker and E. M. Albert). Berkeley: University of California Press.

Foster, G. M. (1979). Fieldwork in Tzintzuntzan: the first thirty years. In *Long-term Field Research in Social Anthropology* (eds G. M. Foster, T. Scudder, E. Colson and R. V. Kemper). Orlando, New York and London: Academic Press.

Foster, G. M., Scudder, T., Colson, E. and Kemper, R. V. (eds) (1979). *Long-term Field Research in Social Anthropology*. Orlando, New York and London: Academic Press.

Foster, G. M., Scudder, T., Colson, E. and Kemper, R. V. (1979a). Conclusion: the long-term study in perspective. In *Long-term Field Research in Social Anthropology* (eds G. M. Foster, T. Scudder, E. Colson and R. V. Kemper). Orlando, New York and London: Academic Press.

Fowler, D. D. (1975). Notes on inquiries in anthropology: a bibliographic essay. In *Towards a Science of Man: Essays in the History of Anthropology* (ed. T. H. H. Thoresen). The Hague: Mouton.

Fowler, D. D. and Fowler, C. S. (1971). *Anthropology of the Numa: John Wesley Powell's Manuscripts on the Numic Peoples of Western North America 1868–1880*. (Smithsonian Contrib. Anthrop. **14**). Washington D.C.: Smithsonian Institute Press.

Fox, Richard (1977). *Urban Anthropology: Cities in Their Cultural Settings*. Englewood Cliffs: Prentice-Hall.

Fox, R. (1982). Principles and pragmatics on Tory Island. In *Belonging: Identity and Social Organisation in British Rural Cultures*. (ed. A. P. Cohen). Manchester: Manchester University Press.

Frake, C. O. (1964). Notes on queries in ethnography. In *Transcultural Studies in Cognition* (eds A. K. Romney and R. Goodwin d'Andrade). *American Anthropologist* **66**, 132–145. Reprinted (1980) in *Language and Cultural Description* (C. O. Frake). Stanford, California: Stanford University Press.

Frankenberg, R. (1957). *Village on the Border*. London: Cohen and West.

Frankenberg, R. (1963). Participant observers. *New Society* 1(23), 22–23. Reprinted

(1982) in *Field Research: A Sourcebook and Field Manual* (ed. R. G. Burgess). London: George Allen and Unwin.

Frankenberg, R. (1966). *Communities in Britain*. Harmondsworth: Penguin Books.

Frazer, J. G. (1887). *Questions on the Customs, Beliefs and Languages of Savages*. Privately printed.

Frazer, J. G. (1905). *Lectures on the Early History of the Kingship*. London: Macmillan.

Frazer, J. G. (1907). *Questions on the Customs . . .* (etc.). Cambridge: Cambridge University Press.

Freedman, M. (1957). *Chinese Family and Marriage in Singapore*. London: H.M.S.O.

Freedman, M. (1963). A Chinese phase in social anthropology. *British Journal of Sociology* **14**, 1–19.

Freeman, D. (1961). On the concept of the kindred. *Journal of the Royal Anthropological Institute* **91**, 192–220.

Freeman, J. M. (1978). Collecting the life history of an Indian untouchable. In *American Studies in the Anthropology of India* (ed. S. Vatuk). New Delhi: Manohar.

Freides, T. (1973). *Literature and Bibliography of the Social Sciences*. Los Angeles: Melville Pub. Co.

Freilich, M. (1963). The natural experiment, ecology and culture. *Southwestern Journal of Anthropology* **19**(1), 21–39.

Freilich, M. (ed.) (1970). *Marginal Natives: Anthropologists at Work*. New York: Harper and Row.

Freilich, M. (1970a). Fieldwork: an introduction. In *Marginal Natives: Anthropologists at Work* (ed. M. Freilich). New York: Harper and Row.

Freilich, M. (1970b). Mohawk heroes and Trinidadian peasants. In *Marginal Natives: Anthropologists at Work* (ed. M. Freilich). New York: Harper and Row.

Freilich, M. (1970c). Toward a formalisation of fieldwork. In *Marginal Natives: Anthropologists at Work* (ed. M. Freilich). New York: Harper and Row.

Friedl, E. (1970). Fieldwork in a Greek village. In *Women in the Field* (ed. P. Golde). Chicago: Aldine.

Galaty, J. G. (1981). Models and metaphors: on the semiotic explanation of segmentary systems. In *The Structure of Folk Models* (eds L. Holy and M. Stuchlik) (A.S.A. Monogr. 20). London, Orlando and New York: Academic Press.

Gallaher, A., Jr. (1964). Plainville: the twice-studied town. In *Reflections on Community Studies* (eds A. J. Vidich, J. Bensman and M. R. Stein). New York: John Wiley and Sons.

Galliher, J. F. (1980). Social scientists' ethical responsibilities to superordinates: looking upward meekly. *Social Problems* **27**(3), 298–308.

Gallin, B. (1959). A case for intervention in the field. *Human Organization* **18**(3), 140–144.

Gallin, B. and Gallin, R. S. (1974). The rural-to-urban migration of an anthropologist in Taiwan. In *Anthropologists in Cities* (eds G. M. Foster and R. V. Kemper). Boston: Little, Brown and Company.

Galtung, J. (1967). *Theory and Methods of Social Research*. London: George Allen and Unwin.

Gans, H. (1968). The participant-observer as a human being. In *Institutions and the Person* (eds H. Becker, B. Geer, D. Riesman and R. Weiss). Chicago: Aldine. Reprinted (1982) in *Field Research: A Sourcebook and Field Manual* (ed. R. G. Burgess). London: George Allen and Unwin.

Garbett, G. K. (1965). A note on a recently introduced card system for processing numerical data. *Man* (O.S.) **106**, 120–121.

Garbett, G. K. (1979). The analysis of social situations. *Man* (N.S.) **5**, 214–227.

Garfinkel, H. (1967). *Studies in Ethnomethodology*. Englewood Cliffs, N.J.: Prentice-Hall.

Gartrell, B. (1979). Is ethnography possible? A critique of 'African Odyssey'. *Journal of Anthropological Research* **35**(4), 426–446.

Gearing, F. O. (1970). *The Face of the Fox*. Chicago: Aldine.

Geer, B. (1964). First days in the field. In *Sociologists at Work* (ed. P. Hammond). New York: Basic Books.

Geertz, C. (1959). *The Religion of Java*. New York: The Free Press.

Geertz, C. (1968). Thinking as a moral act. Ethical dimensions of anthropological fieldwork in the new states. *Antioch Review* **28**(1), 139–158.

Geertz, C. (1975). Thick description: toward an interpretive theory of culture. In *The Interpretation of Cultures: Selected Essays by Clifford Geertz*. London: Hutchinson.

Gellner, E. A. (1970 [1962]). Concepts and society. In *Rationality* (ed. B. R. Wilson). Oxford: Blackwell.

Gellner, E. A. (1973). *Cause and Meaning in the Social Sciences*. London: Routledge and Kegan Paul.

Gerholm, T. and Hannerz, U. (eds) (1982). *The Shaping of National Anthropologies*. Special issue of *Ethnos* **47**(1–2). Stockholm: Etnografiska Museet.

Gibaldi, J. and Achtert, W. S. (1980). *MLA Handbook: For Writers of Research Papers, Theses, and Dissertations* (student edn.). New York: Modern Language Association.

Gibbons, D. C. (1975). Unidentified research sites and fictitious names. *The American Sociologist* **10**(1), 32–36.

Giddens, A. (1976). *New Rules of Sociological Method*. London: Hutchinson.

Gilbert, J. P. and Hammel, E. A. (1966). Computer simulation and analysis of problems in kinship and social structure. *American Anthropologist* **68**, 70–93.

Gillin, J. (1949). Methodological problems in the anthropological study of modern cultures. *American Anthropologist* **51**(3), 392–399.

Gladwin, T. (1960). Petrus Mailo, Chief of Moen (Truk). In *In the Company of Man: Twenty Portraits by Anthropologists* (ed. J. B. Casagrande). New York: Harper and Row.

Glaser, B. G. and Strauss, A. L. (1967). *The Discovery of Grounded Theory: Strategies for Qualitative Research*. New York: Aldine.

Glazer, M. (1972). *The Research Adventure: Promise and Problem of Field Work*. New York: Random House.

Gleason, M. A. (1967). *An Introduction to Descriptive Linguistics.* New York: Holt, Rinehart and Winston.

Glock, C. (ed.) (1967). *Survey Research in the Social Sciences.* New York: Russell Sage Foundation.

Gluckman, M. (1958). *The Analysis of a Social Situation in Modern Zululand.* (Rhodes-Livingstone Paper No. 28). Manchester: Manchester University Press.

Gluckman, M. (1961). Ethnographic data in British social anthropology. *The Sociological Review* 9, 5–17.

Gluckman, M. (ed.) (1964). *Closed Systems and Open Minds: The Limits of Naïvety in Social Anthropology.* Chicago: Aldine.

Gluckman, M. (1965). *Politics, Law and Ritual in Tribal Society.* Oxford: Basil Blackwell.

Gluckman, M. (1967). Introduction. In *The Craft of Social Anthropology* (ed. A. L. Epstein). London: Tavistock Publications.

Goddard, D. (1972). Anthropology: the limits of functionalism. In *Ideology in Social Science: Readings in Critical Social Theory* (ed. R. Blackburn). London: Fontana.

Goffman, E. (1959). *The Presentation of Self in Everyday Life.* Garden City, N.Y.: Doubleday.

Golbeck, A. L. (1980). Quantification in ethnology and its appearance in regional culture trait distribution studies (1888–1939). *Journal of the History of the Behavioural Sciences* 16, 228–240.

Gold, R. L. (1958). Roles in sociological field observations. *Social Forces* 36(3), 217–223.

Goldberg, H. (1967). FBD marriage and demography among Tripolitanian Jews in Israel. *Southwestern Journal of Anthropology* 23, 177–191.

Golde, P. (ed.) (1970). *Women in the Field.* Chicago: Aldine.

Golde, P. (1970a). Odyssey of encounter. In *Women in the Field* (ed. P. Golde). Chicago: Aldine.

Golde, P. (1973). [Letter to F. Redlich on covert research]. Footnote 2, p. 318 in F. Redlich "The Anthropologist as observer: ethical aspects of clinical observations of behaviour". *The Journal of Nervous and Mental Disease* 157(5), 313–319.

Goldkind, V. (1970). Anthropologists, informants and the achievement of power in Chan Kom. *Sociologus* (N.S.) 20(1), 17–41.

Goldstein, K. S. (1964). *A Guide for Fieldworkers in Folklore.* (American Folklore Society Memoir 52). Hatboro, Pennsylvania: Folklore Associates, for the American Folklore Society.

Gonzalez, N. L. S. (1970). Cakchiqueles and Caribs: the social context of fieldwork. In *Marginal Natives: Anthropologists at Work* (ed. M. Freilich), pp. 152–184. New York: Harper and Row.

Goodenough, W. H. (1956). Componential analysis and the study of meaning. *Language* 32, 195–216.

Goodenough, W. H. (1966). Cultural anthropology and linguistics. In *Language in Culture and Society* (ed. D. Hymes). New York: Harper and Row.

Goody, E. N. (ed.) (1978). *Questions and Politeness: Strategies in Social Interaction.* Cambridge: Cambridge University Press.

Goody, J. (1973). From bridewealth to dowry in Africa and Eurasia. In *Bridewealth and Dowry* (eds J. Goody and S. J. Tambiah). (Cambridge papers in Social Anthropology 7). Cambridge: Cambridge University Press.

Goody, J. (1977). *The Domestication of the Savage Mind.* Cambridge: Cambridge University Press.

Gordon, R. L. (1969). *Interviewing Strategy, Techniques and Tactics.* Homewood, Illinois: Dorsey Press.

Gorman, R. A. (1977). *The Dual Vision: Alfred Schutz and the Myth of Phenomenological Social Science.* London: Routledge and Kegan Paul.

Gottdiener, M. (1979). Field research and video tape. *Sociological Inquiry* 49(4), 59–66.

Gould, J. and Kolb, W. L. (eds) (1964). *A Dictionary of the Social Sciences.* London: Tavistock.

Gould, M. A. (1975). Two decades of fieldwork in India: some reflections. In *Encounter and Experience: Personal Accounts of Fieldwork* (eds A. Béteille and T. N. Madan). Delhi: Vikas Publishing House.

Gouldner, A. W. (1970). *The Coming Crisis of Western Sociology.* New York: Basic Books.

Gowers, E. (1965). *Fowler's Modern English Usage.* Oxford: Oxford University Press.

Gowers, E. (1967). *The Complete Plain Words.* Harmondsworth: Penguin.

Great Britain, House of Lords (1982). *Data Protection Bill.* House of Lords Papers and Bills 50. London: H.M.S.O.

Green, J. W. (1978). Problems and consequences of actively helping the host community. In *Ethical Dilemmas in Anthropological Inquiry: A Case Book* (ed. G. N. Appell). Waltham, Mass.: Crossroads Press.

Green, J. (1979). Introduction. In *Zuni: Selected Writings of Frank Hamilton Cushing* (ed. J. Green). Lincoln: University of Nebraska Press.

Griaule, M. (1957). *Methode de l'ethnographie.* Paris: Presses Universitaires de France.

Griffin, J. H. (1960). *Black Like Me.* London: Collins.

Grimshaw, A. D. (1982). Whose privacy? What harm? *Sociological Methods and Research* 11(2), 233–247.

Gronewold, S. (1972). Did Frank Hamilton Cushing go native? In *Crossing Cultural Boundaries* (eds S. T. Kimball and J. B. Watson). San Francisco: Chandler Publishing Company.

Grönfers, M. (1982). From scientific social science to responsible research: the lesson of Finnish gypsies. *Acta Sociologica* 25(3), 249–258.

Gruber, J. W. (1967). Horatio Hale and the development of American anthropology. *Proceedings of the American Philosophical Society* 111, 5–37.

Gruber, J. W. (1970). Ethnographic salvage and the shaping of anthropology. *American Anthropologist* 72, 1289–1299.

Gudschinsky, S. C. (1967). *How to Learn an Unwritten Language.* New York: Holt, Rinehart and Winston.

Guest, R. H. (1960). Categories of events in field observations. In *Human Organization*

Research: Field Relations and Techniques (eds R. N. Adams and J. J. Preiss). Homewood, Illinois: The Dorsey Press, for the Society for Applied Anthropology.

Gullick, J. (1970). Village and city fieldwork in Lebanon. In *Marginal Natives: Anthropologists at Work* (ed. M. Freilich). New York: Harper and Row.

Gulliver, P. H. (1966). *The Family Herds: A Study of Two Pastoral Tribes in East Africa, The Jie and Turkana.* London: Routledge and Kegan Paul.

Gulliver, P. H. (1971). *Neighbours and Networks: The Idiom of Kinship in Social Action Among the Ndendeuli of Tanzania.* Berkeley: University of California Press.

Gupta, K. A. (1979). Travails of a woman fieldworker: a small town in Uttar Pradesh. In *The Fieldworker and the Field* (eds. M. N. Srinivas, A. M. Shah and E. A. Ramaswamy). Delhi: Oxford University Press.

Gutkind, P. C. W. (1967). Comment on Colson. *Current Anthropology* 8, 105–106.

Gutkind, P. C. W. (1969). The social researcher in the context of African national development: reflections on an encounter. In *Stress and Response in Fieldwork* (eds F. Henry and S. Saberwal). New York: Holt, Rinehart and Winston.

Gutkind, P. C. W. and Sankoff, G. (1967). Annotated bibliography on anthropological fieldwork and methods. In *Anthropologists in the Field* (eds D. G. Jongmans and P. C. W. Gutkind). (Studies of Developing Countries 6). Assen: Van Gorcum, Prakke and Prakke; New York: Humanities Press.

Gwaltney, J. L. (1976). On going home again – some reflections of a native anthropologist. *Phylon* 37(3), 236–242.

Gwaltney, J. L. (1981). Common sense and science: urban core black observations. In *Anthropologists at Home in North America: Methods and Issues in the Study of One's Own Society* (ed. D. A. Messerschmidt). Cambridge: Cambridge University Press.

Haddon, A. C. (1890). The ethnography of the western tribe of Torres Strait. *Journal of the Anthropological Institute* 19, 297–440.

Haddon, A. C. (1897). The saving of vanishing knowledge. *Nature* 55, 305–306.

Haddon, A. C. (1898). *The Study of Man.* London: Bliss.

Haddon, A. C. (1902). What the United States is doing for anthropology. *Journal of the Anthropological Institute* 32, 8–24.

Haddon, A. C. (1903). Anthropology: its position and needs. *Journal of the Anthropological Institute* 33, 11–23.

Haddon, A. C. (1905). South African ethnology. *Nature* 72, 471–479.

Haddon, A. C. and Browne, C. R. (1893). The ethnography of the Aran Islands. Co. Galway. *Proceedings of the Royal Irish Academy* 3, 768–830.

Hale, K. (1965). On the use of informants in fieldwork. *The Canadian Journal of Linguistics* 10, 108–119.

Halfpenny, P. (1979). The analysis of qualitative data. *Sociological Review* 27(4), 799–825.

Hall, E. T. (1959). *The Silent Language.* New York: Doubleday.

Hall, R. I. (1972). Agencies of research support: some sociological perspectives. In *The Social Contexts of Research* (eds S. Z. Nagi and R. G. Corwin). New York: John Wiley.

Hallowell, A. I. (1960). The beginnings of anthropology in America. In *Selected Papers from the American Anthropologist 1888–1920* (ed. F. de Laguna). Washington D.C.: American Anthropological Association.

Hallpike, C. R. (1977). *Bloodshed and Vengeance in the Papuan Mountains: The Generation of Conflict in Tanade Society*. Oxford: Clarendon Press; London: Oxford University Press.

Hallpike, C. R. (1978). Accuracy, tact and honesty (Letter). *Man* (N.S.) **13**(3), 477.

Hammel, E. A. and Goldberg, H. (1971). Parallel cousin marriage (letter). *Man* (N.S.) **6**(13), 488–489.

Hammond, P. E. (ed.) (1964). *Sociologists at Work: Essays on the Craft of Social Research*. New York: Basic Books.

Hanna, W. J. (1965). Image-making in field research: some tactical and ethical problems of research in tropical Africa. *American Behavioral Scientist* **8**, 15–20.

Hansen, J. F. (1976). The anthropologist in the field: scientist, friend and voyeur. In *Ethics and Anthropology: Dilemmas in Fieldwork* (eds M. A. Rynkiewich and J. P. Spradley). New York: John Wiley and Sons.

Haring, D. G. (1954). Comment on field techniques in ethnology: illustrated by a survey in the Ryuku Islands. *Southwestern Journal of Anthropology* **10**, 255–267. Reprinted (1956) in *Personal Character and Cultural Milieu* (ed. D. G. Haring). Syracuse: Syracuse University Press.

Harré, R. and Secord, P. F. (1972). *The Explanation of Social Behaviour*. Oxford: Basil Blackwell.

Harrell-Bond, B. (1976). Studying elites: some special problems. In *Ethics and Anthropology: Dilemmas in Fieldwork* (eds M. A. Rynkiewich and J. P. Spradley). New York, Toronto: John Wiley and Sons.

Harris, C. (pseud.) (1974). *Hennage*. New York: Holt, Rinehart and Winston.

Harris, C. (1979). "Participant observation in sociology and social anthropology: a reflection inspired by some of the papers". SSRC Workshop on Participant Observation, Birmingham University, September 24–25.

Harris, M. (1970). Referential ambiguity in the calculus of Brazilian racial identity. *Southwestern Journal of Anthropology* **26**(1), 1–14.

Harris, R. L. (1972). *Prejudice and Tolerance in Ulster: A Study of Neighbours and "Strangers" in a Border Community*. Manchester: Manchester University Press.

Harris, Z. S. and Voegelin, C. F. (1953). Eliciting. *Southwestern Journal of Anthropology* **9**, 59–75.

Hart, C. W. M. (1954). The sons of Turimpi: interpreters and guides. *American Anthropologist* **56**, 242–261.

Hart, C. W. M. (1970). Fieldwork among the Tiwi, 1928–1929. In *Being an Anthropologist* (ed. G. D. Spindler). New York: Holt, Rinehart and Winston.

Hart, H. (1979). *Hart's Rules for Compositors and Readers at the University Press, Oxford* (38th edn.). Oxford: Oxford University Press.

Hastrup, K. (1978). The semantics of biology: virginity. In *Defining Females* (ed. S. Ardener). London: Croom Helm; New York: John Wiley.

Hatfield, C. R. (1973). Fieldwork: toward a model of mutual exploitation. *Anthropological Quarterly* **46**, 15–29. Reprinted (1975) in *Cultural and Social Anthro-*

pology: Introductory Readings in Ethnology (ed. P. B. Hammond), 2nd edn. New York: Macmillan; London: Collier Macmillan.

Hau' ofa, E. (1975). Anthropology and Pacific Islanders. *Oceania* 45(4), 283–289

Healey, A. (1964). *Handling Unsophisticated Linguistic Informants.* (Series A: Occ. Pap. 2). Linguistic Circle of Canberra Publications.

Healey, A. (ed.) (1975). *Language Learner's Field Guide.* Ukarumpa, Papua New Guinea: Summer Institute of Linguistics.

Heider, E. R. (1972). Probabilities, sampling and ethnographic methods: the case of Dani colour names. *Man* (N.S.) 7(3), 448–466.

Helm, J. (1979). Long-term research among the Dogrib and other Dene. In *Long-term Field Research in Social Anthropology* (eds G. M. Foster, T. Scudder, E. Colson and R. V. Kemper). Orlando, New York and London: Academic Press.

Henige, D. (1982). *Oral Historiography.* London: Longman.

Hennigh, L. (1981). The anthropologist as key informant: inside a rural Oregon town. In *Anthropologists at Home in North America: Methods and Issues in the Study of One's Own Society* (ed. D. A. Messerschmidt). Cambridge: Cambridge University Press.

Henriksen, G. (1973). *Hunters in the Barrens: The Naskapi on the Edge of the White Man's World.* (Newfoundland Social and Economic Studies 12). St. John's: Institute of Social and Economic Research.

Henry, F. (1966). The role of the fieldworker in an explosive political situation (with comments and reply). *Current Anthropology* 7(5), 552–559.

Henry, F. (1969). Stress and strategy in three field situations. In *Stress and Response in Fieldwork* (eds F. Henry and S. Saberwal). New York: Holt, Rinehart and Winston.

Henry, F. and Saberwal, S. (eds) (1969). *Stress and Response in Fieldwork.* New York: Holt, Rinehart and Winston.

Henry, J. (1940). A method for learning to talk primitive languages. *American Anthropologist* 42, 635–641.

Henry, J. and Spiro, M. (1953). Psychological techniques: projective tests in fieldwork. In *Anthropology Today, An Encyclopedic Inventory* (ed. A. L. Kroeber). Chicago: Chicago University Press.

Herskovits, M. J. (1948). The ethnographer's laboratory. In *Man and His Works.* New York: A. A. Knopf.

Herskovits, M. J. and Herskovits, F. S. (1934). *Rebel Destiny: Among the Bush Negroes of Dutch Guiana.* New York: McGraw-Hill.

Hiatt, L. R. (1981). *Waiting for Harry.* Film made by Kim McKenzie with L. R. Hiatt. Canberra: Australian Institute of Aboriginal Studies (RAI Film Lib. Cat. No. RA100).

Hickerson, N. P. (1980). *Linguistic Anthropology.* New York: Holt, Rinehart and Winston.

Hicks, G. L. (1977). Informant anonymity and scientific accuracy: the problem of pseudonyms. *Human Organization* 36(2), 214–220.

Hilger, Sister I. (1954). An ethnographic field method. In *Method and Perspective in Anthropology* (ed. R. F. Spencer). Minneapolis: University of Minnesota Press.

Hiller, H. H. (1979). Universality of science and the question of national sociologies. *The American Sociologist* **14**(3), 124–135.

Hinshaw, R. E. (1980). Anthropology, administration and public policy. *Annual Review of Anthropology* **9**, 497–522.

Hinsley, C. (1981). *Savages and Scientists: The Smithsonian Institution and the Development of American Anthropology*. Washington, D.C.: Smithsonian Institution.

Hitchcock, J. T. (1960). Sarat Singh, Head Judge (India). In *In the Company of Man: Twenty Portraits by Anthropologists* (ed. J. B. Casagrande). New York: Harper and Row.

Hitchcock, J. T. (1970). Fieldwork in Ghurka country. In *Being an Anthropologist* (ed. G. D. Spindler). New York: Holt, Rinehart and Winston.

Hockett, C. F. (1957). How to learn Martian. In *Coming Attractions* (ed. M. Greenberg). New York: Gnome Press.

Hockett, C. F. (1958). *A Course in Modern Linguistics*. New York: Macmillan.

Hockey, S. M. (1980). *A Guide to Computer Applications in the Humanities*. London: Duckworth.

Hockings, P. (ed.) (1975). *Principles of Visual Anthropology*. The Hague: Mouton.

Hoijer, H. (1973). History of American linguistics. In *Current Trends in Linguistics* **10**(1): *Linguistics in North America* (ed. T. A. Sebeok). The Hague: Mouton.

Hoinville, G., Jowell, R. and associates (1978). *Survey Research Practice*. London: Heinemann.

Holland, P. W. and Leinhardt, S. (1973). The structural implications of measurement error in sociology. *Journal of Mathematical Sociology* **3**(1), 85–111.

Hollander, A. N. J. Den (1967). Social description: the problem of reliability and validity. In *Anthropologists in the Field* (eds D. G. Jongmans and P. C. W. Gutkind). Assen: Van Gorcum, Prakke and Prakke.

Holleman, J. F. (1958). *African Interlude*. Cape Town: Nasionale Boekhandel.

Hollis, M. (1977). *Models of Man*. Cambridge: Cambridge University Press.

Holmberg, A. (1969). *Nomads of the Longbow: The Siriono of East Bolivia*. Garden City, New York: Natural History Press.

Holzner, B. (1964). Methodology. *A Dictionary of the Social Sciences* (eds J. Gould and W. L. Kolb). London: Tavistock.

Honigan, J. J. (1957). Women in West Pakistan. In *Pakistan Society and Culture* (ed. S. Maran). New Haven: Human Relations Area Files, Yale University Press.

Honigmann, J. J. (1970). Fieldwork in two northern Canadian communities. In *Marginal Natives: Anthropologists at Work* (ed. M. Freilich). New York: Harper and Row.

Honigmann, J. J. (1976). The personal approach in cultural anthropological research. *Current Anthropology* **17**, 243–251.

Horowitz, I. L. (ed.) (1967). *The Rise and Fall of Project Camelot: Studies in the Relationship between Social Science and Practical Politics*. Cambridge, Mass.: M.I.T. Press.

Hostetler, J. A. and Huntington, G. E. (1970). The Hutterites: fieldwork in a north

American communal society. In *Being an Anthropologist* (ed. G. D. Spindler). New York: Holt, Rinehart and Winston.

Howell, S. (1981). Rules not words. In *Indigenous Psychologies* (eds P. Heelas and A. Lock). London, Orlando and New York: Academic Press.

Hsin-Pao Yang (1955). *Fact-finding with Rural People: A Guide to Effective Social Survey*. (Agric. Dev. Paper **52**.) Rome: Food and Agricultural Organisation of the United Nations.

Hsu, F. L. K. (1979). The cultural problem of the cultural anthropologist. *American Anthropologist* **81**, 517–532.

Hubert, J., Forge, A. and Firth, R. (1968). *Methods of Study of Middle-class Kinship in London: A Working Paper on the History of an Anthropological Project, 1960–1965*. Occasional Paper of the Department of Anthropology, London School of Economics and Political Science.

Hughes, D. T. (1980). The responsibilities of anthropologists to Pacific Islanders. *Pacific Studies* **3**(2), 43–51.

Huizer, G. (1973). The a-social role of social scientists in underdeveloped countries: some ethical considerations. *Sociologus* (N.S.) **23**(2), 165–177.

Huizer, G. (1979). Anthropology and politics: from naiveté toward liberation? In *The Politics of Anthropology: From Colonialism and Sexism Toward a View from Below* (eds G. Huizer and B. Mannheim). The Hague, Paris: Mouton.

Huizer, G. and Mannheim, B. (eds) (1979). *The Politics of Anthropology: From Colonialism and Sexism Toward a View from Below*. The Hague: Mouton.

Humphreys, L. (1975). *Tearoom Trade, Impersonal Sex in Public Places*. Chicago: Aldine. (Enlarged edition with a retrospect on ethical issues.)

Hunter, D. E. and Foley, M. A. B. (1976). *Doing Anthropology: A Student-centred Approach to Cultural Anthropology*. London: Harper and Row.

Huxley, F. (1956). *Affable Savages: An Anthropologist Among the Aruba Indians of Brazil*. New York: Viking Press.

Hyman, H. M. with Cobb, W. J., Feldman, J. J., Hart, C. W. and Stember, C. H. (1954). *Interviewing in Social Research*. Chicago: University of Chicago Press.

Hymes, D. H. (1959). Bibliography: fieldwork in linguistics and anthropology. *Studies in Linguistics* **14**(3–4), 82–91.

Hymes, D. H. (ed.) (1964). *Language in Culture and Society*. New York, Evanston and London: Harper and Row.

Hymes, D. H. (ed.) (1965). *The Use of Computers in Anthropology*. London: Mouton.

Hymes, D. H. (ed.) (1972). *Reinventing Anthropology*. New York: Pantheon Books.

Hymes, D. H. (1977). The state of the art in linguistic anthropology. In *Perspectives on Anthropology 1976* (eds A. F. C. Wallace *et al.*,). (Special Publication **10**). Washington, D.C.: American Anthropological Association.

Hymes, D. H. and Fought, J. (1975). American structuralism. In *Current Trends in Linguistics* **13**: *Historiography of Linguistics* (ed. T. A. Sebeok). The Hague: Mouton.

Ifeka-Moller, C. (1976). Changes in property transfer among Greek Cypriot villagers [letter]. *Man* (N.S.) **11**(4), 592–594.

Illustration of the method of recording Indian languages . . . (1879–80). *Annual Report of the Bureau of Ethnology 1879–80.* Washington D.C.: Government Printer.

International Phonetic Association (1949). *The Principles of the International Phonetic Association.* London: University College.

Ishige, N. (ed.) (1980). *The Galela of Halmahera.* (Senri Ethnological Studies 7). Osaka: National Museum of Ethnology.

Jackson, B. and Marsden, D. (1966). *Education and the Working Class.* Harmondsworth: Penguin Books.

Jacobs, S.-E. (1974). Action and advocacy anthropology. *Human Organization* 33(2), 209–215.

Jacobson, D. (1978). The chaste wife: cultural norm and individual experience. In *American Studies in the Anthropology of India* (ed. S. Vatuk). New Delhi: American Institute of Indian Studies and Manohar.

Jacques, E. (1951). *The Changing Culture of a Factory.* London: Tavistock Publications.

Jahoda, M. (1981). To publish or not to publish? *Journal of Social Issues* 37(1), 208–220.

James, H. (1962). The art of fiction (1884). In *The House of Fiction: Essays on the Novel* (ed. L. Edel). London: Mercury Books.

James, W. (1978). Matrifocus on African women. In *Defining Females* (ed. S. Ardener). London: Croom Helm; New York: Wiley.

Janes, R. W. (1961). A note on phases of the community role of the participant observer. *American Sociological Review* 26, 446–450.

Jansen, W. H. II (1973). The applied man's burden: the problem of ethics and applied anthropology. *Human Organization* 32(3), 325–329.

Jarvie, I. C. (1964). *The Revolution in Anthropology.* London: Routledge and Kegan Paul.

Jarvie, I. C. (1967). On the theories of fieldwork and the scientific character of social anthropology. *Philosophy of Science* 24, 223–242.

Jarvie, I. C. (1969). The problem of ethical integrity in participant observation. *Current Anthropology* 10(5), 505–508. In I. C. Jarvie, P. Kloos (and others) (1969).

Jarvie, I. C., Kloos, P. (and others) (1969). Problems of role conflicts in social studies. *Current Anthropology* 10(5), 505–523. See Jarvie (1969); Kloos (1969).

Jaspan, M. A. (1964). Anthropology and commitment to political causes. *Anthropological Forum* 1, 212–219.

Jay, R. (1974). Personal and extrapersonal vision in anthropology. In *Reinventing Anthropology* (ed. D. Hymes). New York: Vintage Books, Random House.

Jayaraman, R. (1979). Problems of entry: Sri Lanka, Zambia and Australia. In *The Fieldworker and the Field: Problems and Challenges in Sociological Investigation* (eds M. N. Srinivas, A. M. Shah and E. A. Ramaswamy). Delhi: Oxford University Press.

Jenkins, R. P. (1979). The value of the evidence [letter]. *Man* (N.S.) 14(1), 161–163.

Johnson, A. W. (1978). *Quantification in Cultural Anthropology: An Introduction to*

Research Design. Stanford University Press (also publ. (1978) as *Research Methods in Social Anthropology.* London: Arnold).

Johnson, D. E. *et al.* (1976). *A Survey of Materials for the Study of Uncommonly Taught Languages.* In 8 fascicles. Arlington, Virginia: Center for Applied Linguistics.

Johnson, J. M. (1975). *Doing Field Research.* London: Collier Macmillan; New York: The Free Press.

Jones, D. (1971). Addendum: social responsibility and the belief in basic research: an example from Thailand. *Current Anthropology* **12**(3), 347–350. [Addendum in J. G. Jorgensen, R. N. Adams (and others). Toward an ethics for anthropologists. *Current Anthropology* **12**(3), 321–356].

Jones, D. (1980). Accountability and the politics of urban research. *Human Organization* **39**(1), 99–104.

Jones, E. L. (1964). The courtesy bias in South-East African surveys. *International Social Science Journal* **16**, 63–69.

Jones, L. McC. and Fischer, C. S. (1978). *Studying Egocentric Networks by Mass Survey.* (Working Paper No. 284). Berkeley: Institute of Urban and Regional Development.

Jongmans, D. G. and Gutkind, P. C. W. (eds) (1967). *Anthropologists in the Field.* (Studies of Developing Countries 6). Assen: Van Gorcum, Prakke and Prakke; New York: Humanities Press.

Jordan, D. K. (1981). The ethnographic enterprise and the bureaucratization of ethics: the problem of human subjects legislation. *Journal of Anthropological Research* **37**(4), 415–419.

Jorgensen, J. G. (1971). On ethics and anthropology. *Current Anthropology* **12**(3), 321–334.

Jorgensen, J. G., Adams, R. N. (and others) (1971). Toward an ethics for anthropologists. *Current Anthropology* **12**(3), 321–356. [With comments and a reply].

Joshi, P. C. (1979). Fieldwork experience: relived and reconsidered: rural Uttar Pradesh. In *The Fieldworker and the Field: Problems and Challenges in Sociological Investigation* (eds M. N. Srinivas, A. M. Shah and E. A. Ramaswamy). New Delhi: Oxford University Press.

Journal of Social Issues (1977a). Privacy as a behavioral phenomenon (ed. S. T. Margulis). *Journal of Social Issues* **33**(3), 1–195.

Journal of Social Issues (1977b). Research among racial and cultural minorities: problems, prospects, and pitfalls (eds D. Montero and G. E. Levine). *Journal of Social Issues* **33**(4), 1–178.

Jowell, R. (1982). Ethical concerns in data collection. In *Data Protection and Privacy: Proceedings of a Conference* (ed. C. D. Raab). London: Social Research Association.

Jules, H. and Spiro, M. E. (1953). Psychological techniques: projective tests in fieldwork. In *Anthropology Today* (ed. A. L. Kroeber). Chicago: University of Chicago Press.

Junker, B. H. (1960). *Fieldwork: An Introduction to the Social Sciences.* Chicago: University of Chicago Press.

Junod, H. A. (1898). *Les Ba-Ronga. Étude ethnographique sur les indigènes de la baie de Delagoa.* Neuchâtal.

Kaberry, P. H. (1957). Malinowski's contribution to field-work methods and the writing of ethnography. In *Man and Culture: An Evaluation of the Work of Bronislaw Malinowski* (ed. R. Firth). London: Routledge and Kegan Paul.

Kahn, R. L. and Cannell, C. F. (1967). *The Dynamics of Interviewing: Theory, Technique, and Case.* New York: John Wiley.

Kane, E. (1982a). "Ballybran". [Letter] *RAIN (Royal Anthropological Institute News)* **52** (October), 14–15.

Kane, E. (1982b). Cui Bono? Do Aon Duine. (To whose advantage? Nobody's). *RAIN (Royal Anthropological Institute News)* **50** (June), 1–2.

Kanitkar, H. A. (1979). A guide to bibliographies and reference works related to anthropology and sociology of South Asia. In *South Asian Bibliography: A Handbook and Guide* (ed. J. D. Pearson). Sussex: Harvester Press.

Kapferer, B. (1969). Norms and the manipulation of relationships in a work context. In *Social Networks in Urban Situations: Analyses of Personal Relationships in Central African Towns* (ed. J. Clyde Mitchell).Manchester: Manchester University Press, for Institute of African Studies.

Kapferer, B. (1972). *Strategy and Transaction in an African Factory: African Workers and Indian Management in a Zambian Town.* Manchester: Manchester University Press.

Kaplan, A. (1964). *The Conduct of Inquiry: Methodology for Behavioral Science.* San Francisco: Chandler Publishing Company.

Kasakoff, A. B. and Adams, J. W. (1977). Spatial location and social organisation: an analysis of Tikopian patterns. *Man* (N.S.) **12**, 48–64.

Keat, R. and Urry, J. (1975). *Social Theory as Science.* London: Routledge and Kegan Paul.

Keesing, F. M. (1959). *Field Guide to Oceania.* (Publication 701, Field guide series 1). Washington D.C.: National Academy of Sciences/National Research Council.

Keesing, R. M. (1979). Anthropology in Melanesia: retrospect and prospect. In *The Politics of Anthropology: From Colonialism and Sexism Toward a View from Below* (eds G. Huizer and B. Mannheim). The Hague: Mouton.

Keiser, R. L. (1970). Fieldwork among the Vice Lords of Chicago. In *Being an Anthropologist: Fieldwork in Eleven Cultures* (ed. G. D. Spindler). New York: Holt, Rinehart and Winston.

Kelman, H. C. (1982). Ethical issues in different social science methods. In *Ethical Issues in Social Science Research* (eds T. L. Beauchamp, R. R. Faden, R. J. Wallace, Jr. and L. Walters). Baltimore: The Johns Hopkins University Press.

Kemper, R. V. (1979). Fieldwork among Tzintzuntzan migrants in Mexico City: retrospect and prospect. In *Long-term Field Research in Social Anthropology* (eds G. M. Foster, T. Scudder, E. Colson and R. V. Kemper). Orlando, New York and London: Academic Press.

Kesby, J. (1982). The Preface. *Progress and the Past Among the Rangi of Tanzania.* New Haven, Connecticut: Human Relations Area Files.

Keyes, C. F. (1977). *The Golden Peninsula: Culture and Adaptation in Mainland Southeast Asia*. London: Collier Macmillan.

Kimball, S. T. (1955). Problems of studying American culture. *American Anthropologist* 57, 1131–1142.

Kimball, S. T. and Partridge, W. L. (1979). *The Craft of Community Study*. Gainesville, Florida: University Presses of Florida.

Kimball, S. T. and Watson, J. B. (eds) (1972). *Crossing Cultural Boundaries: The Anthropological Experience*. San Francisco: Chandler Publishing Company.

King, M., King, F. and Martodipoero, S. (1978). *Primary Child Care: A Manual for Health Workers*. Oxford: Oxford University Press.

Klineberg, O. (1980). Historical perspectives: cross cultural psychology before 1960. In *Handbook of Cross Cultural Psychology 1: Perspectives* (eds H. C. Triandis and W. W. Lambert). Boston: Allyn and Bacon.

Klockars, C. B. (1979). Dirty hands and deviant subjects. In *Deviance and Decency: The Ethics of Research with Human Subjects* (eds C. B. Klockars and F. W. O'Connor). Beverly Hills and London: Sage Publications.

Klockars, C. B. and O'Connor, F. W. (eds) (1979). *Deviance and Decency: The Ethics of Research with Human Subjects*. Beverly Hills and London: Sage Publications.

Kloos, P. (1969). Role conflicts in social fieldwork. *Current Anthropology* 10, 509–512.

Kloos, P. and Claessen, W. (1981). *Current Issues in Anthropology in the Netherlands*. Rotterdam: Anthropological Branch of the Netherlands Sociological and Anthropological Society.

Kluckhohn, C. (1943). The personal document in anthropological science. In *The Use of Personal Documents in History, Anthropology and Sociology* (eds L. Gottschalk, C. Kluckhohn and R. Angell). United States of America, Social Science Research Council, Committee on Appraisal of Research.

Kluckhohn, C. (1960). A Navaho politician. In *In the Company of Man: Twenty Portraits by Anthropologists* (ed. J. B. Casagrande). New York: Harper and Row.

Kluckhohn, F. R. (1940). The participant–observer technique in small communities. *American Journal of Sociology* 46, 331–343.

Köbben, A. J. F. (1967). Participation and quantification: fieldwork among the Djuka (bush negroes of Surinam). In *Anthropologists in the Field* (eds D. G. Jongmans and P. C. W. Gutkind). Assen: Van Gorcum, Prakke and Prakke.

Koepping, K.-P. (1976). On the epistemology of participant observation and the generating of paradigms. In *Occasional Papers in Anthropology* 6 (ed. P. K. Lauer). University of Queensland Anthropology Museum.

Koestler, A. (1969). *The Act of Creation* (revised Danube edn.). London: Hutchinson.

Korn, S. (1978). The formal analysis of visual systems as exemplified by a study of Abelam (Papua New Guinea) paintings. In *Art in Society: Studies in Style, Culture and Aesthetic* (eds M. Greenhalgh and J. V. S. Megaw). London: Duckworth.

Krashen, S. D. (1981). *Second Language Acquisition and Second Language Learning*. Oxford: Pergamon Press.

Krathwohl, D. (1976). *How to Prepare a Research Proposal*.

Kress, G. R. and Hodge, R. (1979). *Language as Ideology*. London: Routledge and Kegan Paul.

Krige, E. J. (1960). Agnes Winifred Hoernlé: an appreciation. *African Studies* 19(3), 138–144.

Kroeber, A. L. (1943). Franz Boas: the man. In *Franz Boas 1858–1942* (eds A. L. Kroeber *et al.,*). (Memoir of the American Anthropological Association 61).

Kuhn, T. S. (1970). *The Structure of Scientific Revolutions* (2nd edn.). Chicago: University of Chicago Press.

Kuper, A. (1973). *Anthropologists and Anthropology: The British School 1922–72*. London: Allen Lane.

Kuper, H. (1947). *An African Aristocracy: Rank Among the Swazi*. London: Oxford University Press, for International African Institute.

Lakoff, R. (1975). Linguistic theory and the real world. *Language Learning* 25(2), 309–338.

Lamphere, L. (1977). Review essay: Anthropology. *Signs* 2(3), 612–627.

Lamphere, L. (1979). The long-term study among the Navaho. In *Long-term Field Research in Social Anthropology* (eds G. M. Foster, T. Scudder, E. Colson and R. V. Kemper). Orlando, New York and London: Academic Press.

Lancy, D. F. and Strathern, A. J. (1981). 'Making twos': pairing as an alternative to the taxonomic mode of representation. *American Anthropologist* 83, 773–795.

Landers, R. (1970). A woman anthropologist in Brazil. In *Women in the Field: Anthropological Experiences* (ed. P. Golde). Chicago: Aldine Publishing Company.

Langacker, R. W. (1972). *Fundamentals of Linguistic Analysis*. New York: Harcourt Brace Jovanovich.

Langacker, R. W. (1973). *Language and its Structure: Some Fundamental Linguistic Concepts*. New York: Harcourt Brace Jovanovich.

Langham, I. (1981). *The Building of British Social Anthropology: W. H. R. Rivers and his Cambridge Disciples in the Development of Kinship Studies*. Dordrecht: D. Reidel.

Langness, L. L. (1965). *The Life History in Anthropological Science*. (Studies in Anthropological method). London: Holt, Rinehart and Winston.

Laracy, H. (1976). Malinowski at war, 1914–18. *Mankind* 10, 264–268.

Larson, D. N. (1964). Making use of anthropological fieldnotes. *Practical Anthropology* 11(3), 143–144.

Leach, E. R. (1957). The epistemological background to Malinowski's empiricism. In *Man and Culture: An Evaluation of the Work of Bronislaw Malinowski* (ed. R. Firth). London: Routledge and Kegan Paul.

Leach, E. R. (1961). *Pul Eliya. A Village in Ceylon. A Study of Land Tenure and Kinship*. Cambridge: Cambridge University Press.

Leach, E. R. (1967). An anthropologist's reflections on a social survey. In *Anthropologists in the Field* (eds D. G. Jongmans and P. C. W. Gutkind). Assen: Van Gorcum. (First published 1958 in *Ceylon Journal of History and Social Studies* I(1), 9–20).

Leach, E. R. (1968 [1961]). *Rethinking Anthropology*. London: The Athlone Press.

Leacock, E. B. (1981). *Myths of Male Dominance*. New York: Monthly Review Press.

LeClair, E. E., Jr. (1960). Problems of large-scale anthropological research. In *Human Organization Research* (eds R. N. Adams and J. J. Preiss). Homewood, Ill.: The Dorsey Press.

Lee, R. B. (1979). Hunter-gatherers in process: the Kalahari Research Project, 1963–1976. In *Long-term Field Research in Social Anthropology* (eds G. M. Foster, T. Scudder, E. Colson and R. V. Kemper). Orlando, New York and London: Academic Press.

Leeds, A. (1969). Ethics report criticized. American Anthropological Association *Newsletter* **10**(6), 3–4.

Leggett, G. H., Mead, C. D. and Charvat, W. (1974). *Prentice-Hall Handbook for Authors.* Englewood Cliffs: Prentice-Hall.

Leibowitz, L. (1980). Perspectives in the evaluation of sex-differences. In *Towards an Anthropology of Women* (ed. R. Reiter). New York: Monthly Review Press.

Lesser, A. (1981). Franz Boas. In *Totems and Teachers: Perspectives on the History of Anthropology* (ed. S. Silverman). New York: Columbia University Press.

Lévi-Strauss, C. (1955). *Tristes tropiques.* Paris: Libraire Plon.

Lévi-Strauss, C. (1960). On manipulated sociological models. *Bijdragen tot de Taal-, Land- en Volkenkunde* **112**, 45–54.

Lévi-Strauss, C. (1969). The story of Asdiwal. In *The Structural Study of Myth and Totemism* (ed. E. R. Leach). London: Tavistock.

Lévi-Strauss, C. (1973). *Tristes tropiques.* (English translation by J. Weightman and D. Weightman). London: Jonathan Cape.

Lévi-Strauss, C. (1981). *A World on the Wane.* (Translation of *Tristes tropiques* by J. Russell). London: Hutchinson.

Lewis, D. (1973). Anthropology and colonialism. *Current Anthropology* **14**(5), 581–602 [with comments and a reply].

Lewis, I. (1973). *The Anthropologist's Muse: An Inaugural Lecture.* London: The London School of Economics and Political Science.

Lewis, I. (1976). *Social Anthropology in Perspective: The Relevance of Social Anthropology.* Harmondsworth: Penguin.

Lewis, O. (1951). *Life in a Mexican Village: Tepoztlan Restudied.* Urbana: University of Illinois Press.

Lewis, O. (1959). *Five Families: Mexican Case Studies in the Culture of Poverty.* New York: Basic Books.

Lewis, O. (1961). *The Children of Sanchez: Autobiography of a Mexican Family.* New York: Random House.

Lewis, O, (1964). *Pedro Martinez: A Mexican Peasant and His Family.* London: Secker and Warburg.

Lewis, O. (1965). *La Vida: A Puerto Rican Family in the Culture of Poverty—San Juan and New York.* London: Secker and Warburg.

Lewis, O. (1968). *A Study of Slum Culture: Backgrounds for La Vida.* New York: Random House.

Lewis, O. (1969). *A Death in the Sanchez Family.* London: Secker and Warburg.

Lewis, O., Lewis, R. M. and Rigdon, S. M. (1977). *Living the Revolution: An Oral History of Contemporary Cuba.* London: University of Illinois Press.

Li, T. (1980). *Social Science Reference Sources: A Practical Guide*. Westport, Conn.: Greenwood Press.

Lieberman, D. and Dressler, W. (1977). Bilingualism and cognition of St. Lucian disease terms. *Medical Anthropology* 1, 81–110.

Liebow, E. (1967). *Tally's Corner, A Study of Negro Street-Corner Men*. London: Routledge and Kegan Paul; Boston: Little, Brown and Co.

Light, L. and Kleiber, N. (1981). Interactive research in a feminist setting: the Vancouver Women's Health Collective. In *Anthropologists At Home in North America: Methods and Issues in the Study of One's Own Society* (ed. D. A. Messerschmidt). Cambridge: Cambridge University Press.

Lofland, J. (1971). *Analysing Social Settings: A Guide to Qualitative Observation and Analysis*. Belmont, California: Wadsworth.

Loizos, P. (1975). *The Greek Gift: Politics in a Cypriot Village*. Oxford: Blackwell.

Loizos, P. (1975a). Changes in property transfer among Greek Cypriot villages. *Man* (N.S.) 10, 503–523.

Loizos, P. (1981). *The Heart Grown Bitter: A Chronicle of Cypriot War Refugees*. Cambridge: Cambridge University Press.

Lomnitz, L. A. (1977). *Networks and Marginality: Life in a Mexican Shantytown*. Orlando, New York and London: Academic Press.

Long, G. L. and Dorn, D. S. (1982). An assessment of the ASA Code of Ethics and Committee of Ethics. *The American Sociologist* 17 (May), 80–86.

Lowie, R. H. (1933). Queries. *American Anthropologist* 35, 288–296.

Lowie, R. H. (1937). *The History of Ethnological Theory*. New York: Rinehart.

Lowie, R. H. (1940). Native languages as ethnographic tools. *American Anthropologist* 42, 81–89.

Lowie, R. H. (1959). *Robert H. Lowie Ethnologist: A Personal Record*. Berkeley: University of California Press.

Lowie, R. H. (1960). My Crow interpreter. In *In the Company of Man: Twenty Portraits by Anthropologists* (ed. J. B. Casagrande). New York: Harper and Row.

Lowry, J. (1981). Theorising 'Observation'. *Communication and Cognition* 14(1), 7–23.

Lukes, S. (1967). Some problems about rationality. *Archives Européennes de Sociologie* 8, 247–264

Lundberg, C. C. (1968). A transactional conception of fieldwork. *Human Organization* 27(1), 45–49.

Lundsgaarde, H. P. (1970–71). Privacy: an anthropological perspective on the right to be let alone. *Houston Law Review* 8, 858–875.

Luszki, M. B. (1957). Team research in social science: major consequences of a growing trend. *Human Organization* 16(1), 21–24. Reprinted (1960) in *Human Organization Research: Field Relations and Techniques* (eds R. N. Adams and J. J. Preiss). Homewood, Ill.: The Dorsey Press, for the Society for Applied Anthropology.

Lynch, F. R. (1977). Field research and future history: problems posed for ethnographic sociologists by the "Doomsday Cult" making good. *The American Sociologist* 12 (April), 80–88.

Lynd, R. S. and Lynd, H. M. (1929). *Middletown: A Study in Contemporary American Culture*. New York: Harcourt and Brace.

Lyons, J. (1981). *Language and Linguistics: An Introduction*. Cambridge: Cambridge University Press.

MacCormack, C. and Strathern, M. (eds) (1980). *Nature, Culture and Gender*. Cambridge: Cambridge University Press.

Macdonald, O. J. S. (1980). *The Preservation of Personal Health in Warm Climates*. London: The Ross Institute of Tropical Medicine and E. G. Berryman.

MacFarlane, A. (1977). *Reconstructing Historical Communities*. Cambridge: Cambridge University Press.

Mackinnon, K. (1976). *Language, Education and Social Processes in a Gaelic Community*. London: Routledge and Kegan Paul.

McArthur, N. (1961). *Introducing Population Statistics*. Melbourne: Oxford University Press.

McCall, G. J. and Simmons, J. L. (eds) (1969). *Issues in Participant Observation: A Text and a Reader*. Reading, Mass.: Addison-Wesley.

McCurdy, D. W. (1976). The medicine man. In *Ethics and Anthropology: Dilemmas in Fieldwork* (eds. M. A. Rynkiewich and J. P. Spradley). New York: John Wiley.

Madan, T. N. (1969). Political pressures and ethical constraints upon Indian sociologists. In *Ethics, Politics and Social Research* (ed. G. Sjoberg). London: Routledge and Kegan Paul.

Madge, J. (1957). *The Tools of Social Science*. London: Longmans.

Madge, J. (1963). *The Origins of Scientific Sociology*. London: Tavistock.

Mafeje, A. (1976). The problem of anthropology in historical perspective: an inquiry into the growth of social sciences. *Canadian Journal of African Studies* **10**(2), 307–333.

Maines, D. R., Shaffir, W. and Turowetz, A. (1980). Leaving the field in ethnographic research: reflections on the entrance-exit hypothesis. In *Fieldwork Experience: Qualitative Approaches to Social Research* (eds W. B. Shaffir, R. A. Stebbins and A. Turowetz). New York: St. Martin's Press.

Malinowski, B. (1915). The natives of Mailu: preliminary results of the Robert Mond research work in British New Guinea. *Transactions of the Royal Society of South Australia* **39**, 494–706.

Malinowski, B. (1920). Classificatory particles in the language of Kiriwina. *Bulletin of the School of Oriental Studies* **1**, 33–78.

Malinowski, B. (1922). *Argonauts of the Western Pacific: An Account of Native Enterprise and Adventure in the Archipelagoes of Melanesian New Guinea*. London: Routledge and Kegan Paul.

Malinowski, B. (1922a). Ethnology and the study of society. *Economica* **2**, 208–219.

Malinowski, B. (1923). The problem of meaning in primitive languages. In *The Meaning of Meaning* (eds C. K. Ogden and I. A. Richards). London: Routledge.

Malinowski, B. (1930). The rationalization of anthropology and administration. *Africa* **3**, 405–429.

Malinowski, B. (1932). *The Sexual Life of Savages in North-Western Melanesia* (2nd edn.), London: Routledge.

Malinowski, B. (1935). *Coral Gardens and Their Magic,* Vol. 1. *Soil Tilling and Agricultural Rites in the Trobriand Islands.* London: Allen and Unwin.

Malinowski, B. (1967). *A Diary in the Strict Sense of the Term* (with an introduction by R. Firth). London: Routledge and Kegan Paul; New York: Harcourt, Brace and World.

Mamak, A. and McCall, G. (eds) (1978). *Paradise Postponed: Essays on Research and Development in the South Pacific.* Rushcutters Bay, New South Wales: Pergamon Press.

Man, E. H. (1885). *On the Aboriginal Inhabitants of the Andaman Islands.* London.

Mandelbaum, D. G. (1960). A reformer of his people (South India). In *In the Company of Man: Twenty Portraits by Anthropologists* (ed. J. B. Casagrande). New York: Harper and Row.

Mandelbaum, D. G. (1973). The study of life history: Ghandi. *Current Anthropology* **14**, 177–206.

Mandelbaum, D. G. (1980). *The Todas* in time perspective. *Reviews in Anthropology* **7**, 279–302.

Mandelbaum, D. G., Lasker, G. W. and Albert, E. M. (eds) (1963). *The Teaching of Anthropology.* Berkeley: University of California Press.

Mangin, W. (1979). Thoughts on twenty-four years of work in Peru: The Vicos project and me. In *Long-term Field Research in Social Anthropology* (eds G. M. Foster, T. Scudder, E. Colson and R. V. Kemper). Orlando, New York and London: Academic Press.

Mann, P. H. (1976). *Methods of Sociological Enquiry.* Oxford: Blackwell.

Manners, R. A. (1956). Functionalism, realpolitik, and anthropology in under-developed areas. *America Indigena* **16**, 7–33.

Manners, R. A. and Kaplan, D. (1968). *Theory in Anthropology: A Source Book.* London: Routledge and Kegan Paul.

Manning, P. K. (1967). Problems in interpreting interview data. *Sociology and Social Research* **51**(3), 302–316.

Manning, P. K. and Fabrega, M., Jr. (1976). Fieldwork and the 'New Ethnography'. *Man* (N.S.) **11**, 39–52.

Maquet, J. J. (1964). Objectivity in anthropology. *Current Anthropology* **5**(1), 47–55.

Marcus, G. E. (1979). Ethnographic research among elites in the Kingdom of Tonga: some methodological considerations. *Anthropological Quarterly* **52**(3), 135–151.

Marcus, G. E. (1980). Rhetoric and the ethnographic genre in anthropological research. *Current Anthropology* **21**(4), 507–510.

Marett, R. R. and Penniman, T. K. (1932). *Spencer's Scientific Correspondence with Sir J. G. Frazer and Others.* Oxford: Clarendon Press.

Mark, J. (1980). *Four Anthropologists: An American Science in its Early Years.* New York: Science History Publications.

Marriott, A. (1962). *Greener Fields: Experiences Among the American Indians.* New York: Doubleday (Dolphin Books).

Marsh, C. (1982). *The Survey Method: the contribution of surveys to sociological explanation.* (Contemporary Social Research 6.) London: Allen and Unwin.

Marshall, G. (1970). In a world of women: field work in a Yoruba community. In *Women in the Field* (ed. P. Golde). Chicago: Aldine.

Marwick, M. G. (1947). The study of social attitudes. *Human Problems in British Central Africa* (*Rhodes-Livingstone Journal* 5, 44–47).

Marwick, M. G. (1967). The study of witchcraft. In *The Craft of Social Anthropology* (ed. A. L. Epstein). London: Tavistock.

Mauss, M. (1947). *Manuel d'Ethnographie*. Paris: Payot.

Maxwell, R. J. (1970). A comparison of field research in Canada and Polynesia. In *Marginal Natives: Anthropologists at Work* (ed. M. Freilich). New York: Harper and Row.

May, W. (1980). Doing ethics: the bearing of ethical theories on fieldwork. *Social Problems* 27(3), 358–370.

Maybury-Lewis, D. (1965). *The Savage and the Innocent*. London, New York: Evans Bros.

Maybury-Lewis, D. (1967). *Akwē-Shavante Society*. Oxford: Clarendon Press.

Mayer, A. C. (1968). The significance of quasi-groups in the study of complex societies. In *The Social Anthropology of Complex Societies* (ed. M. Banton). London: Tavistock.

Mayer, I. (1975). The patriarchal image: routine dissociation in Gusii families. *African Studies* 34(4), 259–281.

Mayer, P. (ed.) (1972). *Socialization: The Approach from Social Anthropology*. London: Tavistock.

Maynard, E. (1974). The growing negative image of the anthropologist among American Indians. *Human Organization* 33(4), 402–404.

Mayntz, R., Holm, K. and Huebner, R. (1976). *Introduction to Empirical Sociology* (Trans. A. Hammond, H. Davis, and D. Shapira). Harmondsworth: Penguin.

Mead, M. (1933). More comprehensive field methods. *American Anthropologist* 35, 1–15.

Mead, M. (1939). Native languages as field-work tools. *American Anthropologist* 41, 189–205. Reprinted (1964) in *Anthropology, A Human Science*. Princeton: New Jersey: Van Nostrand.

Mead, M. (1942). *And Keep Your Powder Dry: An Anthropologist Looks at America*. New York: Morrow.

Mead, M. (1959). *An Anthropologist at Work: Writings of Ruth Benedict*. London: Secker and Warburg.

Mead, M. (1960). Weaver of the border (New Britain). In *In the Company of Man: Twenty Portraits by Anthropologists* (ed. J. B. Casagrande). New York: Harper and Row.

Mead, M. (1964). *Anthropology, A Human Science: Selected Papers, 1939–1960*. Princeton: New Jersey: Van Nostrand.

Mead, M. (1965). *Anthropologists and What They Do*. New York: Franklin and Watts.

Mead, M. (1969). Research with human beings: a model derived from anthropological field practice. *Daedalus, Proceedings of the American Academy of Arts and Sciences* 98(2), 361–386.

Mead, M. (1970). Fieldwork in the Pacific Islands, 1925–1967. In *Women in the Field* (ed. P. Golde). Chicago: Aldine.

Mead, M. (1972). *Blackberry Winter: My Early Years*. London: Angus and Robertson; New York: William Morrow.

Mead, M. (1977). *Letters From the Field 1925–1975*. New York: Harper and Row.

Mead, M. (1978). The evolving ethics of applied anthropology. In *Applied Anthropology in America* (eds E. M. Eddy and W. L. Partridge). New York: Columbia University Press.

Mead, M., Chapple, E. D. and Brown, G. (1949). Report of the committee on ethics. *Human Organization* **8**(2), 20–21.

Meggitt, M. J. (1979). Reflections occasioned by continuing anthropological field research among the Enga of Papua New Guinea. In *Long-term Field Research in Social Anthropology* (eds G. M. Foster, T. Scudder, E. Colson and R. V. Kemper). Orlando, New York and London: Academic Press.

Mencher, J. P. (1975). Viewing hierarchy from the bottom up. In *Encounter and Experience* (eds A. Béteille and T. N. Madan). Delhi: Vikas Publishing House.

Merton, R. K. (1956). *The Focussed Interview: A Manual of Problems and Procedures*. Glencoe, Illinois: Free Press.

Messerschmidt, D. A. (ed.) (1981). *Anthropologists at Home in North America: Methods and Issues in the Study of One's Own Society*. Cambridge: Cambridge University Press.

Methods of Study of Culture Contact in Africa [L. P. Mair (ed.)] (1938). Oxford: Oxford University Press, for the International African Institute (International African Institute Memorandum **1**).

Metraux, R. (1969). Appendix II: Study program in human health and the ecology of Man – China: measures taken for the protection of confidentiality in an inter-disciplinary study of health and cultural adaptation based on retrospective life histories. *Daedalus* **98**(2), 379–381. [In Mead (1969) q.v.]

Middleton, J. (1970). *The Study of the Lugbara: Expectation and Paradox in Anthropological Research*. (Case Studies in Anthropological Methods). New York: Holt, Rinehart and Winston.

Miller, S. M. (1952). The participant-observer and 'over-rapport'. *American Sociological Review* **17**(1), 97–99.

Milton, K. (1979). Male bias in anthropology? *Man* (N.S.) **14**(1), 40–54.

Mitchell, J. C. (1949). The collection and treatment of family budgets in primitive communities as a field problem. *Rhodes-Livingstone Journal* **8**, 50–56.

Mitchell, J. C. (1956). *The Yao Village*. Manchester: Manchester University Press.

Mitchell, J. C. (1957). *The Kalela Dance*. (Rhodes-Livingstone Paper **27**). Manchester: Manchester University Press.

Mitchell, J. C. (1966). Theoretical orientations in African urban studies. In *The Social Anthropology of Complex Societies* (ed. M. Banton). (A.S.A. Monogr. **4**). London: Tavistock.

Mitchell, J. C. (1967). On quantification in social anthropology. In *The Craft of Social Anthropology* (ed. A. L. Epstein). London: Tavistock.

Mitchell, J. C. (1969). The concept and use of social networks. In *Social Networks in Urban Situations: Analyses of Personal Relationships in Central African Towns* (ed. J. C. Mitchell). Manchester: Manchester University Press, for Institute of African Studies.

Mitchell, J. C. (1974). Social networks. *The Annual Review of Anthropology* 3, 279–299.

Mitchell, J. C. (1979). Networks, algorithms and analysis. In *Perspectives on Social Network Research* (eds P. W. Holland and S. Leinhardt). Orlando, New York and London: Academic Press.

Mitchell, J. C. (1983). Case and situation analysis. *Sociological Review* (NS) 31(2), 187–211.

Mitchell, R. E. (1965). Survey materials collected in the developing countries: sampling, measurement, and interviewing obstacles to intra- and international comparisons. *International Social Science Journal* 17(4), 665–685.

Monberg, T. (1975). Informants fire back: a micro-study in anthropological methods. *Journal of the Polynesian Society* 84, 218–224.

Money, A. S. and Smallwood, R. L. (1978). *M.H.R.A. Style Book: Notes for Authors, Editors and Writers of Dissertations.* London: Modern Humanities Research Association.

Monkhouse, F. J. and Wilkinson, H. R. (1971). *Maps and Diagrams: Their Compilation and Construction.* London: Methuen.

Moore, D. S. (1979). *Statistics, Concepts and Controversies.* San Francisco: Freeman.

Moore, F. C. T. (1969). Translator's introduction. In *The Observation of Savage Peoples* (J.-M. Degérando). London: Routledge and Kegan Paul.

Moore, R. (1974). *Pit-men, Preachers and Politics.* Cambridge: Cambridge University Press.

Morgan, D. H. J. (1972). The British Association scandal: the effect of publicity on a sociological investigation. *Sociological Review* (N.S.) 20, 185–206.

Morgan, L. H. (1862). Circular in reference to the degree of relationship among different nations. *Smithsonian miscellaneous collection* 2, No. 10. Washington: Smithsonian Institution.

Morgan, L. H. (1971 [1870]). *Systems of Consanguinity and Affinity of the Human Family.* Washington: Smithsonian Institution.

Moser, C. and Kalton, G. (1971). *Survey Methods in Social Investigation* (2nd edn.). London: Heinemann.

Mulvaney, D. J. (1970). The anthropologist as tribal elder. *Mankind* 7, 207–217.

Mulvaney, D. J. (1971). The ascent of Aborginal man: Howitt as anthropologist. In *Come Wind, Come Weather: A Biography of Alfred Howitt* (ed. M. H. Walker). Carlton: Melbourne University Press.

Mulvaney, D. J. (1981). Gum leaves on the golden bough: Australia's palaeolithic survivals discovered. In *Antiquity and Man: Essays in Honour of Glyn Daniel* (eds J. D. Evans *et al.*,). London: Thames and Hudson.

Murdock, G. P. (1975). *Outline of World Cultures* (5th edn.). New Haven: Human Relations Area Files.

Murdock, G. P., Ford, C. S., Hudson, A. E., Kennedy, R., Simmons, L. W. and Whiting, J. W. M. (1967). *Outline of Cultural Materials*. (Behaviour Science Outlines 1) (4th edn.). New Haven: Human Relations Area Files

Murray, W. B. and Buckingham, R. W. (1976). Implications of participant observation in medical studies. *Canadian Medical Association Journal* **115**, 1187–1190.

Myers, C. S. (1940). Obituary: Doctor Alfred Cort Haddon, FRS. *Geographical Journal* **96**, 230–231.

Myers, J. E. (1969). Unleashing the untrained: some observations on student ethnographers. *Human Organization* **28**(2), 155–159.

Myres, J. L. (1929). The science of man in the service of the state. *Journal of the Royal Anthropological Institute* **59**, 19–52.

Myres, J. L. (1951). The origin of Man. *Man* **51**, 1–2.

Nachmias, D. and Nachmias, C. (1976). *Research Methods in the Social Sciences*. London: Edward Arnold.

Nadel, S. F. (1937). Experiments on culture psychology. *Africa* **10**, 421–435.

Nadel, S. F. (1937a). A field experiment in racial psychology. *British Journal of Psychology* **28**, 195–211.

Nadel, S. F. (1939). The application of intelligence tests in the anthropological field. In *The Study of Society: Methods and Problems* (eds F. C. Bartlett *et al.*,). London: Kegan Paul.

Nadel, S. F. (1939a). The interview technique in social anthropology. In *The Study of Society: Methods and Problems* (eds F. C. Bartlett *et al.*,). London: Kegan Paul.

Nadel, S. F. (1953). *The Foundations of Social Anthropology*. New York: Free Press.

Nader, L. (1970). From anguish to exultation. In *Women in the Field: Anthropological Experiences* (ed. P. Golde). Chicago: Aldine Publishing Co.

Nader, L. (1972). Up the anthropologist – perspectives gained from studying up. In *Reinventing Anthropology* (ed. Dell Hymes). New York: Pantheon.

Nakhleh, K. (1979). On being a native anthropologist. In *The Politics of Anthropology: From Colonialism and Sexism Toward a View from Below* (eds G. Huizer and B. Mannheim). The Hague: Mouton.

Naroll, R. and Cohen, R. (1970). *A Handbook of Method in Cultural Anthropology*. New York and London: Columbia University Press.

Nash, D. (1963). The ethnologist as stranger: an essay in the sociology of knowledge. *Southwestern Journal of Anthropology* **19**, 149–167.

Nash, D. and Wintrob, R. (1972). The emergence of self-consciousness in ethnography. *Current Anthropology* **13**, 527–542.

Nash, J. (1974). Ethics and politics in social science research. *Transactions of the New York Academy of Sciences* (series 2) **36**, 497–511.

Nash, J. (1975). Nationalism and fieldwork. *Annual Review of Anthropology* **4**, 225–245.

Nash, J. (1981). Ethnographic aspects of the world capitalist system. *Annual Review of Anthropology* **10**, 393–423.

Needham, R. (1971). Introduction. In *Rethinking Kinship and Marriage* (ed. R. Needham). (A.S.A. Monogr. 10). London: Tavistock.

Needham, R. (1974). Surmise, discovery and rhetoric. In *Remarks and Inventions: Sceptical Essays about Kinship*. London: Tavistock.

Nejelski, P. (ed.) (1976). *Social Research in Conflict with Law and Ethics*. Cambridge, Mass.: Ballinger Publishing Company. Originally published (1976) as: *Forschung im Konflikt mit Recht und Ethik*. Stuttgart: Enke.

Newby, H. (1979). *The Deferential Worker*. Harmondsworth: Penguin.

Newmark, P. (1981). *Approaches to Translation*. Oxford: Pergamon Press.

Nicholson, H. (1957). *A Dictionary of American–English Usage; Based on Fowler's Modern English Usage*. New York: Oxford University Press.

Nida, E. A. (1947). *Bible Translating: An Analysis of Principles and Procedures with Special Reference to Aboriginal Languages*. New York: American Bible Society.

Nida, E. A. (1957). *Learning a Foreign Language*. Ann Arbor, Michigan: Friendship Press.

Nida, E. A. (1964). *Toward a Scientific Theory of Translation*. Leiden: E. J. Brill.

Nie, N. H., Hull, C. H., Jenkins, J. G., Steinbrenner, K. and Bent, D. H. (1979). *Statistical Package for the Social Sciences* (2nd edn.). New York: McGraw Hill.

Norbeck, E. (1970). Changing Japan: field research. In *Being an Anthropologist* (ed. G. D. Spindler). New York: Holt, Rinehart and Winston.

Nukunya, G. K. (1969). *Kinship and Marriage Among the Anlo Ewe*. New York: Humanities Press.

Øyen, E. (1972). The impact of prolonged observation on the role of the 'neutral observer' in small groups. *Acta Sociologica* 15(3), 254–266.

Okely, J. (1975). The self and scientism. *Journal of the Anthropological Society of Oxford* 6(3), 171–188.

Okely, J. (1979). "Participant observation". Unpublished paper presented to S.S.R.C. Workshop on Participant Observation. University of Birmingham, September 24–25.

Okely, J. (1983). *The Traveller Gypsies*. Cambridge: Cambridge University Press.

Olesen, V. and Whittaker, E. (1967). Role-making in participant observation: processes in the researcher-actor relationship. *Human Organization* 26, 273–281.

Oppenheim, A. N. (1966). *Questionnaire Design and Attitude Measurement*. London: Heinemann.

Orenstein, A. and Phillips, W. R. F. (1978). *Understanding Social Research*. Boston: Allyn and Bacon.

Orlans, H. (1969). Ethical problems in the relations of research sponsors and investigators. In *Ethics, Politics and Social Research* (ed. G. Sjoberg). London: Routledge and Kegan Paul.

Orlans, H. (1973). *Contracting for Knowledge*. San Francisco: Jossey-Bass.

Osgood, C. (1964). Semantic differential technique in the comparative study of cultures. In *Transcultural Studies in Cognition* (eds A. K. Romney and R. G. D'Andrade). *American Anthropologist* 66(3), 171–200.

Owen-Hughes, D. (1978). From bridewealth to dowry in mediterranean Europe. *Journal of Family History* 3, 262–296.

Owusu, M. (1978). The ethnography of Africa: the usefulness of the useless. *American Anthropologist* 80, 310–334.

Owusu, M. (1979). Colonial and post-colonial anthropology of Africa: scholarship or sentiment? In *The Politics of Anthropology: From Colonialism and Sexism Toward a View from Below* (eds G. Huizer and B. Mannheim). The Hague: Mouton.

Paddock, J. (1965). Private lives and anthropological publications. *Mesoamerican Notes* **6**, 59–65.

Paluch, A. K. (1981). The Polish background to Malinowski's work. *Man* (N.S.) **16**, 276–285.

Pandey, T. N. (1972). Anthropologists at Zuni. *Proceedings of the American Philosophical Society* **116**(4), 321–337.

Pandey, T. N. (1975). 'India man' among American Indians. In *Encounter and Experience* (eds A. Beteille and T. N. Madan). Delhi: Vikas Publishing House.

Pandey, T. N. (1979). The anthropologist-informant relationship: the Navajo and Zuni in America and the Tharu in India. In *The Fieldworker and the Field: Problems and Challenges in Sociological Investigation* (eds. M. N. Srinivas, A. M. Shah and E. A. Ramaswamy). Delhi: Oxford University Press.

Papanek, H. (1964). The woman fieldworker in a purdah society. *Human Organization* **23**, 160–163.

Parkin, D. (ed.) (1982). *Semantic Anthropology* (A.S.A. Monogr. **22**). London, Orlando and New York: Academic Press.

Pastner, C. McC. (1982). Rethinking the role of the woman fieldworker in purdah societies. *Human Organization* **4**(3), 262–264.

Patrick, J. (pseud.) (1973). *A Glasgow Gang Observed*. London: Eyre Methuen.

Patterson, G. J. (1978). Problems in urban ethnic research. In *Ethical Dilemmas in Anthropological Inquiry: A Case Book* (ed. G. N. Appell). Waltham, Mass.: Crossroads Press.

Patwardhan, S. (1979). Making my way through caste images: Untouchables in Poona. In *The Fieldworker and the Field* (eds M. N. Srinivas, A. M. Shah and E. A. Ramaswamy). Delhi: Oxford University Press.

Paul, B. D. (1953). Interview techniques and field relationships. In *Anthropology Today: An Encyclopedic Inventory* (ed. A. L. Kroeber). Chicago: Chicago University Press.

Payne, S. Le B. (1965 [1951]). *The Art of Asking Questions*. Princeton: Princeton University Press.

Pelto, P. J. (1970). Research in individualistic societies. In *Marginal Natives: Anthropologists at Work* (ed. M. Freilich). New York: Harper and Row.

Pelto, P. J. and Pelto, G. H. (1973). Ethnography: the fieldwork enterprise. In *Handbook of Social and Cultural Anthropology* (ed. J. J. Honigman). Chicago: Rand McNally.

Pelto, P. J. and Pelto, G. H. (1978). *Anthropological Research: The Structure of Inquiry* (2nd edn). Cambridge: Cambridge University Press.

Pepinsky, H. E. (1980). A sociologist on police patrol. In *Fieldwork Experience: Qualitative Approaches to Social Research* (eds W. B. Shaffir, R. A. Stebbins and. A. Turowetz). New York: St. Martin's Press.

Perlman, M. L. (1970). Intensive fieldwork and scope sampling: methods for studying the same problem at different levels. In *Marginal Natives: Anthropologists at Work* (ed. M. Freilich). New York: Harper and Row.

Peters, E. (1963). Aspects of rank and status among Muslims in a Lebanese village. In *Mediterranean Countrymen* (ed. J. Pitt-Rivers). Paris: Mouton.

Peters, E. (1970). The proliferation of segments in the lineages of the Bedouin of Cyrenaica. In *Peoples and Cultures of the Middle East* (ed. L. M. Sweet), Vol. 1. New York: Natural History Press.

Peters, E. (1972). Shifts in power in a Lebanese village. In *Rural Politics and Social Change in the Middle East* (eds R. T. Autoun and I. Hayik). Bloomington: Indiana University Press.

Pettigrew, J. (1981). Reminiscences of fieldwork among the Sikhs. In *Doing Feminist Research* (ed. H. Roberts). London: Routledge and Kegan Paul.

Phillips, B. S. (1971). *Social Research Strategy and Tactics*. New York: Macmillan; London: Collier-Macmillan.

Phillips, D. L. (1971). *Knowledge From What?: Theories and Methods in Social Research*. Chicago: Rand McNally.

Phillips, D. L. (1973). *Abandoning Method*. San Francisco: Jossey-Bass.

Phillips, H. P. (1959). Problems of translation and meaning in field work. *Human Organization* 18(4), 184–192. Reprinted (1960) in *Human Organization Research. Field Relations and Techniques* (eds. R. N. Adams and J. J. Preiss). Homewood, Illinois: The Dorsey Press, for the Society for Applied Anthropology.

Phillipson, M. (1972). Phenomenological philosophy and sociology. In *New Directions in Sociological Theory* (eds P. Filmer, M. Phillipson, D. Silverman and D. Walsh). London: Collier-Macmillan.

Philpott, S. B. (1973). *West Indian Migration: The Montserrat Case*. London: Athlone Press.

Piddington, R. (1957). *An Introduction to Social Anthropology*, Vol. 2. London: Oliver and Boyd.

Pike, K. L. (1947). *Phonemics: A Technique for Reducing Languages to Writing*. Ann Arbor: University of Michigan Press.

Pinxten, R. (1981). Observation in anthropology: positivism and subjectivism combined. *Communication and Cognition* 14(1), 57–83.

Pitt-Rivers, J. A. (1971). *The People of the Sierra* (2nd edn.). Chicago: University of Chicago Press.

Pitt-Rivers, J. A. (1978). The value of the evidence [letter]. *Man* (N.S.) 13(2), 319–322.

Pitt-Rivers, J. A. (1979). The value of the evidence [letter]. *Man* (N.S.) 14(1), 160.

Platt, J. (1976). *The Realities of Social Research: An Empirical Study of British Sociologists*. London: Sussex University Press.

Polanyi, M. (1958). *Personal Knowledge: Towards a Post-critical Philosophy*. New York: Harper and Row.

Polsky, N. (1967). Research method, morality and criminology. In *Hustlers, Beats and Others*. Chicago: Aldine.

Polunin, I. (1970). Visual and sound recording apparatus in ethnographic fieldwork. *Current Anthropology* 11(1), 3–22.

Pons, V. (1969). *Stanleyville: An African Urban Community Under Belgian Administration*. Oxford University Press, for the International African Institute.

Posner, J. (1980). Urban anthropology: fieldwork in semifamiliar settings. In *Fieldwork Experience: Qualitative Approaches to Social Research* (eds W. B. Shaffir, R. A. Stebbins and A. Turowetz). New York: St. Martin's Press.

Powdermaker, H. (1962). *Coppertown; Changing Africa, the Human Situation on the Rhodesian Copperbelt*. New York: Harper and Row.

Powdermaker, H. (1966). *Stranger and Friend: The Way of an Anthropologist*. New York: W. W. Norton. Published (1967) in London by Secker and Warburg.

Powdermaker, H. (1968). Field work. In *International Encyclopaedia of the Social Sciences* (ed. D. Sills), **51**, 418–424. New York: Macmillan and Free Press.

Powell, H. A. (1969). Genealogy, residence and kinship in Kiriwina. *Man* (N.S.) **4**, 177–202.

Preston, R. J. (1966). Edward Sapir's anthropology: style, structure, and method. *American Anthropologist* **68**, 1105–1128.

Pryce, K. (1979). *Endless Pressure*. Harmondsworth: Penguin.

Putnam, H. (1975). Language and reality. In *Mind, Language and Reality* (H. Putnam). (Philosophical Papers **2**, 272–290). Cambridge: Cambridge University Press.

Putnam, H. (1978). *Meaning and the Moral Sciences*. London: Routledge and Kegan Paul.

Pym, B. (1955). *Less than Angels*. London: Jonathan Cape.

Quiggin, A. H. (1942). *Haddon the Headhunter: A Short Sketch of the Life of A. C. Haddon*. Cambridge: Cambridge University Press.

Quinn, N. (1977). Anthropological studies on women's status. *Annual Review of Anthropology* **6**, 181–225.

Raab, C. D. (1982). *Data Protection and Privacy: Proceedings of a Conference*. London: SRA Publications. [Social Research Association].

Rabinow, P. (1977). *Reflections on Fieldwork in Morocco*. Berkeley: University of California Press.

(Radcliffe-) Brown, A. R. (1922). *The Andaman Islanders*. Cambridge: Cambridge University Press.

Radcliffe-Brown, A. R. (1952). *Structure and Function in Primitive Society*. New York: The Free Press.

Radin, P. (1926). *Crashing Thunder: The Autobiography of an American Indian*. New York: Appleton.

Radin, P. (1927). *Primitive Man as Philosopher*. New York: Appleton.

Radin, P. (1965 [1933]). *The Method and Theory of Ethnology: An Essay in Criticism*. New York: Basic Books.

RAIN (1982). Fiftieth issue of RAIN – with a special section on ethnographic film. *RAIN (Royal Anthropological Institute News)* (June) No. 50.

Raine, K. (1973). *Farewell Happy Fields: Memories of Childhood*. London: Hamish Hamilton.

Rainwater, L. and Pittman, D. J. (1967). Ethical problems in studying a politically sensitive and deviant community. *Social Problems* **14**(4), 357–366.

Raman Unni, K. (1979). On tracks and tracts in my fieldwork: rural Kerala. In *The Fieldworker and the Field: Problems and Challenges in Sociological Investigation* (eds M. N. Srinivas, A. M. Shah and E. A. Ramaswamy). Delhi: Oxford University Press.

Ramaswamy, E. A. (1979). Being where the action was: trade unions in Coimbatore. In *The Fieldworker and the Field: Problems and Challenges in Sociological Investigation* (eds M. N. Srinivas, A. M. Shah and E. A. Ramaswamy). Delhi: Oxford University Press.

Rapp, R. (see also, Reiter, R.) (1979). Review essay: Anthropology. *Signs* **4**(3), 497–513.

Read, K. E. (1965). *The High Valley.* New York: Scribner; London: George Allen and Unwin.

Read, M. W. (1980). *Ethnographic Field-notes and Interview Transcripts. Some Preliminary Observations on the Computer Management of Text.* (SRU Working Paper **8**). Cardiff: Sociology Research Unit.

Record, J. C. (1969). The research institute and the pressure group. In *Ethics, Politics and Social Research* (ed. G. Sjoberg). London: Routledge and Kegan Paul.

Redfield, R. (1930). *Tepoztlan, A Mexican Village: A Study of Folk Life.* Chicago: Chicago University Press.

Redfield, R., Linton, R. and Herskovits, M. J. (1936). Memorandum for the study of acculturation. *American Anthropologist* **38**, 149–152.

Redlich, F. (1973). The anthropologist as observer: ethical aspects of clinical observations of behaviour. *The Journal of Nervous and Mental Disease* **157**(5), 313–319.

Reiss, A. J., Jr. (1979). Governmental regulation of scientific inquiry: some paradoxical consequences. In *Deviance and Decency: The Ethics of Research with Human Subjects* (eds C. B. Klockars and F. W. O'Connor). Beverly Hills, London: Sage Publications.

Reiter, R. (see also, Rapp, R.) (1980). *Towards an Anthropology of Women.* New York: Monthly Review Press.

Reynolds, P. D. (1975). Value dilemmas in the professional conduct of social science. *International Social Science Journal* **27**(4), 563–611.

Reynolds, P. D. (1982). *Ethics and Social Science Research.* Englewood Cliffs, N.J.: Prentice-Hall. [A condensed and reorganized version of his *Ethical Dilemmas and Social Science Research*, Jossey Bass, 1979].

Rich, J. (1968). *Interviewing Children and Adolescents.* London: Macmillan.

Richards, A. I. (1939). The development of fieldwork methods in social anthropology. In *The Study of Society: Methods and Problems* (eds F. C. Bartlett *et al.*). London: Kegan Paul.

Richards, A. I. (1967). African systems of thought: an Anglo-French dialogue (review article). *Man* (N.S.) **2**(2), 286–298.

Richards, A. I. (1980). Foreword. In *Elmdon: Continuity and Change in a North-west Essex Village 1861–1964* (ed. Jean Robin). Cambridge: Cambridge University Press.

Richards, I. A. (1932). *Mencius on the Mind.* London: Kegan Paul, Trench Frubner.

Riesman, D. and Watson, J. (1964). The Sociability Project: a chronicle of frustration and achievement. In *Sociologists at Work: Essays on the Craft of Social Research* (ed. P. E. Hammond). New York, London: Basic Books.

Riggs, F. W. (1982). Terminological problems: a proposed solution. *International Union of Anthropological and Ethnological Sciences Newsletter* 4 (December).

Rimmer, P. J., Drakakis-Smith, D. W. and McGee, T. G. (eds) (1978). *Food, Shelter and Transport in Southeast Asia and the Pacific.* Canberra: Australian National University.

Rink, H. J. (1866–71). *Eskimoiske Eventyr og Sagn.* Copenhagen.

Rivers, W. H. R. (1900). A genealogical method of collecting social and vital statistics. *Journal of the Anthropological Institute* 30, 74–82.

Rivers, W. H. R. (1906). *The Todas.* London: Macmillan.

Rivers, W. H. R. (1913). Anthropological research outside America. In *Reports of the Present Condition and Future Needs of the Science of Anthropology.* (Publ. 200). Washington: Carnegie Institute.

Rivers, W. H. R. (1914). *The History of Melanesian Society.* Cambridge: Cambridge University Press.

Rivers, W. H. R. (ed.) (1922). *Essays on the Depopulation of Melanesia.* Cambridge: Cambridge University Press.

Roberts, J. (1956). *Zuni Daily Life Notebook 3.* Laboratory of Anthropology: University of Nebraska.

Robertson, A. F. (1970). *Community of Strangers: A Journal of Discovery in Uganda.* London: Scolar Press.

Robin, J. (1980). *Elmdon, Continuity and Change in a North-west Essex Village 1861–1964.* Cambridge: Cambridge University Press.

Robinson, W. S. (1951). The logical structure of analytic induction. *American Sociological Review* 16, 812–818.

Rogers, S. C. (1978). Woman's place: a critical review of anthropological theory. *Comparative Studies in Society and History* 20(1), 123–162.

Rohner, R. (1966). Franz Boas: ethnographer of the Northwest coast. In *Pioneers of American Anthropology* (ed. J. Helm). Seattle: University of Washington Press.

Rohner, R. and Rohner, E. (1969). Franz Boas and the development of north American ethnology and ethnography. In *The Ethnography of Franz Boas* (ed. R. Rohner). Chicago: University of Chicago Press.

Rohrlich-Leavitt, R. (ed.) (1975). *Women Cross-culturally: Change and Challenge.* The Hague: Mouton.

Rohrlich-Leavitt, R., Sykes, B. and Weatherford, E. (1975). Aboriginal woman: male and female anthropological perspectives. In *Women Cross-culturally: Change and Challenge* (ed. R. Rohrlich-Leavitt). The Hague: Mouton.

Rooksby, R. L. (1971). W. H. R. Rivers and the Todas. *South Asia* 1, 109–121.

Rosaldo, M. Z. (1980). The use and abuse of anthropology: reflections on feminism and cross-cultural understanding. *Signs* 5(3), 389–417.

Rose, F. G. C. (1960). *Classification of Kin, Age Structure and Marriage amongst the Groote Eylandt Aborigines: A Study in Method and a Theory of Australian Kinship.*

Berlin: Akademie-Verlag, Deutsche Akademie de Wissenschaften zu Berlin 3: Sektion für Völkerkundliche Forschungen.

Roth, J. (1962). Comments on "secret observation". *Social Problems* 9, 283–284.

Rothschild, Lord (1982). *An Enquiry into the Social Science Research Council.* London: Her Majesty's Stationery Office, Cmnd. 8554.

Rowe, J. H. (1953). Technical aids in anthropology: a historical survey. In *Anthropology Today: An Encyclopedic Inventory.* Chicago: Chicago University Press.

Rowe, J. H. (1965). The Renaissance foundations of anthropology. *American Anthropologist* 67, 1–20.

Royal Anthropological Institute (1951). *Notes and Queries on Anthropology* (6th edn., 4th impression). London: Routledge and Kegan Paul.

Ryan, A. (1970). *The Philosophy of the Social Sciences.* London: Macmillan.

Ryan, A. (ed.) (1973). *The Philosophy of Social Explanation.* Oxford: Oxford University Press.

Rynkiewich, M. A. and Spradley, J. P. (eds) (1976). *Ethics and Anthropology: Dilemmas in Fieldwork.* New York: John Wiley.

Saberwal, S. (1968). The problem. *Seminar* 112, 12–13. Reprinted (1973) in *To See Ourselves: Anthropology and Modern Social Issues* (ed. T. Weaver). Glenview, Illinois: Scott, Foresman.

Saberwal, S. (1969). Rapport and resistance among the Embu of Central Kenya (1963–64). In *Stress and Response in Fieldwork* (eds F. Henry and S. Saberwal). New York: Holt, Rinehart and Winston.

Sagarin, E. and Moneymaker, J. (1979). The dilemma of researcher immunity. In *Deviance and Decency: The Ethics of Research with Human Subjects* (eds C. B. Klockars and F. W. O'Connor). Beverly Hills, London: Sage Publications.

Salamone, F. (1977). The methodological significance of the lying informant. *Anthropological Quarterly* 50(3), 117–124.

Salamone, F. (1979). Epistemological implications of fieldwork and their consequences. *American Anthropologist* 81, 46–60.

Samarin, W. J. (1967). *Field Linguistics: A Guide to Linguistic Fieldwork.* New York: Holt, Rinehart and Winston.

Samson, B. L. (1980). *The Camp at Wallaby Cross: Aboriginal Fringe Dwellers in Darwin.* Canberra: Australian Institute of Aboriginal Studies.

Santos, B. de S. (1981). Science and politics: doing research in Rio's squatter settlements. In *Law and Social Enquiry: Case Studies of Research* (ed. R. Luckham). Uppsala: Scandinavian Institute of African Studies; New York: International Center for Law in Development.

Schatzman, L. and Strauss, A. (1973). *Field Research: Strategies for a Natural Sociology.* Englewood Cliffs, N.J.: Prentice-Hall.

Scheffler, I. (1972). Vision and revolution: a postscript on Kuhn. *Philosophy of Science* 39, 366–374.

Schensul, S. L. and Schensul, J. L. (1978). Advocacy and applied anthropology. In *Social Scientists as Advocates* (eds G. H. Weber and G. J. McCall). Beverly Hills, California: Sage Publications.

Scheper-Hughes, N. (1981). Cui Bonum – for whose good?: a dialogue with Sir Raymond Firth. *Human Organization* **40**(4), 371–372.

Scheper-Hughes, N. (1982). 'Ballybran' [Letter]. *RAIN (Royal Anthropological Institute News)* (August), **51**, 12–13.

Scholte, B. (1971). Discontents in anthropology. *Social Research* **38**(4), 777–807.

Scholte, B. (1980). Anthropological traditions. Their definition. In *Anthropology: Ancestors and Heirs* (ed. S. Diamond). The Hague: Mouton.

Schon, D. A. (1969). *Invention and the Evolution of Ideas.* London: Tavistock.

Schoolcraft, H. R. (1886). Plan for American ethnological investigation (written in 1846). *Annual Report . . . Smithsonian Institution for 1885,* 907–914.

Schutz, A. (1972). *The Phenomenology of the Social World* (Trans. G. Walsh and F. Lehnertog). London: Heinemann Educational.

Schwab, W. B. (1965). Comparative field techniques in two African towns. *Human Organization* **24**(4), 373–380.

Schwab, W. B. (1970). Comparative field techniques in urban research in Africa. In *Marginal Natives: Anthropologists at Work* (ed. M. Freilich). New York: Harper and Row.

Schwartz, M. S. and Schwartz, C. G. (1955). Problems in participant observation. *American Journal of Sociology* **60**(4), 343–353.

Schwartz, T. (1962). The Paliau movement in the Admiralty Islands, 1946–1954. *Anthropological Papers of the American Museum of Natural History, New York* **49**(2).

Schwartz-Barcott, D. (1981). Review of G. H. Herdt. *Guardians of the Flutes: Idioms of Masculinity, a Study of Ritualized Homosexual Behaviour.* (New York: McGraw-Hill, 1981.) *American Anthropologist* **83**(3), 703–704.

Scudder, T. and Colson, E. (1979). Long-term research in Gwembe Valley, Zambia. In *Long-term Field Research in Social Anthropology* (eds G. M. Foster, T. Scudder, E. Colson and R. V. Kemper). Orlando, New York and London: Academic Press.

Seitel, P. (1980). *See so That We may See: Performances and Interpretations of Traditional Tales from Tanzania.* Bloomington: Indiana University Press.

Seligman, C. G. (1910). *The Melanesians of British New Guinea.* Cambridge: Cambridge University Press.

Seligman, C. G. and Seligman, B. Z. (1911). *The Veddas.* Cambridge: Cambridge University Press.

Seligman, C. G. and Seligman, B. Z. (1932). *Pagan Tribes of the Nilotic Sudan.* London: Routledge.

Selltiz, C., Jahoda, M., Deutsch, M. and Cook, S. W. (1966). *Research Methods in Social Relations.* New York: Holt, Rinehart and Winston.

Shaffir, W. B., Stebbins, R. A. and Turowetz, A. (eds) (1980). *Fieldwork Experience: Qualitative Approaches to Social Research.* New York: St. Martin's Press.

Sharma, U. (1981). Male bias in anthropology. *South Asian Review* **2**, 34–39.

Shaw, B. (1980). Life history writing in anthropology: a methodological review. *Mankind* **12**(3), 226–233.

Shelton, A. J. (1963). The 'Miss Ophelia' syndrome as a problem in African field research. *Practical Anthropology* **11**, 259–265.

Shils, E. A. (1959). Social enquiry and the autonomy of the individual. In *The Human Meaning of the Social Sciences* (ed. D. Lerner). Cleveland and New York: Meridian.

Shokeid, M. (1971). Fieldwork as predicament rather than spectacle. *Archives Européennes de Sociologie* 12(1), 111–122.

Sieber, S. D. (1982). The integration of fieldwork and survey methods. In *Field Research: A Source Book and Field Manual* (ed. R. G. Burgess). London: Allen and Unwin. First published (1973) in *American Journal of Sociology* 78(6), 1335–1359.

Silberman, L. (1947). Logic and problems of sampling. *Human Problems in British Central Africa* (*Rhodes-Livingstone Journal* 5, 1–17).

Sjoberg, G. (ed.) (1969a [1967]). *Ethics, Politics and Social Research*. London: Routledge and Kegan Paul.

Sjoberg, G. (1969b). Project Camelot: selected reactions and personal reflections. In *Ethics, Politics and Social Research* (ed. G. Sjoberg). London: Routledge and Kegan Paul.

Sjoberg, G. and Nett, R. (1968). *A Methodology for Social Research*. New York: Harper and Row.

Skeat, W. W. and Blagden, C. O. (1906). *Pagan Races of the Malay Peninsula*. London: Macmillan.

Slater, M. K. (1976). *African Odyssey: An Anthropological Adventure*. New York: Anchor Books.

Slobodin, R. (1978). *W. H. R. Rivers*. New York: Columbia University Press.

Slocum, S. (1975). Woman the gatherer: male bias in anthropology. In *Toward an Anthropology of Women* (ed. R. R. Reiter). New York and London: Monthly Review Press.

Smith, M. (1959). Boas' "natural history" approach to field method. In *The Anthropology of Franz Boas: Essays on the Centennial of his Birth* (ed. W. Goldschmidt). *Mem. Amer. Anthrop. Assoc.* 89.

Social Problems (1973). Social control of social research (eds G. Sjoberg and W. B. Littrell). *Social Problems* 21(1), 1–143.

Social Problems (1980). Ethical problems of fieldwork (eds J. Cassell and M. L. Wax). *Social Problems* 27(3), 259–378.

Social Research Association (1980). *Terms and Conditions of Social Research Funding in Britain. Report of the Working Group*. London: Social Research Association.

Social Science Research Council (1968). *Research in Social Anthropology*. London: Heinemann.

Sociological Review (1979). (Papers from symposium on the handling of Social Science Research Council qualitative data, Warwick, 1977), 27(4).

Société pour l'Étude des Langues Africaines (1971). *Enquête et description des Langues à tradition orale* (ER 74 du CNRS). Volumes 1–5. Paris: SELAF.

Society for Applied Anthropology (1963). Statement on the ethics of the Society for Applied Anthropology. *Human Organization* 22, 237.

Society for Applied Anthropology (1974). Statement on professional and ethical responsibilities. *Human Organization* 33(3), opposite p. 219.

Solomon Islands, Ministry of Education and Cultural Affairs (1976). Information for research workers in the Solomon Islands. *Current Anthropology* 17(1), 163–164.

Soloway, I. and Walters, J. (1977). Workin' the corner: the ethics and legality of ethnographic fieldwork among active heroin addicts. In *Street Ethnography: Selected Studies of Crime and Drug Use in Natural Settings* (ed. R. S. Weppner). (Sage Annual Reviews of Drug and Alcohol Abuse 1). Beverly Hills, London: Sage Publications.

Sommer, R. (1971). Some costs and pitfalls in field research. *Social Problems* **19**(2), 162–166.

Southwold, M. (1982). Personal communication to John Blacking.

Spier, R. F. G. (1970). *Surveying and Mapping: A Manual of Simplified Techniques.* New York: Holt, Rinehart and Winston.

Spillius, J. (1957). Natural disorder and political crisis in a Polynesian society: an exploration of operational research. *Human Relations* **10**, 3–27, 113–125.

Spindler, G. D. (1970). *Being an Anthropologist.* New York: Holt, Rinehart and Winston.

Spindler, G. D. and Spindler, L. (1970). Fieldwork among the Menomini. In *Being an Anthropologist* (ed. G. D. Spindler). New York: Holt, Rinehart and Winston.

Spoehr, A. (1981). Lewis Henry Morgan and his Pacific collaborators: a nineteenth-century chapter in the history of anthropological research. *Proceedings of the American Philosophical Society* **125**, 449–459.

Spradley, J. P. (1979). *The Ethnographic Interview.* New York: Holt, Rinehart and Winston.

Spradley, J. P. (1980). *Participant Observation.* New York: Holt, Rinehart and Winston.

Srinivas, M. N. (1966). *Social Change in Modern India* (Chapter 5: Some Thoughts on the Study of One's Own Society). Cambridge: Cambridge University Press.

Srinivas, M. N., Shah, A. M. and Ramaswamy, E. A. (eds) (1979). *The Fieldworker and the Field: Problems and Challenges in Sociological Investigation.* Delhi: Oxford University Press.

Stacey, M. (1969). *Methods of Social Research.* London: Pergamon Press.

Stanley, N. S. (1981). "The extra dimension": a study and assessment of the methods employed by Mass Observation in its first period, 1937–1940. Ph.D. thesis: CNAA, Birmingham Polytechnic.

Stanner, W. E. H. (1960). Durmugam, A Nangiomeri (Australia). In *In the Company of Man: Twenty Portraits by Anthropologists* (ed. J. B. Casagrande). New York: Harper and Row.

Stavenhagen, R. (1971). Decolonizing applied social science. *Human Organization* **30**(4), 333–357.

Stephenson, J. B. and Greer, L. S. (1981). Ethnographers in their own cultures: two Appalachian cases. *Human Organization* **40**(2), 123–130.

Stephenson, R. M. (1978). The CIA and the professor: a personal account. *The American Sociologist* **13**, 128–133 (see also comments pp. 153–177).

Stern, B. J. (ed.) (1930). Selections from the letters of Lorimer Fison and A. W. Howitt to Lewis Henry Morgan. *American Anthropologist* **37**, 257–279, 419–453.

Stocking, G. W., Jr. (1968). *Race, Culture and Evolution: Essays in the History of Anthropology.* New York: The Free Press; London: Collier-Macmillan.

Stocking, G. W. (1973). From chronology to ethnology: James Cowles Pritchard and British anthropology 1800–1850. In *Researches into the Physical History of Man* (J. C. Pritchard). Chicago: Chicago University Press.

Stocking, G. W. (ed.) (1974). *The Shaping of American Anthropology 1883–1911: A Franz Boas Reader*. New York: Basic Books.

Stocking, G. W. (1976). Ideas and institutions in American anthropology: thoughts toward a history of the interwar years. In *Selected Papers from the American Anthropologists 1921–1945* (ed. G. W. Stocking). Washington: American Anthropological Association.

Stocking, G. W. (1979). "The intensive study of limited areas": toward an ethnographic context for the Malinowskian innovation. *History of Anthropology Newsletter* 7(2), 9–12.

Strathern, A. (1978). Review of C. R. Hallpike, *Bloodshed and Vengeance in the Papuan Mountains* (Oxford, 1977). *Man* (N.S.) **13**(1), 150–151.

Strathern, A. (1979). Accuracy, tact and honesty [letter]. *Man* (N.S.) **14**(2), 354.

Strathern, A. (1979a). Anthropology, "snooping", and commitment: a view from Papua New Guinea. In *The Politics of Anthropology: From Colonialism and Sexism Toward a View from Below* (eds G. Huizer and B. Mannheim). The Hague: Mouton.

Strathern, A. (1979b). "Participant observation". Unpublished paper presented to S.S.R.C. Workshop on Participant Observation, University of Birmingham, September 24–25.

Strathern, A. (1981). *Fieldwork and Theory in Social Anthropology*. An Inaugural Lecture delivered at University College London. London: University College London.

Strathern, A. (1983). Research in Papua New Guinea: cross-currents of conflict. *RAIN (Royal Anthropological Institute News)* 58 (October), 4–10.

Strathern, M. (1981). Culture in a netbag: the manufacture of a subdiscipline in anthropology. *Man* (N.S.) **16**(1), 665–688.

Strathern, M. (1981a). *Kinship at the Core*. Cambridge: Cambridge University Press.

Stretton, W. G. (1893). Customs, rites and superstitions of the Aboriginal tribes of the Gulf of Carpentaria. *Transactions and Proceedings of the Royal Society of South Australia* **17**, 227–253.

Strunk, W. (1972). *The Elements of Style* (with revisions, an introduction and a chapter on writing by E. B. White), 2nd edn. New York: MacMillan.

Stryker, S. (1980). *Symbolic Interactionism: A Social Structural Version*. Menlo Park, California: The Benjamin/Cummings Publishing Co.

Sturtevant, W. C. (1960). A Seminole medicine maker. In *In the Company of Man: Twenty Portraits by Anthropologists* (ed. J. B. Casagrande). New York: Harper and Row.

Taplin, G. (1879). Questions on Aboriginal folklore, etc. In *The Folklore, Manners and Customs of the South Australian Aborigines* (G. Taplin). Adelaide: E. Spiller.

Taylor, C. (1971). Interpretation and the sciences of man. *The Review of Metaphysics* **25**, 3–51.

Tebape, O. P. (1978). The issuing of research permits in Botswana. In *Proceedings of the Workshop on Outlining the Botswana Research Landscape* held at University College of Botswana, 29–30 May (ed. B. Weimer). Gaberone: National Institute for Development and Cultural Research, Documentation Unit.

Thomas, E. M. (1959). *The Harmless People.* New York: Knopf.

Thomas, W. I. and Znaniecki, F. (1958 [1920]). *The Polish Peasant in Europe and America.* New York: Dover.

Thompson, L. (1970). Exploring American Indian communities in depth. In *Women in the Field: Anthropological Experiences* (ed. P. Golde). Chicago: Aldine.

Thompson, P. (1978). *The Voice of the Past: Oral History.* Oxford: Oxford University Press.

Thorne, B. (1980). "You still takin' notes?" Fieldwork and problems of informed consent. *Social Problems* **27**(3), 284–297.

Thouless, R. H. (1939). Problems of terminology in the social sciences. In *The Study of Society: Methods and Problems* (eds F. C. Bartlett *et al.*,). London: Kegan Paul.

Thurnwald, R. (1912). Probleme der ethno-psychologischen Forschung. Zur Praxis der ethno-psychologischen Ermittelungen besonders durch sprachliche Forschungen. *Zeitschrift für augewandte Psychologie* **5**, 1–27, 117–124.

Tiffany, S. W. (1978). Models and the social anthropology of women: a preliminary assessment. *Man* (N.S.) **13**(1), 34–51.

Tiryakian, E. A. (ed.) (1971). *The Phenomenon of Sociology: A Reader in the Sociology of Sociology.* New York: Appleton-Century-Crofts.

Toit, B. M. du (1980). Ethics, informed consent, and fieldwork. *Journal of Anthropological Research* **36**(3), 274–286.

Tonkin, J. E. A. (1971). The use of ethnography. *Journal of the Anthropological Society of Oxford* **2**, 134–136.

Tozzer Library Index to Anthropological Subject Headings (1981). 2nd edn. Boston, Mass.: G. K. Hall.

Tremblay, M. A (1957). The key informant technique: a non-ethnographic application. *American Anthropologist* **59**, 688–701. Reprinted (1982) in *Field Research: A Sourcebook and Field Manual* (ed. R. G. Burgess). London: George Allen and Unwin.

Trend, M. G. (1978). Freedom, confidentiality, and regulation: a dissent. *Human Organization* **37**(1), 88–89.

Trend, M. G. (1980). Applied social research and the government: notes on the limits of confidentiality. *Social Problems* **27**(3), 342–349.

Trist, E. L. and Bamforth, K. W. (1951). Some social and psychological consequences of the Longwall method of coal-getting. *Human Relations* **4**(1), 3–38.

Tugby, D. J. (1964). Towards a code of ethics for applied anthropology. *Anthropological Forum* **1** (Nov.), 220–231.

Turabian, K. L. (1982). *A Manual for Writers of Research Papers, Theses and Dissertations.* London: Heinemann.

Turnbull, C. (1961). *The Forest People.* London: Jonathan Cape. Published in paperback by Picador in 1976.

Turnbull, C. (1972). *The Mountain People*. New York: Simon and Schuster; London: Jonathan Cape (1973). Published in paperback by Picador 1974.

Turner, R. H. (1953). The quest for universals in sociological research. *American Sociological Review* **18**, 604–611.

Turner, R. (ed.) (1974). *Ethnomethodology*. Harmondsworth: Penguin.

Turner, V. W. (1957). *Schism and Continuity in an African Society: A Study of Ndembu Village Life*. Manchester: Manchester University Press.

Turner, V. W. (1960). Muchana the hornet, interpreter of religion (Northern Rhodesia). In *In the Company of Man: Twenty Portraits by Anthropologists* (ed. J. B. Casagrande). New York: Harper and Row.

Tuzin, D. F. and Schwartz, T. (1980). Margaret Mead and New Guinea: an appreciation. *Oceania* **50**, 241–247.

Tyler, S. A. (1969). *Cognitive Anthropology*. New York: Holt, Rinehart and Winston.

University of Chicago Press (1969). *A Manual of Style for Authors, Editors and Copywriters*. Chicago: University of Chicago Press.

Urry, J. (1972). *Notes and Queries on Anthropology* and the development of field methods in British anthropology, 1870–1920. *Proceedings of the Royal Anthropological Institute for 1972*, 45–57.

Urry, J. (1977). The literature and sources of social anthropology. In *Use of Social Science Literature* (ed. N. Roberts). London: Butterworths.

Urry, J. (1982). From ethnology to zoology: A. C. Haddon's conversion to anthropology. *Canberra Anthropology* **5**(2), 58–85.

Van Maanen, J. (1981). Notes on the production of ethnographic data in an American police agency. In *Law and Social Enquiry: Case Studies of Research* (ed. R. Luckham). (Studies of Law in Social Change and Development 5). Uppsala: Scandinavian Institute of African Studies; New York: International Center for Law in Development.

Vansina, J. (1969). *Oral Tradition*. London: Routledge and Kegan Paul.

Velsen, J. van (1964). *The Politics of Kinship*. Manchester: Manchester University Press.

Velsen, J. van (1967). The extended-case method and situational analysis. In *The Craft of Social Anthropology* (ed. A. L. Epstein). London: Tavistock Publications.

Veth, P. J. (1881–97). *Midden Sumatra: Reizen en Onderzoekingen der Sumatra Expeditie* (Vols 1–9). Leiden: E. J. Brill.

Vidich, A. J. (1955). Participant observation and the collection and interpretation of data. *American Journal of Sociology* **60**(4), 354–360.

Vidich, A. J. (1974). Ideological themes in American anthropology. *Social Research* **41**(4), 719–745.

Vidich, A. J. and Bensman, J. (1954). The validity of field data. *Human Organization* **13**, 20–27. Reprinted (1960) in *Human Organization Research* (eds R. N. Adams and J. J. Preiss). Homewood, Illinois: Dorsey Press.

Vidich, A. J. and Bensman, J. (1964). The Springdale case: academic bureaucrats and sensitive townspeople. In *Reflections on Community Studies* (eds A. J. Vidich, J. Bensman and M. R. Stein). New York: John Wiley.

Vidich, A. J., Bensman, J. and Stein, M. R. (eds) (1964). *Reflections on Community Studies*. New York: John Wiley.

Voegelin, C. F. (1959). Guide for transcribing unwritten languages in field work. *Anthropological Linguistics* 1(6), 1–28.

Vogt, E. Z. (1974). *Aerial Photography in Anthropological Field Research*. Cambridge, Mass.: Harvard University Press.

Vogt, E. Z. (1979). The Harvard Chiapas project: 1957–1975. In *Long-term Field Research in Social Anthropology* (eds G. M. Foster, T. Scudder, E. Colson and R. V. Kemper). Orlando, New York and London: Academic Press.

Wacaster, C. T. and Firestone, W. A. (1978). The promise and problems of long-term, continuous fieldwork. *Human Organization* 37(3), 269–275.

Wadel, C. (1973). *Now, Whose Fault is That?* St. John's, Newfoundland: Institute of Social and Economic Research.

Wagley, C. (1960). Champukin of the village of the tapirs (Brazil). In *In the Company of Man: Twenty Portraits by Anthropologists* (ed. J. B. Casagrande). New York: Harper and Row.

Wagner, J. (1980). *Images of Information: Still Photography in the Social Sciences*. London: Sage.

Wakeford, J. (comp. and ed.) (1979). *Research Methods Syllabuses in Sociology Departments in the U.K.* British Sociological Association Methodology Conference, January 1979.

Walford, A. J. (1982). *Walford's Guide to Reference Materials. Vol. 2: Social and Historical Sciences, Philosophy and Religion* (4th edn.). London: The Library Association.

Walker, A. L. and Lidz, C. W. (1977). Methodological notes on the employment of indigenous observers. In *Street Ethnography: Selected Studies of Crime and Drug Use in Natural Settings* (ed. R. S. Weppner). Beverly Hills, London: Sage Publications.

Walker, M. H. (1971). *Come Wind, Come Weather: A Biography of Alfred Howitt*. Carlton: Melbourne University Press.

Wallis, W. D. (1912). The methods of English anthropologists. *American Anthropologist* 14, 178–186.

Wallman, S. (1978). Epistemologies of sex. In *Female Hierarchies* (eds L. Tiger and H. T. Fowler). Chicago: Beresford Book Service.

Wallman, S. (1983). Rethinking inner London. *London Journal* (in press).

Wallman, S. and associates (1982). *Living in South London*. London: London School of Economics; Gower Press.

Walter, J. A. (1975). 'Delinquents' in a 'Treatment' Situation. The Processing of Boys in a List D School. Ph.D. thesis, Aberdeen University.

Warren, C. A. B. (1977). Fieldwork in the gay world: issues in phenomenological method. *Journal of Social Issues* 33(4), 93–107.

Warren, C. A. B. (1980). Data presentation and the audience: responses, ethics, and effects. *Urban Life* 9(3), 282–308.

Warren, C. A. B. and Rasmussen, P. K. (1977). Sex and gender in field research. *Urban Life* 6(3), 349–369.

Warwick, D. P. (1980). *The Teaching of Ethics in the Social Sciences.* Hastings-on-Hudson, New York: The Hastings Center.

Warwick, D. P. (1982). Types of harm in social research. In *Ethical Issues in Social Science Research* (eds T. L. Beauchamp, Ruth R. Faden, R. J. Wallace, Jr. and L. Walters). Baltimore: The Johns Hopkins University Press.

Watson, J. B. (1960). A New Guinea 'Opening Man'. In *In the Company of Man: Twenty Portraits by Anthropologists* (ed. J. B. Casagrande). New York: Harper and Row.

Watson-Franke, M.-B. (1980). Bias: male and female [letter]. *Man* (N.S.) **15**(2), 377–380.

Wax, M. L. (1977). On fieldwork and those exposed to fieldwork: Federal regulations and moral issues. *Human Organization* **36**, 321–328.

Wax, M. L. (1979a). On the presentation of the self in fieldwork: the dialectic of mutual deception and disclosure. *Humanity and Society* **3**(4), 248–259.

Wax, M. L. (1979b). The reluctant Merlins of Camelot: ethics and politics of overseas research. In *Federal Regulations: Ethical Issues and Social Research* (eds M. L. Wax and Joan Cassell). (AAAS Selected Symposium **36**). Boulder, Colorado: Westview Press.

Wax, M. L. (1980). Paradoxes of 'consent' to the practice of fieldwork. *Social Problems* **27**(3), 272–283.

Wax, M. L. and Cassell, J. (eds) (1979). *Federal Regulations: Ethical Issues and Social Research.* (AAAS Selected Symposium **36**). Boulder, Colorado: Westview Press.

Wax, M. L. and Cassell, J. (1981). From regulation to reflection: ethics in social research. *The American Sociologist* **16**(4), 224–229.

Wax, R. H. (1952). Reciprocity as a field technique. *Human Organization* **11**(3), 34–37. Reprinted (1960) with the title 'Reciprocity in fieldwork' in *Human Organization Research: Field Relations and Techniques* (eds R. N. Adams and J. J. Preiss). Homewood, Illinois: Dorsey Press for the Society for Applied Anthropology.

Wax, R. H. (1960). Twelve years later: an analysis of field experience. In *Human Organization Research* (eds. R. N. Adams and J. J. Preiss). Homewood, Illinois: Dorsey Press, for the Society for Applied Anthropology.

Wax, R. H. (1968). Participant observation. *International Encyclopedia of the Social Sciences* **11**, 238–241.

Wax, R. H. (1971). *Doing Fieldwork: Warnings and Advice.* Chicago: The University of Chicago Press.

Wax, R. H. (1979). Gender and age in fieldwork and fieldwork education: no good thing is done by any man alone. *Social Problems* **26**(5), 509–522 (with comment by Jeanne Guillemin, p. 523).

Weaver, S. M. (1978). A challenge to anthropological inquiry on an Indian Reserve. In *Ethical Dilemmas in Anthropological Inquiry: A Case Book* (ed. G. N. Appell). Waltham, Mass.: Crossroads Press.

Weaver, T. (ed.) (1973). *To See Ourselves: Anthropology and Modern Social Issues.* Glenview, Illinois: Scott, Foresman.

Webb, E. J., Campbell, D. T. , Schwartz, R. D. and Sechrest, L. (1966). *Unobtrusive Measures: Non-reactive Research in the Social Sciences.* Chicago: Rand McNally.

Weidman, H. H. (1970). On ambivalence and the field. In *Women and the Field* (ed. P. Golde). Chicago: Aldine.

Weiner, A. B. (1976). *Women of Value, Men of Renown.* Austin and London: University of Texas Press.

Weiner, A. B. (1979). Trobriand kinship from another view: the reproductive power of women and men. *Man* (N.S.) **14**, 328–348.

Weiner, J. and Lourie, J. (1969). *Human Biology: A Guide to Field Methods.* Oxford: Blackwell.

Weiss, M. S. (1977). The research experience in a Chinese-American community. *Journal of Social Issues* **33**(4), 120–132.

Wellman, B. and Crump, B. (1977). *Interview Schedule/Aide-memoire: East York Social Network Project. Phase IV.* Toronto: Centre for Urban and Community Studies.

Went, A. E. J. (1972). Four late nineteenth-century expeditions organised by the Royal Irish Academy. In *Proceedings of the Second International Conference on the History of Oceanography: Challenger Expedition Centenary* **1**, 305–309 (*Proceedings of the Royal Society of Edinburgh* Sec. **B**: 72).

Weppner, R. S. (1977). Street ethnography: problems and prospects. In *Street Ethnography: Selected Studies of Crime and Drug Use in Natural Settings* (ed. Robert S. Weppner). (Sage Annual Reviews of Drug and Alcohol Abuse, Vol. 1). Beverly Hills, London: Sage Publications.

Werner, D. (1980). *Where There is no Doctor.* London: Macmillan (paperback).

Werner, O. and Campbell, D. T. (1970). Translating, working through interpreters, and the problem of decentering. In *A Handbook of Method in Cultural Anthropology* (eds R. Naroll and R. Cohen). New York and London: Columbia University Press.

White, G. W. (1958). The Human Relations Area Files. *College and Research Libraries* **19**(2), 111–117.

White, L. (1966). *The Social Organization of Ethnological Thought.* (Rice University Monographs on Cultural Anthropology **25**(4).) Houston, Texas.

Whiting, B. and Whiting, J. (1970). Methods of observing and recording. In *A Handbook of Method in Cultural Anthropology* (eds R. Naroll and R. Cohen). New York and London: Columbia University Press.

Whitten, N. E. (1970). Network analysis and processes of adaptation among Ecuadorian and Nova Scotian Negroes. In *Marginal Natives: Anthropologists at Work* (ed. M. Freilich). New York: Harper and Row.

Whyte, W. F. (1955). *Street Corner Society.* Chicago: University of Chicago Press.

Whyte, W. F. (1958). Freedom and responsibility in research: the "Springdale" case. *Human Organization* **17**(2), 1–2.

Whyte, W. F. (1960). Interviewing in field research. In *Human Organization Research: Field Relations and Techniques* (eds R. N. Adams and J. J. Preiss). Homewood, Illinois: The Dorsey Press. Reprinted (1982) in *Field Research: A Sourcebook and Field Manual* (ed. R. G. Burgess). London: George Allen and Unwin.

Whyte, W. F. (1964). The slum: On the evolution of Street Corner Society. In *Reflections on Community Studies* (eds A. J. Vidich, J. Bensman and M. R. Stein). New York: Harper Torchbooks.

Whyte, W. F. (1979). On making the most of participant observation. *The American Sociologist* **14** (February), 56–66.

Wichmann, A. (1912). *Entdeckungsgeschichte von Neu-Guinea (1885 bis 1912).* (Nova Guinea **2**(2)). Leiden: E. J. Brill.

Wild, R. (1978). The background to *Bradstow*. In *Inside the Whale: Ten Personal Accounts of Social Research* (eds C. Bell and S. Encel). Rushcutters Bay, New South Wales: Pergamon Press.

Wilkins, P. (1982). Regaining the Golden Stool: social anthropology and applied research. *Journal of the Anthropological Society of Oxford* **13**(1), 112–120.

Willer, D. and Willer, J. (1973). *Systematic Empiricism: Critique of a Pseudoscience.* Englewood Cliffs, N.J.: Prentice-Hall.

Williams, R. G. A. (1981). Logical analysis as a qualitative method. *Sociology of Health and Illness* **3**, 140–187.

Williams, T. R. (1967). *Field Methods in the Study of Culture.* New York: Holt, Rinehart and Winston.

Willis, P. (1976). The man in the iron cage: notes on method. *Working Papers in Cultural Studies* **9**, 135–143.

Wilson, B. R. (ed.) (1970). *Rationality.* Oxford: Basil Blackwell.

Wilson, P. J., McCall, G. and Geddes, W. R. (1975). More thoughts on the Ik and anthropology (with a reply by Colin Turnbull). *Current Anthropology* **16**(3), 343–358 (with comments).

Wilson, T. P. (1970). Conceptions of interaction and forms of sociological explanation. *American Sociological Review* **35**, 697–710.

Winch, P. (1958). *The Idea of Social Science and Its Relation to Philosophy.* London: Routledge and Kegan Paul.

Wintrob, R. M. (1969). An inward focus: a consideration of psychological stress in fieldwork. In *Stress and Response in Fieldwork* (eds F. Henry and Satish Saberwal). New York: Holt, Rinehart and Winston.

Wolcott, H. F. (1975). Feedback influences on fieldwork, or: A funny thing happened on the way to the beer garden. In *Urban Man in Southern Africa* (eds C. Kileff and W. C. Pendleton). Gwelo: Mambo Press.

Wolcott, H. F. (1981). Home and away: personal accounts in ethnographic style. In *Anthropologists at Home in North America: Methods and Issues in the Study of One's Own Society* (ed. D. A. Messerschmidt). Cambridge: Cambridge University Press.

Wolf, E. R. (1968). Kinship, friendship and patron-client relations in complex societies. In *The Social Anthropology of Complex Societies* (ed. M. Banton). (A.S.A. Monogr. **4**). London: Tavistock.

Wolf, E. R. and Jorgensen, J. G. (1970). Anthropology on the warpath in Thailand. *New York Review of Books* 19 November, 26–35.

Wolff, H. (1959). Intelligibility and inter-ethnic attitudes. *Anthropological Linguistics* **1**(3), 34–41. Reprinted (1964) in *Language in Culture and Society* (ed. D. Hymes). New York, Evanston and London: Harper and Row.

Wolff, K. H. (1960). The collection and organization of field materials: a research report. In *Human Organization Research* (eds R. N. Adams and J. J. Preiss). Homewood, Illinois: The Dorsey Press.

Wolfgang, M. E. (1981). Confidentiality in criminological research and other ethical issues. *The Journal of Criminal Law and Criminology* **72**(1), 345–361.

Worrall, J. (1978). Against too much method. Review of P. K. Feyerabend: Against method. *Erkenntnis* **13**, 279–295.

Worsley, P. (1982). Barriers to ethnographic fieldwork [letter]. *RAIN (Royal Anthropological Institute News)* (December) **53**, 12.

Wright, G. H. von. (1971). *Explanation and Understanding.* Ithaca: Cornell University Press.

Yengoyan, A. A. (1970). Open networks and native formalism: the Mandaya and Pitjandjara cases. In *Marginal Natives: Anthropologists at Work* (ed. M. Freilich). New York, Evanston and London: Harper and Row.

Young, K., Wolkowitz, C. and McCullagh, R. (eds) (1981). *Of Marriage and the Market: Women's Subordination in International Perspective.* London: S.C.E. Books.

Young, M. W. (ed.) (1979). *The Ethnography of Malinowski: The Trobriand Islands 1915–1918.* London: Routledge and Kegan Paul.

Zinn, M. B. (1979). Field research in minority communities: ethical, methodological and political considerations by an insider. *Social Problems* **7**(2), 209–219.

Znaniecki, F. (1934). *The Methods of Sociology.* New York: Rinehart.

Author index

Subject Index